Cooling towers

Coal blending system

Absorbers for flue gas desulphurization (scrubbers)

3 Generating units with coal-fired boilers

A MODERN COAL-FIRED GENERATING PLANT
Photograph courtesy of Allegheny Power Systems

Computer-Aided Power Systems Analysis

Second Edition

Computer-Aided Power Systems Analysis

Second Edition

Dr. George Kusic

University of Pittsburgh
Pittsburgh, Pennsylvania, U.S.A.

CRC Press
Taylor & Francis Group
Boca Raton   London   New York

CRC Press is an imprint of the
Taylor & Francis Group, an **informa** business

MATLAB® is a trademark of The MathWorks, Inc. and is used with permission. The MathWorks does not warrant the accuracy of the text or exercises in this book. This book's use or discussion of MATLAB® software or related products does not constitute endorsement or sponsorship by The MathWorks of a particular pedagogical approach or particular use of the MATLAB® software.

CRC Press
Taylor & Francis Group
6000 Broken Sound Parkway NW, Suite 300
Boca Raton, FL 33487-2742

© 2009 by Taylor & Francis Group, LLC
CRC Press is an imprint of Taylor & Francis Group, an Informa business

No claim to original U.S. Government works
Printed in the United States of America on acid-free paper
10 9 8 7 6 5 4 3 2 1

International Standard Book Number-13: 978-1-4200-6106-2 (Hardcover)

Visit the Taylor & Francis Web site at
http://www.taylorandfrancis.com

and the CRC Press Web site at
http://www.crcpress.com

Dedication

To the Serbian father

of power engineering—Nikola Tesla

Contents

Preface

This textbook presents basic principles of power systems from the point of view of the central control facility. Each chapter of the book has accompanying computer software that is integrated with power system concepts and extends the student's understanding beyond hand calculations. The software is representative of computer programs used to control and monitor large scale interconnections of power systems.

Chapter 1 discusses central digital computers used to process real time power system data, present select data to an operator, and allow the operator to implement commands to equipment in the field. All electric utilities, from the largest to the smallest rural electrification cooperatives, have progressively increased the amount of instrumentation delivering information to the central computer and the control authority of the central facility. The central point of view emphasizes emergency power to/from interconnection neighbors and the economic exchange of power. This chapter also demonstrates the droop property of generation that allows power systems to operate without a central control facility.

Chapter 2 develops the analytical characteristics of the major equipment used in power systems. These characteristics are transformed into the symmetrical component frame to simplify system analysis. Software performs the transformation for unbalanced networks and for transmission lines with specified dimensions and conductor type. This chapter, except for symmetrical components and the per unit system, can be covered cursorily by a reader with a prior course in power equipment.

Chapter 3 introduces computer-aided methods to construct and invert the important bus impedance and bus admittance matrices. Software programs are used by the students for networks consisting of more than several busses and lines. The matrices are used for short circuit calculations and to describe networks for power flow. Sparse matrix techniques are introduced since they are essential to programs used for huge interconnections such as the eastern United States. Several matrix iterative methods are presented in the chapter because they remain in use for special applications in large-scale power systems.

Chapter 4 is devoted to linearized network equations to calculate short circuit currents for balanced and unbalanced faults, and contingency cases. Contingency cases are modifications of network flow conditions due to line switching, generation changes, or load demand changes. Several basic contingency methods which employ sparse matrix techniques are presented.

Chapter 5 contains the Gauss-Seidel, Newton-Raphson, and Fast-Decoupled methods to iteratively solve the non-linear equations that describe network power flow. Network reactive compensation and adjustment of operating conditions are discussed. A software program is provided that uses a Newton-Raphson method and has a common input data format for other programs.

Chapter 6 calculates the minimum cost base power setting for generators on the network. A unit commitment program shows economic dispatch over a time interval with a constrained variable number of units on-line. The B coefficients used to approximate transmission line power losses are derived by a network reduction program. Congestion on networks is discussed. The Jacobian from a power flow program is used to adjust generator power setting as modified by transmission line losses. Economic exchange of power between areas is discussed.

Chapter 7 has state estimation to calculate the voltage and phase angle of all busses on the system from real-time measurements on the system. Both the line flow and general method for state estimation are presented. The chapter demonstrates selection of measurements required to determine the state of the power system. Normalized residuals are used to detect the most probable error among measurements. The accompanying state estimator software program is flexible for the user to select bus injections, line flows, a "bad data" threshold, and set the measurement data weighting factors. An accompanying program allows the user to corrupt measurements with Gaussian white noise.

A fundamental concept in the book is that computer applications yield more insight into system behavior than is possible from hand calculations on system elements. The book develops methods which are used for full-scale networks, so that the student can understand and expand computer control of systems. The problems at the end of each chapter apply the methods and generally require the accompanying computer software to calculate the solutions.

The format of the accompanying software is such that a student can save files of all problems he runs, along with their own documentation. For example, a common network file of bus interconnections, bus injections, and line flows can be used in five different chapters of the book, with minor modifications for particular cases.

All textbook software is in the form of self-contained executable programs which are independent of the computer operating system. Input data file formats and names are carefully defined. All programs run with or without the Powerpoint presentation. The only program dependent upon other software is the use of MATLAB to compute dynamic response for several problems in Chapter 1. These Chapter 1 problems can be run on any dynamic simulation software.

The author has selected the material in this textbook based upon interaction with utility and industrial colleagues, teaching power systems for a number of years, and many publications in the power field. The references at the end of each chapter are often the first of publications on the subject and serve as an introduction to extensive literature in the power field. There are many contributors to our present state of technology in power systems.

The author expresses my appreciation to my wife, Alexandra, and the University of Pittsburgh Department of Electrical and Computer Engineering that encouraged me to write this book.

1

Central Operation and Control of Power Systems

1.1 General

Electric power systems came into service in the 1880s and since that time have grown enormously in size and complexity. Their power generation, transmission, and distribution methods and equipment have consistently improved in performance and reliability. The power industry provided the first large-scale application of nuclear energy, was among the first to use analog controls for turbine generators, and introduced the use of on-line digital control computers. These advances were often mandatory because of the enormous growth in megawatt requirements of the power systems.

As power systems increased in size, so did the number of lines, substations, transformers, switchgear, and so on. Their operation and interactions became more complex; therefore, it became essential to monitor this information simultaneously for the total system at a focal point, which is now called an *energy control center*. A fundamental design feature of energy control centers is that they increase system reliability and economic feasibility, but system operation is possible *without* the energy control center. This is essentially a fail-safe type of operation. In practice, all communication links between equipment and the control center could be interrupted and the electric service maintained. For example, most systems maintain local as opposed to centralized control of protective switchgear that must function within several cycles of the 60 Hz line upon detection of an overload. The same switchgear may be operated on a slower basis from the energy control center. As another example, a generating plant in the system remains synchronized to the transmission network and maintains its existing power output level even without signals received from the control center.

An energy control center fulfills the function of coordinating the response of the system elements in both normal operations and emergency conditions. The burden of repetitious control in normal situations is delegated to the digital computer, and selective monitoring is performed by human operators. Essentially, the digital computer is used to process the incoming stream of data, detect abnormalities, and then alarm the human operator via lights, buzzers, or graphical interface (i.e., computer screens) presentations. Many lower-level or less serious cases of exceeding normal limits are routinely handled by the digital computer. For example, if an increase in generation must be allocated to all machines of the system and a generator is already at its high power limit, its share would be distributed among the remaining units. A more serious abnormality detected by the digital computer may cause suspension of normal control functions.

In extreme emergencies, such as loss of a major generator or excess power demands by a neighboring utility on the tie lines, many alarms would be detected and the system would enter an emergency state. The operator may be flooded with information and no diagnostics. The present state of the art is that the digital computer programs perform the first attempts at diagnosing the source of several simultaneous alarms, but remedial or corrective actions remain the responsibility of the human operator, usually called a *dispatcher*.

1.2 Control Center of a Power System

Figure 1.1 is a photograph of the interior of an energy control center. An operator or dispatcher is shown seated at one of several consoles that command the system simultaneously. The interfaces and equipment available to the operator are termed a *man–machine interface* and often consist of the following:

1. Color light-emitting diode (LED) or cathode ray tube (CRT) graphical and numerical presentations of data from transmission networks, substations, distribution networks, system performance indices, and more. The selection of screen presentations is by means of mouse controls on menus, dedicated function keys on the keyboard, paging buttons, and even voice command interpreters.

2. Editing keyboards that allow authorized consoles (or password-protected users) to change system operating conditions or system parameters, by entering data into the digital computer software code.

3. Special function keyboards that allow the dispatcher to raise and lower transformer tap settings, switch line capacitors, and change transmission or distribution network operation.

4. Mouse or pointer controls to open or close circuit breakers or switches and to activate data presentation or other operations directly on the display screens.

FIGURE 1.1
Interior view of an energy control center. Dispatcher sits at a console containing computer LED interface, keyboard to enter control commands, and telephone lines. Wall shows one-line diagrams of the power system with appropriate lights for warning and indications. (Photograph courtesy of PJM interconnection. Copyright and use by permission of PJM)

Critical functions often require confirmation before they are implemented (e.g., open/close transmission line circuit breakers).

5. Alarm lights, audible alarms, dedicated telephone communications with system generating stations and transmission substations, and conventional telephone links to neighboring utilities, which are available at every console.

Large data displays such as shown in Figure 1.1 are usually characterized by a fixed background and numerical data fields, or status points, which are periodically updated with measured data. The numerical data or device fields are fixed in size and location until the computer source code is edited. New LED displays can be created and entered into the computer system, and the colors, symbols, and blinking or nonblinking formats may be changed from the console through display compilers.

Figure 1.2 shows a dispatcher seated at a control console. The operator points a stylus or light pen to a sensitive area of the screen display, then the computer software prompts him to enter the value via the keyboard. Using the pen or a mouse cursor, an operator may typically do the following:

1. Position the mouse pointer to an active point associated with the device and click the mouse key. In the case of a circuit breaker on a power line, the status of the device is represented by a separate symbol, adjacent to the graphical point. In the case of a tap-changing transformer, the display field is numeric, adjacent to the transformer.

FIGURE 1.2

An operator selecting a numerical value to be changed in the database. Behind the operator is a large display, called a mimic board, of a one-line diagram of the power system. (Photograph courtesy of Westinghouse IED, Pittsburgh, PA)

2. After the computer software senses the cursor position, the activated display blinks and the pointer moves to a home position to give a clear view of the selection. In the case of a device such as a circuit breaker, the status is displayed by lighted areas for ON, OFF, TRIP, or LOCKOUT. For a tap-changing transformer, either the "raise" or "lower" lights are displayed and the numerical field for its ratio is updated.

3. After the hardware in the field has performed the breaker action, the breaker symbol changes either (or both) symbol or color. The symbols and colors (though frequently the same) are selectable on a per-field basis. In this case, as with all successful operations, a message documenting the action is entered on a log record, which may be later or simultaneously printed.

4. Each type of electrical device in the field that is controllable has a time limit (usually in terms of computer scan cycles) within which a normal remote operation should be completed. An OPERATION FAILED is shown on the LED display and logged should the time interval be exceeded.

Some of the other control features available to the operator are:

1. *System commands*: May select the mode by which the digital computer controls the entire power system (explained in Section 1.4).

2. *Units*: Selects the manual or base loading mode, and so on, for each generator supplying power to the system.

3. *Automatic generation control (AGC) data entry*: Permits the operator to set ACE operating modes, inadvertent payback, and other parameters.

4. *Data/entry readout*: Values may be entered on the keyboard for subsequent introduction to the digital computer program.

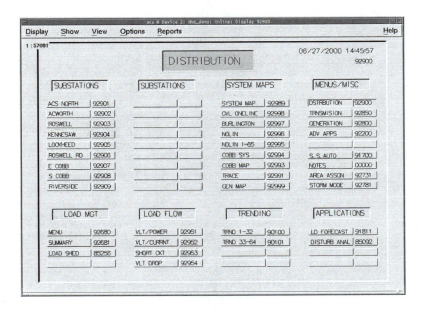

FIGURE 1.3
Main menu for a dispatcher to select a display of generation, transmission networks, distribution networks, or computational programs such as power flow or load forecast. (Display courtesy of Advanced Control Systems, Atlanta, GA)

5. *Alarms:* Allows the operator to find the source of the alarm in the system, then correct the source of the alarm.
6. *Plant/substation* select: Allows the dispatcher or operator to bring forth presentations on generating plants or substation power distribution points.
7. *Special functions:* Allow entry or retrieval of data used to control the power system.
8. *Readout control:* Directs output to different LED consoles or line printers.
9. *CPU control:* Switches the data acquisition and command operations to either of the redundant computers to be used on-line, and the other is applied to nonessential programs.

Figure 1.3 shows a typical main menu for a dispatcher to select displays.

1.3 Digital Computer Configuration

Consistent with principles of high reliability and fail-safe features, electric utilities have almost universally applied redundancy for data acquisition from equipment in the field, and for computer control of the power system. The interface to the field equipment is called SCADA, which is an abbreviation for supervisory control and data acquisition. The critical parts of the system transfer from one redundant system to the other automatically in case of failure. Figure 1.4 shows typical functions that are carried out by the equipment called an energy management system (EMS).

The incoming/outgoing information to/from the computers through the SCADA interface to the field equipment is stored in hot standby, so upon a fail-over or switch in status command, the information is stored and made available to the backup system. The maximum age of the stored data to transfer is constantly reduced as the speed of computers increases. Figure 1.5 is a typical block diagram of the redundant computer configuration and indicates some of the necessary digital equipment.

For the typical EMS of Figure 1.5, there are five major operation areas indicated by groups of computers—transmission, distribution, interchange, engineering/technician, and storm emergency—which are often authorization centers. The capability and responsibility to operate equipment are usually divided among these groups. For example, the distribution consoles, such as shown in Figure 1.6, can only monitor, switch, connect, and disconnect circuits 69 kV and below, while the transmission consoles have the authority for all lines above 69 kV. A chief dispatcher at the storm console may be the only computer set that has control over all elements of the power system.

The data acquisition and implementation equipment in the field, the remote terminal units (RTUs), are interfaced to the computer system through input–output microprocessors that have been programmed to communicate as well as preprocess the analog information, check for limits, convert to another system of units, and so on. The microprocessors can transfer data in and out of the main database memory without interrupting the operation of any computer terminal. Often, the microprocessors are also redundant, in that equipment interfaces may be switched to spare units upon detecting a malfunction.

As a result of redundancy and fail-over precautions, for all critical hardware functions there is often a guaranteed 99.8% or more availability. Software (computer programs) also allow for multilevel hardware failures and initialization of application programs if failures occur.

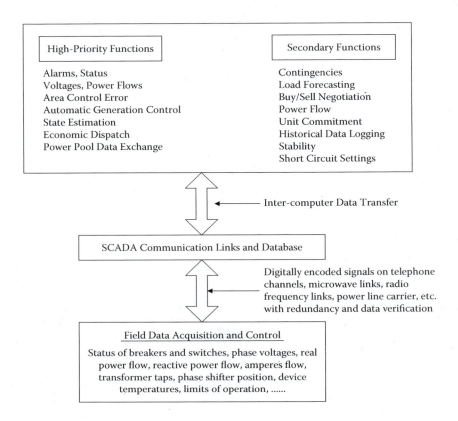

FIGURE 1.4
Typical digital computer control and monitoring for power systems.

Another feature of the computer system is that critical operating functions are maintained during either preventive or corrective maintenance. Besides hardware, new digital code to control the system may be compiled and tested in the engineer/technician consoles without interrupting normal operations, then switched to become the on-line software.

The digital computers are usually employed in a fixed-cycle operating mode with priority interrupts wherein the computer periodically performs a list of operations. The most critical functions have the fastest scan cycle. Typically the following categories are scanned as often as every second:

- All status points, such as switchgear position (open or closed), substation loads and voltages, transformer tap positions, and capacitor banks
- Transmission line and distribution line flows and voltages
- Tie-line flows and interchange schedules
- Generator loads, voltage, operating limits, and boiler capacity
- Telemetry verification to detect failures and errors in the bilateral communication links between the digital computer and the remote equipment

The turbine generators are often commanded to new power levels every 2–4 sec, sharing the load adjustment based on each unit's response capability in MW/min. The absolute

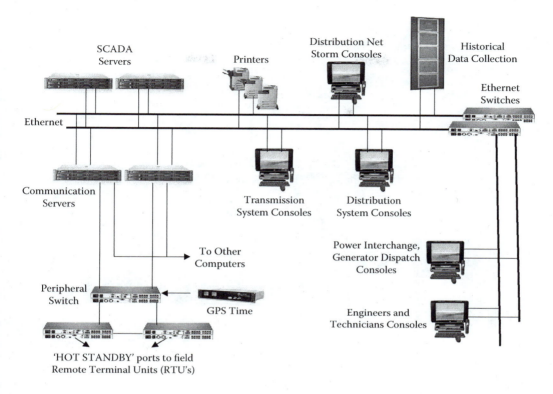

FIGURE 1.5
Typical redundant computer control for an energy management center.

power output of each unit is typically adjusted every 5 min by the computer executing an economic dispatch program to determine the base power settings. Many other system operations, such as the recording of load, forecasting of load, and determination of which generators to start up or to stop, are considered noncritical, so the computer executes these programs on a less frequent basis.

Most secondary-priority programs (those run less frequently) may be executed on demand by the operator for study purposes or to initialize the power system. From an authorized terminal, an operator may also alter the digital computer code in the execution if a parameter changes in the system. For example, the MW/min capability of a generating unit may change if one of its throttle valves is temporarily removed for maintenance, so the unit's share of regulating power must accordingly be decreased by the code. The authorized terminal may change this limit. The computer software compilers and data handlers are designed to be versatile and readily accept operator inputs.

1.4   Automatic Generation Control for a Power System

Automatic generation control and economic load dispatch are two principal areas of concern for generation control on large, interconnected power systems. The role and aim of each is quite different, but both act on the generator through controlling the prime mover, and both vary the generation relatively slowly.

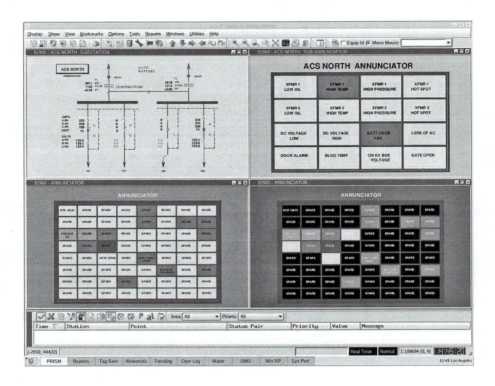

FIGURE 1.6
Dispatcher display for a substation in split screen form shows real-time flows and voltages at the substation as well as monitors on equipment in the substation. (Photograph courtesy of Advanced Control Systems, Atlanta, GA)

Automatic generation control (AGC) is an on-line computer control that maintains the overall system frequency and the net tie-line load exchange between the power companies in the interconnection. The common practice is to carry out generation control on a decentralized basis; that is, each individual area [1] tries to maintain its scheduled interchange of power.

Economic load dispatch, described in Chapter 6, is also an on-line computer control, whose function is to supply the existing system load demand from all currently operating generators in the most economical manner in terms of minimal fuel cost. A variety of conditions, such as the presence of hydro along with fossil-fuel power stations, a multiarea structure for the interconnected power system, or even pollution control, can be features of an economic dispatch operation [2].

Usually, neighboring power companies are interconnected by one or more transmission lines called *tie lines*, as depicted graphically in Figure 1.7. The electrical areas shown in Figure 1.7 are *separate power systems* under the control of an AGC in a central digital computer. The boundaries of an area are the points on the tie lines where a utility's ownership, maintenance, and loss accounting ends and those of its neighbors begin. There are very few isolated power systems that are not connected to neighbors by means of tie lines. The power systems employ tie lines for the following reasons:

1. Tie lines allow a local or "pool" exchange and sale of power between the power companies on a predetermined schedule.

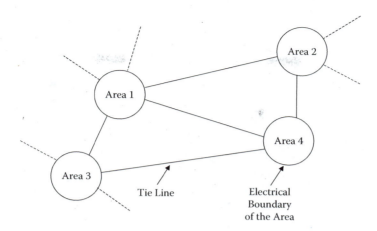

FIGURE 1.7
Typical interconnected power systems.

2. Tie lines allow areas experiencing disturbances to draw on other areas for help.
3. Tie lines provide a long-distance transmission line for the sale and transfer of power (e.g., on an interstate or international basis).

Interconnections are made so that operating areas can share generation and load. This sharing is normally on a scheduled basis as forced by the AGC. However, during times of disturbance, when an area is unable to meet its own regulating requirements, unscheduled sharing occurs, as dictated by generator governing responses and by the contributory function of the AGC frequency bias of each area. Such unscheduled interchanges persist until either the disturbed area can itself fully respond to its local requirement, or normal schedules are reset so that the contribution is taken out of the unscheduled class and put into the scheduled class.

1.4.1 Area Control Error

To maintain a net interchange of power with its area neighbors, an AGC uses real power flow measurements of all tie lines emanating from the area and subtracts the scheduled interchange to calculate an error value. The net power interchange, together with a gain, B (MW/0.1 Hz), called the frequency bias, as a multiplier on the frequency deviation, is called the *area control error* (ACE) and is given by

$$\text{ACE} = \sum_{k=1}^{k} P_k - P_s + 10B(f_{\text{act}} - f_0) \quad \text{MW} \tag{1.1}$$

Here P_k is an MW tie flow defined as positive *out* of the area, P_s is the scheduled MW interchange, and f_0 is the scheduled base frequency. When a system is not interconnected, only the frequency term is used. The interchange power P_s is generally scheduled for periods of the day and is changed as blocks of MWh are bought or sold to neighboring utilities. A *positive* ACE or positive net exchange of power represents a flow *out* of the area. Figure 1.8

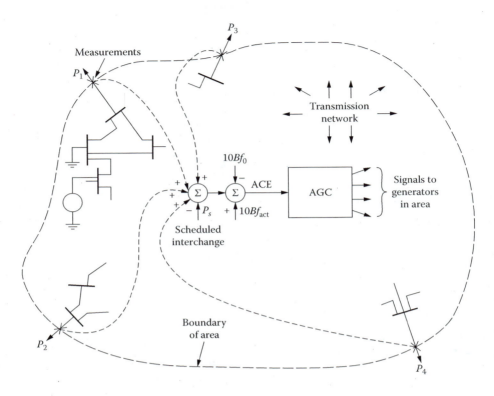

FIGURE 1.8
Using ACE as an input to the AGC. (Tie-line power flow is measured and telemetered to the central control computer. Four tie lines are shown.)

schematically shows the ACE signal used by the central computer to control generation within the electrical boundary encompassing the system's transmission network.

The frequency bias term, B (MW/0.1 Hz), is usually on the order of 1% of the power system's estimated peak power demand. Thus, a 23,000 MW system has a frequency bias of $B = 236$. The bias could be lower for a system with significant nuclear generation for which it is desired to operate the reactors at constant temperature, hence constant output power of the generators. Alternately, the bias could be higher when an area supplies regulation for a neighbor.

The tie lines are generally connected to the transmission network at locations where their specific power flow must be established by adjusting or shifting the power output of generators in order to achieve a desired flow value. Notice that the real power summation of the ACE calculation loses information regarding the flow on individual tie lines; thus, the *AGC sensing only ACE does not control the flow on the individual tie lines* but is concerned with area net generation. Often, the tie lines transfer power through the area from one neighbor to the next, called *wheeling power*. The wheeling power cancels algebraically in the ACE. However, this is exactly the process whereby one area purchases or sells blocks of power (MWh) with nonneighbor utilities. For example, consider the four-area power pool shown in Figure 1.9, which is operating at 60 Hz, so the frequency error is zero. If area A desires to sell a block of power, p, for 1 h beginning at 1 A.M., it would introduce p into its own ACE as a scheduled interchange (with sign change), causing its tie lines to export

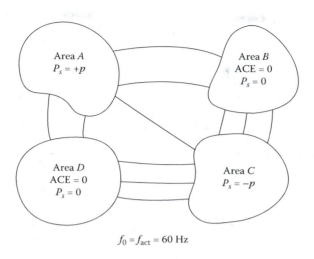

$f_0 = f_{act} = 60$ Hz

FIGURE 1.9
Sale of power from area A to area C.

power until its AGC forces Equation 1.1 to zero. Simultaneously, area C introduces $-p$ into its ACE (with sign change) so that power flows into area C until its own AGC reduces Equation 1.1 to zero. Notice that areas B and D usually participate in this interchange because of the interarea tie lines, so they must be informed of the transaction to determine possible tie-line overloads. Generally, the characteristic patterns of area B and D network flows are calculated or known *a priori* and stored as tie distribution factors for the case when neighbors A and C are exchanging power, so critical situations are avoided. It is common practice in the United States for neighboring areas to use after-the-fact accounting on a monthly basis for the use of their facilities for wheeling power, with the charge based on a percentage cost of the power exchanged. The economic sale or purchase of power and the use of the distribution factors in these transactions are discussed in Chapter 6.

The minimum requirements of the AGC on controlling the interchange of power and frequency have been established by the North American Electric Reliability Council (NERC), which is comprised of representatives of the major operating power pools. This committee had previously specified that in normal operation the ACE should be zero at least once every 10 min, and there was an average limit on ACE for all 10 min periods. After January 1998, NERC established more statistical criteria [3, 4] for both normal and disturbance conditions (labeled CPS1 and CPS2). The goal of the normal control is to limit frequency errors to

$$Root - Mean - Square\,(\Delta f_1) \leq \varepsilon_1$$

where Δf_1 is the 1 min average of the deviation from specified frequency (normally 60 Hz) and ε_1 is a 1 min target bound for a 12-month period. The value of ε_1 is calculated by NERC based upon yearly averages of frequency measurements, and is intended to be progressively reduced each year to improve regulation. The frequency deviation is used in the following AGC performance measures.

1.4.1.1   CPS1, 1 Minute Average

$$CF_1 = \frac{1}{\varepsilon_1^2}\left[\left(\frac{ACE}{-10B}\right)_1 \times \Delta f_1\right]$$

B is the frequency bias term, MW/0.1 Hz.

The quantity CF_1 is a compliance factor, and in each term the subscript 1 means a 1 min average. For the ACE signal, only the periodic AGC cycles are averaged. The CF_1 term is used in $CPS1 = (2 - AVG\ (CF_1)) \times 100\%$, where the average AVG is taken over 1 month, but for reporting purposes it is a 12-month average.

1.4.1.2   CPS2, 10 Minute Average

$$CF2 = \frac{\overline{ACE_{10}}}{L_{10}} = \frac{\overline{ACE_{10}}}{1.65\varepsilon_{10}\sqrt{(10B)(10B_S)}}$$

In the last equation, CF2 is a 10 min compliance factor. B_S is the bias for the entire interconnection, $\overline{ACE_{10}}$ is a 10 min average, and ε_{10} is again an rms target for frequency control. If CF2 > 1 for a 10 min period, the area control is out of compliance. The number of 10 min intervals that CF2 > 1 is counted for a ratio:

$$R = \frac{Intervals\_CF2 > 1}{Total\_Intervals}$$

R is used in the percentage measure:

$$CPS2 = (1 - R) \times 100\%$$

The allowable limit, L_{10}, of the average ACE for power systems (averaged over 10 min) is

$$L_{10} = 1.65\varepsilon_{10}\sqrt{(10B)(10B_S)} \quad \text{MW}$$

where the computed value L_{10} replaces a prior value based upon load pickup, i.e., the greatest hourly change in the net system (native) load of the control area on the day of maximum peak load. Figure 1.10 shows the daily load characteristics of an eastern U.S. power system. The value L_{10} is determined annually. Typical shares of two utilities in the eastern U.S. interconnection for the year 2005 were L_{10} = 101 MW for a 13,000 MW peak power system and L_{10} = 117 MW for a 23,000 MW peak power system. All power systems in the United States and Canada are interconnected into four groups that have a common 60 Hz frequency (see Figure 3.1a and b). The regulation goals for these areas as defined by NERC are shown in Table 1.1.

A typical ACE signal of an ACG that meets minimum performance is shown in Figure 1.11, where the average value is less than 100% of the limit allowed, and the peak excursions are acceptable.

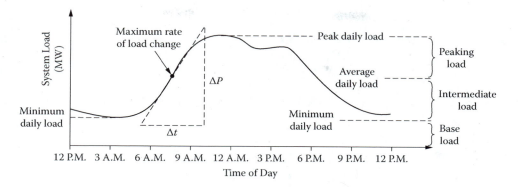

FIGURE 1.10
Daily load cycle for a power system. (Maximum rate of load change is indicated.)

TABLE 1.1

Frequency Control Goals for Major U.S.–Canada Interconnections

Interconnection	ε_1 (Hz)	ε_{10} (Hz)
Eastern (East of Rocky Mountains)	0.0180	0.0057
Hydro Quebec (Eastern Canada)	0.0210	0.0249
Western (West of Rocky Mountains)	0.0228	0.0073
ERCOT (Texas)	0.0300	0.0073

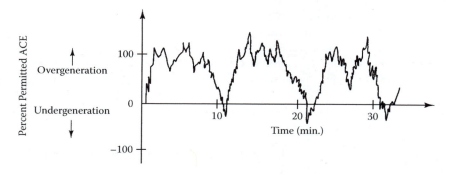

FIGURE 1.11
Example of ACE close to permissible limit.

1.4.1.3   Disturbance Conditions

During a disturbance, the AGC controls cannot maintain ACE within the above criteria for normal load variations. However, an area is expected to activate operating reserve to recover normal conditions for ACE within 15 min. A disturbance is considered to be an event such as a greater than 80% loss of the largest generation plant on the system, or a power transfer loss of more than 80% capacity of the system's largest transmission line. These two conditions, called contingency cases, are treated in Chapter 4.

1.5   Operation without Central Computers or AGC

It must be emphasized that power systems are entirely capable of operating without a central computer or AGC. This is a result of turbine generator speed controls built into generating stations and natural load regulation. These characteristics force generators within an area to share load and cause interconnected power areas to share load.

To understand the basic concepts involved in the operation of an interconnected power system, the hypothetical model shown in Figure 1.12 will be analyzed. First assume that breaker T is open and there is no tie-line flow between areas A and D. Area D represents an operating area of the interconnection in which a sudden load or generation change occurs. Area A is considered a single operating area representing the remainder of the

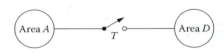

FIGURE 1.12
Representation of a simple interconnected system.

interconnection. This implies a basic assumption in regard to the interconnection; the many operating entities that comprise area A share a composite action. Furthermore, it is assumed that areas share a disturbance in proportion to their generating-capacity size and operating characteristics. Both of these fundamental assumptions are based on operating experience. Let the area A overall generation–frequency characteristic be represented by curve GG in Figure 1.13, which is a composite response curve from all the generators in area A. Figure 1.13 essentially summarizes how the rotating shaft speed of steam turbines, gas turbines, diesel engines, or other prime movers varies with the electrical load on the generator. The shaft speed—and consequently the electrical line frequency—changes with load reflected onto the prime mover.

The generation-frequency characteristic curve has a negative slope, or droop, with frequency because each turbine generator control is a type 0 control. Typically, the prime mover has approximately a 5% speed droop from no load to full load. For the purposes of the present study, the assumption is made that over the restricted frequency and loading range under consideration, the generation characteristic has a constant incremental slope. Each turbine generator control may have nonlinearities and deadband regions, but considering a number of such generating situations in the area, a composite would be as in Figure 1.13. The area connected load is defined by curve LL, as shown in Figure 1.14, which may be intuitively envisioned as the increase in load when the rotating machinery in the area is forced to increase speed.

Although difficult to verify by experimental data, these models are generally accepted. Therefore, the basic equations describing generation and load are

$$G_A = G_0 + 10\beta_1 \left(f_{\text{act}} - f_0 \right) \quad \text{MW} \tag{1.2}$$

$$L_A = L_0 + 10\beta_2 \left(f_{\text{act}} - f_0 \right) \quad \text{MW} \tag{1.3}$$

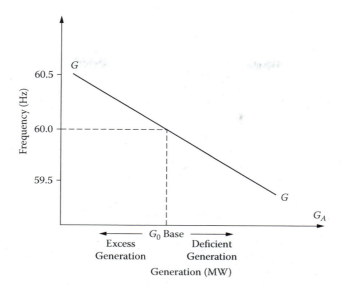

FIGURE 1.13
Area A generation-frequency characteristic curve GG.

where G_A is the total generation on system A (MW); G_0 the base generation on the system (MW at 60 Hz); L_A is the total load on system A (MW); L_0 is the base load on the system (MW at 60 Hz); f_{act} is the system frequency (Hz); f_0 is the base frequency (60 Hz), often a scheduled quantity; β_1 is the cotangent of generation–frequency characteristic in MW/0.1 Hz (this quantity is negative in sign and is called the natural generation governing), $\beta_1 < 0$; and β_2 is the cotangent of load-frequency characteristic in MW/0.1 Hz, $\beta_2 > 0$.

For a steady-state frequency, total generation must equal total effective load, and prevailing frequency is defined by the point of intersection I_0 of the GG and LL curves, shown at 60 Hz in Figure 1.15.

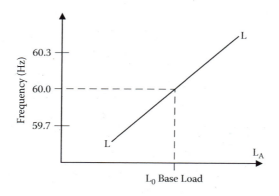

FIGURE 1.14
Effective area A load-frequency characteristic curve LL.

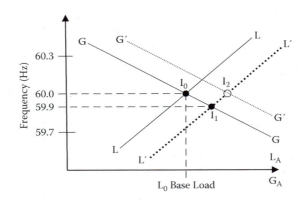

FIGURE 1.15
Isolated operation in area A in response to a load change.

The generation characteristic and the load characteristic can be added algebraically to obtain the combined area analytic expression. The composite generation–load-frequency expression is

$$G_A - L_A = G_0 + 10\beta_1 (f_{act} - f_0) - L_0 - 10\beta_2 (f_{act} - f_0) \tag{1.4}$$

Now assume that there is a load increase in area A of magnitude to move the load-frequency characteristic to position $L'L'$. The new system frequency will be defined by the intersection of the GG generation line and the new load line $L'L'$. This point of intersection, labeled I_1, is shown in Figure 1.15 at 59.9 Hz. If it is desired to return system frequency to 60 Hz, this is possible by means of offsetting the automatic generation control (AGC) in area A, shifting generation curve GG to the new position $G'G'$, which intersects the prevailing $L'L'$ load line at 60.0 Hz, point I_2. However, if there is no isochronous (constant frequency) sensing in AGC to perform corrective action, but only individual generators with droop, the frequency does not return to 60 Hz.

Equation 1.4 may be written in terms of increments about a base as

$$\Delta_A = G_A - G_0 + L_0 - L_A = 10\beta_1 (f_{act} - f_0) - 10\beta_2 (f_{act} - f_0)$$
$$= 10B_A X_A (f_{act} - f_0) \tag{1.5}$$
$$= 10B_A X_A \Delta f \quad \text{MW}$$

where $B_A < 0$ is the natural regulation characteristic of area A expressed in percent of generation per 0.1 Hz, and the X_A is the generating capacity of area A in megawatts. Thus, should the load increase by $+\Delta_A$ MW or generation decrease in area A, the resulting frequency deviation would be

$$\Delta f = \frac{\Delta_A}{10B_A X_A} \quad \text{Hz} \tag{1.6}$$

which is a negative quantity because B_A is negative. The system speed (ac line frequency) decreases due to added load.

From Figure 1.12, define a real power line flow, ΔT_L, as a positive quantity out of the area A. The combined effect on frequency for a load increase (or a generation decrease) and positive tie flow on area A is then

$$\Delta f = \frac{\Delta_A + \Delta T_L}{10 B_A X_A} \quad \text{Hz} \tag{1.6a}$$

where $\Delta_A + \Delta T_L$ is the net megawatt change.

To understand the effect of tie-line flow when systems A and D are interconnected, assume that the breaker T of Figure 1.12 is closed with generation and load equal at 60 Hz in both areas. There is initially no tie-line power interchange between the two areas. Now a disturbance occurs in area D that causes the common system frequency to drop to 59.9 Hz. Assuming that a supplemental control of AGC does not act, the area A generation increases and the power to the load decreases in area A. The resulting tie line flow is between intercepts of the GG and LL curves with the 59.9 Hz line, as shown in Figure 1.16. The difference between generation and effective load, or the net excess power in the area, flows out of area A over the tie lines as a contribution to the disturbance in area D. The contributory effect is comprised of two components: $-\Delta L$, which represents a decrease in load power in area A, and ΔG, which represents the increase of generation in area A with decrease in frequency. The tie-line flow between A and D is then

$$\Delta T_L = \Delta G_A - \Delta L_A \quad \text{MW} \tag{1.7}$$

where ΔT_L is the net change in tie-line power flow from initial conditions, which is a positive value directed from A toward D.

If area A has an AGC that applies frequency *bias*, B, the effect of the bias is to increase the MW/Hz response of the generation. The $G'G'$ and $G''G''$ lines of Figure 1.16 represent

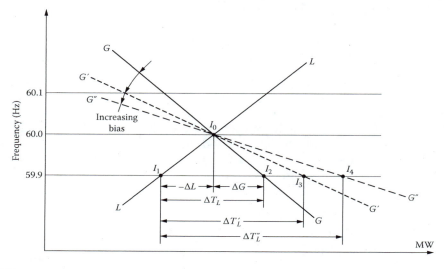

FIGURE 1.16
Governing and regulating characteristics of area A in interconnected operation.

increasing amounts of bias applied by the AGC. Corresponding tie flows $\Delta T'_L$ and $\Delta T''_L$ are larger. Bias is discussed in Section 1.7.

For area D, ΔT_L is the tie-line power flow directed from A to D. Power flow into the area appears as a load decrease, so the frequency change due to a disturbance Δ_D and tie-line flow from A to D is

$$\Delta f = \frac{\Delta_D - \Delta T_L}{10 B_D X_D} \quad \text{Hz} \tag{1.8}$$

Let $\Delta_{AD} = \Delta_D$ be the magnitude of the disturbance that occurs in area D and $\Delta_A = 0$.

Since the frequency is common to both systems,

$$\Delta f = \frac{\Delta T_L}{10 B_A X_A} = \frac{\Delta_{AD} - \Delta T_L}{10 B_D X_D} \quad \text{Hz} \tag{1.9}$$

Solving Equation 1.9 for the tie-line flow, we obtain

$$\Delta T_L = \frac{(10 B_A X_A) \Delta_{AD}}{10 B_A X_A + 10 B_D X_D} \quad \text{MW} \tag{1.10}$$

The net power change in area D is

$$\Delta_{AD} - \Delta T_L = \frac{(10 B_D X_D) \Delta_{AD}}{10 B_A X_A + 10 B_D X_D} \quad \text{MW} \tag{1.11}$$

From this derivation it is seen that the interconnected system comprised of areas A and D *participates in sharing disturbances* as weighted by their generating capacity:

$$\Delta_{AD} = 10 B_A X_A \, \Delta f + 10 B_D X_D \, \Delta f = (10 B_A X_A + 10 B_D X_D) \, \Delta f \tag{1.12}$$

Thus, by interconnecting the systems, the frequency fluctuations due to disturbances are reduced and system performance is improved by means of tie-line flows. Equation 1.12 is rewritten in transfer function to show this:

$$\frac{\Delta f}{\Delta_{AD}} = \frac{1}{10 B_A X_A + 10 B_D X_D} \quad \text{Hz/MW} \tag{1.12a}$$

For more than one tie line between areas A and D, Equation 1.10 describes the *net* flow between the areas. The power flow on each of the multiple tie lines connecting areas is dependent on transmission line impedances and the location of generation and load, so it is calculated by the methods of Chapter 5. If two or more areas are interconnected, only the net flow into or out of each area is estimated by means of generalizing Equations 1.10 and 1.11.

Example 1.1

Two areas are interconnected (see Figure E1.1). The generating capacity of area A is 36,000 MW and its regulating characteristic is 1.5% of capacity per 0.1 Hz. Area D has a generating capacity of 4,000 MW and its regulating characteristic is 1% of capacity per 0.1 Hz. Find each area's share of a +400 MW disturbance (load increase) occurring in area D and the resulting tie-line flow.

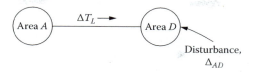

FIGURE E1.1

Solution

Using Equation 1.12a, the change in frequency is

$$\Delta f = \frac{\Delta_{AD}}{10B_A X_A + 10B_D X_D}$$

$$= \frac{400}{-10(0.015)(36,000) - 10(0.01)(4000)}$$

$$= -0.068964 \text{ Hz}$$

The tie-line flow is given by Equation 1.10:

$$\Delta T_L = \frac{(10B_A X_A)\Delta_{AD}}{10B_A X_A + 10B_D X_D} = \frac{(5400)400}{5800} = 372.4 \text{ MW}$$

Through the interconnection, the larger system considerably aids the other system. The smaller system needs only to absorb a 27.6 MW load. The frequency regulation is much better; for if area D were alone, its frequency would decrease $1 \text{ Hz} = \Delta_{AD}/10B_D X_D$ due to the increased load.

1.6 Parallel Operation of Generators

The tie-line flows and frequency droop described for interconnected power areas are composite characteristics based on parallel operation of generators. That areas must have speed or frequency droop as opposed to isochronous (constant-speed) operation is obvious, for if each area could maintain its speed $\omega = 2\pi f$ despite synchronizing torques, then a load common to both areas, by superposition, would have the terminal voltage

$$V_{\text{load}} = V_1 \sin \omega_1 t + V_2 \sin \omega_2 t \tag{1.13}$$

where subscripts 1 and 2 refer to the areas and t is time in seconds. Combining the terms of Equation 1.13 results in line frequencies that are the sum and difference of f_1 and f_2, which is objectionable. Although it is possible to use a reference frequency for both areas, in principle both areas as well as generating units must be capable of independent operation should communication links be interrupted.

A generator speed versus load characteristic is a function of the type of governor used on the prime mover (type 0 for a speed droop system, type 1 for an isochronous system, etc.) as the capacity of the generator. Consider an extreme case where generator 1, of limited capacity, is paralleled to an infinite bus of constant frequency, as shown in Figure 1.17.

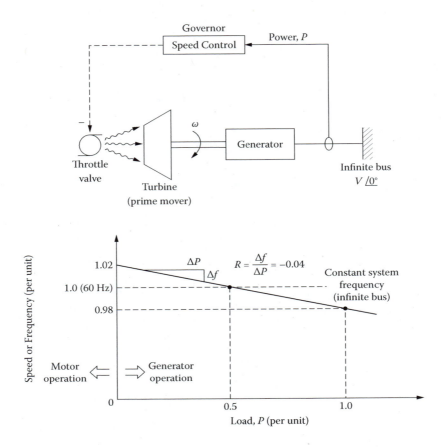

FIGURE 1.17
Parallel operation of a droop generator with an infinite bus.

(An infinite bus can absorb or supply unlimited power at constant voltage and constant frequency.) In the figure, the generator droop characteristic is such that it is loaded to 50% of its capacity when paralleled to the bus. The regulation of the unit with an implicit algebraic sign is defined as

$$\text{unit speed regulation} = R = \frac{\Delta f(\text{p.u.})}{\Delta P(\text{p.u.})} = \frac{\Delta f(\text{Hz})/60(\text{Hz})}{\Delta P(\text{MW})/P_{\text{rate}}(\text{MW})} \tag{1.14}$$

where P_{rate} is the megawatt rating of the generator and p.u. represents "per unit." The regulation is assumed to be constant for the range of interest here. The governor shown in Figure 1.17 has a steady-state regulation of 4%. If it is desired to increase the power delivered by the generator, the prime-mover torque is increased, which results in a shift of the speed droop curve as shown in Figure 1.18.

By means of adjusting the prime-mover torque, the power output of the generator is set to the desired level, including motor operation. The shifts in generator output are performed by means of momentary shaft speed changes with respect to the infinite bus at constant frequency. Thus, Figure 1.18 is equivalent to changing the shaft angle δ_1 of the synchronous machine with respect to the reference $\delta_2 = 0$ shown in Figure 1.19. For a

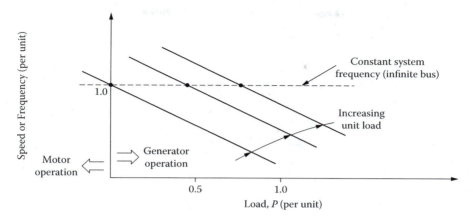

FIGURE 1.18
Adjusting prime-mover torque to load a generator.

simplified, cylindrical rotor machine, the real power flow is given by

$$P = \frac{V_1 V_2}{X} \sin(\delta_1 - \delta_2) \tag{1.15}$$

where X is the synchronous reactance and the voltages are expressed as phasors. The phasors and reactances are discussed in Chapter 2.

Steady-state output power changes for the generator of Figure 1.18 and Figure 1.19 are due to prime-mover steady-state changes, and no description is given here of the transients necessary to reach this operating point. The transients will be a function of the generator inductances and resistances, the voltage regulators, and the prime-mover dynamic characteristics.

Generally, two generating units that are paralleled both have different governor–speed droop characteristics, or the characteristics may vary with load. When parallel, the power exchange between machines forces them to synchronize at a common frequency, as the coupling impedance between machines (e.g., impedance of the transmission lines) is small compared to the load equivalent impedances. Consider the case of two parallel units of

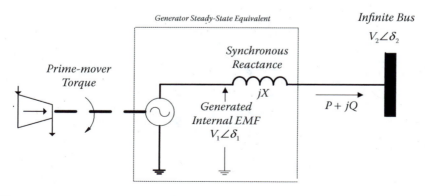

FIGURE 1.19
Synchronous generator connected to an infinite bus.

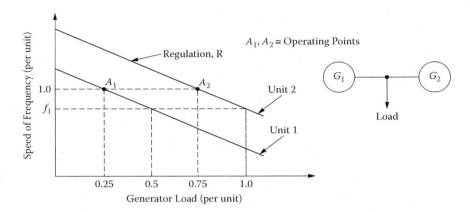

FIGURE 1.20
Parallel operation of identical units with different speed settings.

equal capacity that have equal regulation and are initially operating at 1.0 base speed, as shown in Figure 1.20.

When the paralleled system is operated at base speed, unit 1 at point A_1 satisfies 25% of the total load, and unit 2 at point A_2 supplies 75%. If the total load is increased to 150%, the frequency decreases to f_1. Since the droop curves are linear, unit 1 will increase its load to 50% of rating and unit 2 will reach 100% of rating. Further increases in system load will cause unit 2 to be overloaded.

The case when two units of different capacity and regulation characteristics are operated in parallel is shown schematically in Figure 1.21. For these two different units in parallel, their regulation characteristics are

$$R_1 = \frac{\Delta f(\text{p.u.})}{\Delta P_1(\text{p.u.})} = \frac{-\Delta f/60}{\Delta P_1/P_{1\text{rate}}} \quad \text{p.u.} \tag{1.16a}$$

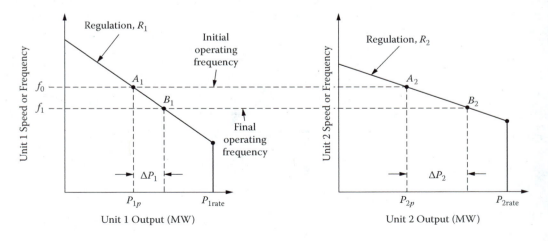

FIGURE 1.21
Two different generators operated in parallel.

$$R_2 = \frac{-\Delta f/60}{\Delta P_2/P_{2\,\text{rate}}} \quad \text{p.u.} \tag{1.16b}$$

Since the frequency is common to both units, they will share according to the ratio

$$\frac{\Delta P_1}{\Delta P_2} = \frac{R_2 P_{1\,\text{rate}}}{R_1 P_{2\,\text{rate}}} \quad \text{p.u.} \tag{1.17}$$

If the initial load is $P_{1,0} + P_{2,0}$, a change in load is satisfied by

$$\Delta L = \Delta P_1 + \Delta P_2 = \frac{-\Delta f\, P_{1\,\text{rate}}}{60 R_1} - \frac{\Delta f\, P_{2\,\text{rate}}}{60 R_2} \quad \text{MW} \tag{1.18}$$

Hence, the equivalent regulation of the paralleled system is

$$\bar{R}_{\text{system}} = \frac{\Delta f}{60 \Delta L} = \frac{-1}{\dfrac{P_{1\,\text{rate}}}{R_1} + \dfrac{P_{2\,\text{rate}}}{R_2}} \quad \text{1/MW} \tag{1.19}$$

or in terms of per unit, with the system capacity as base, this is

$$R_{\text{system}} = \bar{R}_{\text{system}}\,(P_{1\,\text{rate}} + P_{2\,\text{rate}}) \quad \text{p.u.} \tag{1.19a}$$

such that the regulating performance of the system is a capacity-weighted combination of the generators. A change in load ΔL is satisfied by the two units in shifting the operating points from A_1, A_2 to B_1, B_2 in Figure 1.21, with the ratio of the increments given by Equation 1.17.

Notice that if $10 B_i = 1/R_i$, the parallel operation of two generators expressed by Equation 1.19 is exactly analogous to Equation 1.12a, which describes interconnected power systems.

Example 1.2

Two synchronous generators are initially supplying a common load (see Figure E1.2) at 1 p.u. frequency (60 Hz). The rating of unit 1 is 337 MW and has 0.03 p.u. droop built into its governor. Unit 2 is rated at 420 MW and has 0.05 p.u. droop.

Find each unit's share of a 0.10 p.u. (75.7 MW or 10% of the total generation) increase in the load demand. Also find the new line frequency.

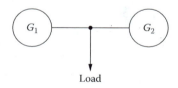

Load

FIGURE E1.2

Solution

The equivalent system regulation is a weighted average of both units.

$$R_{\text{system}} = \frac{\Delta f}{\Delta L} = \frac{-1}{\dfrac{P_{1\,\text{rate}}}{R_1} + \dfrac{P_{2\,\text{rate}}}{R_2}}\,(P_{1\,\text{rate}} + P_{2\,\text{rate}})$$

$$= \frac{-1}{\dfrac{337}{0.03} + \dfrac{420}{0.05}}\,(337 + 420)$$

$$= -0.0386 \quad \text{p.u.}$$

The new system frequency due to a load increase is

$$\Delta f(\text{p.u.}) = R_{\text{system}} \, \Delta L(\text{p.u.}) = -0.0386 \times 0.10 = -0.00386 \ \text{p.u.}$$

which is a decrease of 0.231 Hz. Both generating units have the same frequency. Their share of the load is determined by their regulation:

$$\Delta P_1 = \frac{\Delta f(\text{p.u.})}{R_1} = \frac{-0.00386}{-0.03} = 0.129 \ \text{p.u.} \rightarrow + 43.3 \quad \text{MW}$$

$$\Delta P_2 = \frac{\Delta f(\text{p.u.})}{R_2} = \frac{-0.00386}{-0.05} = 0.0771 \ \text{p.u.} \rightarrow + 32.4 \ \text{MW}$$

1.7   Network Power Flows

Power flow is *simultaneous* in time on all interconnected transmission lines and networks. Regardless if the power flow is on interconnected high-voltage, three-phase transmission lines, or through lower-voltage lines connected through transformers, the current flow between any two electrically connected points, or nodes/buses, depends upon the difference in instantaneous voltage at the points. Currents and voltages obey Kirchhoff's basic circuit laws. Engineering units of kilovolts (kV), kiloamps (kA), ohms (Ω), and megavolt-amperes reactive (Mvar) are used to express transmission line electrical quantities. The earth below transmission lines is usually the neutral potential, or ground potential voltage of 0.0 V. Often the earth is considered to be an ideal conductor of zero resistivity, or an infinitely conductive ground plane, beneath transmission lines. This assumption leads to writing Kirchhoff's current and voltage laws in nodal form for power networks with all voltages defined with respect to zero ground potential.

For the two transmission lines shown in Figure 1.22, the injected phase A current and power, respectively, may be written using phasors (Chapter 2) in nodal form as

$$I_{1A} = I_1 + I_2 = \frac{V_{1A}\angle\delta_1 - V_{2A}\angle\delta_2}{R_1 + jX_1} + \frac{V_{1A}\angle\delta_1 - V_{3A}\angle\delta_3}{R_2 + jX_2} \tag{1.20}$$

$$S_{1A} = P_{1A} + jQ_{1A} = V_{1A}\angle\delta_1(I_{1A})^*$$

$$= V_{1A}\angle\delta_1 \left[\frac{V_{1A}\angle-\delta_1 - V_{2A}\angle-\delta_2}{R_1 - jX_1} + \frac{V_{1A}\angle-\delta_1 - V_{3A}\angle-\delta_3}{R_2 - jX_2} \right]$$

$$= \frac{V_{1A}^2 - V_{1A}V_{2A}\cos(\delta_1 - \delta_2) - jV_{1A}V_{2A}\sin(\delta_1 - \delta_2)}{R_1 - jX_1} \tag{1.21}$$

$$+ \frac{V_{1A}^2 - V_{1A}V_{3A}\cos(\delta_1 - \delta_3) - jV_{1A}V_{3A}\sin(\delta_1 - \delta_3)}{R_2 - jX_2}$$

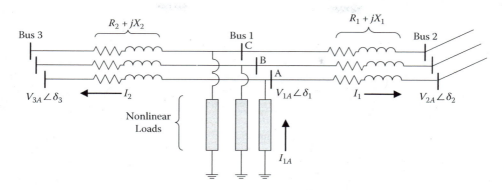

FIGURE 1.22
Notation for nodal equations on three-phase transmission lines above perfectly conducting earth.

In a similar manner, phases B and C can be treated. Lines connected to buses 2 and 3 also have currents written in terms of injections at these nodes (buses) and voltages with respect to the earth reference. This process of adding connected buses continues to add simultaneous equations for current and power flow on the entire interconnection, extending through tie lines into neighboring power systems. In fact, the entire eastern United States and eastern Canada comprise a huge interconnection of generation introducing power into the network and load points to where power is delivered (power consumption). The continental United States and Canada have four such interconnections with distinctly different 60 Hz power linked only by back-to-back ac/dc/ac converters or high-voltage direct-current (HVDC) lines (see Figure 3.1a and b).

A significant problem is how or where on a network to limit the computation of simultaneous nodal equations in order to obtain a reasonable size and solvable set of equations. In most cases, a utility lumps the neighboring power areas into a fixed load or generation on their tie lines, and holds these as constant for its calculations. A computational problem that arises is that loads on the network do not behave as a linear resistance or impedance, but as constant real and reactive power demand $P + jQ$. This destroys a linear solution of Kirchhoff's equations, because as the voltage across a load increases, the current decreases to keep the power constant. The nonlinear computation problem is treated by power flow methods of Chapter 5. A consequence of real and reactive power requirements is that the simultaneous equations are overdetermined, and one node (or bus) must be designated as a reference.

1.7.1 Oversimplified Power Flow (dc Power Flow)

A number of oversimplifications are performed on power flow Equation 1.21 in order to demonstrate electrical coupling of buses on the network, and selection of one bus as the reference or slack bus. The coupling links the entire interconnection with simultaneous flow equations. The simplified formulation is called a dc power flow.

In Chapter 2, balanced three-phase power flow is defined using only the phase A equivalent. The first simplification is that all nodal voltages on the network are maintained at 1.0 per-unit voltage by reactive injections, and the phase angle, δ, is very small and its difference is small with respect to all other buses. These regulated voltage and small-angle assumptions imply

$$V_1 V_2 \cos(\delta_1 - \delta_2) \approx 1.0 \ \text{(p.u.)}^2$$

and

$$V_1 V_2 \sin(\delta_1 - \delta_2) \approx \delta_1 - \delta_2 \ (\text{p.u.})^2$$

Another idealizing assumption is that the transmission lines are lossless and inductive reactive only, $R + jX \approx jX$, without any capacitive effects. These assumptions lead to approximating Equation 1.21 for power flow into bus 1, connected to buses 2 and 3, in per-unit (p.u.) form as

$$S_{1A} \approx P_{1A} = P_{12} + P_{13} = (\delta_1 - \delta_2)/X_1 + (\delta_1 - \delta_3)/X_2 \ (\text{p.u.}) \qquad (1.21a)$$

The reactive power flow terms (j or imaginary terms) are zero because of small-angle and unity voltage approximations. Injections at any bus are the sum of line flows to other connected buses. Idealized and lossless power flow, which is not true for real networks, is demonstrated in the following example.

Example 1.3

Do the following:

(a) Apply the dc power flow to compute the phase angles and real power flow on lines of the three-bus radial network shown in Figure E1.3a.
(b) Apply the dc power flow to compute inter-area power flow on the interconnection of areas A and D shown in Figure E1.3b.
(c) Compute MW flows on the network of Figure E1.3b if the capacity of area A is 36,000 MW and area D is 4,000 MW, as defined in Example 1.1.

Solution

(a) For the three-bus, two-line radial network of Figure E1.3a, bus 1 is selected to be the slack or reference bus at 1.0 per-unit voltage. Because only differences of phase angles are used on the network, any bus may be selected as a reference, so $\delta_1 = 0.0$ radians is arbitrarily selected as the phase reference.

In the lossless network of Figure E1.3a, whatever power goes into buses 2 and 3 must come out of bus 1, $P_1 = -(P_2 + P_3)$. Therefore, only power flow into buses 2 and 3 is described by power flow approximations. The injected power at buses 2 and 3 may be written as

$$P_2 = 2 = \frac{\delta_2 - \delta_1}{X_1} + \frac{\delta_2 - \delta_3}{X_2} \quad (\text{p.u.}) \qquad P_3 = 3 = \frac{\delta_3 - \delta_2}{X_2} \quad (\text{p.u.})$$

Bus voltages are implicit in the above p.u. equations. The impedances X_1 and X_2 are also in per unit (Chapter 2), so that power flow is dimensionally correct. The phase angles at buses 2 and 3 are calculated (with $\delta_1 = 0$) from the simultaneous equations

$$\begin{bmatrix} P_2 \\ P_3 \end{bmatrix} = \begin{bmatrix} 2 \\ 3 \end{bmatrix} = \begin{bmatrix} \dfrac{1}{X_1} + \dfrac{1}{X_2} & \dfrac{-1}{X_2} \\ \dfrac{-1}{X_2} & \dfrac{1}{X_2} \end{bmatrix} \begin{bmatrix} \delta_2 \\ \delta_3 \end{bmatrix} \qquad (\text{p.u.})$$

Numerical values are inserted, the 2×2 matrix inverted, and the phase angles calculated as $\delta_2 = 0.15\,rad$, $\delta_3 = 0.27\,rad$. From these phase angles, the line flows are

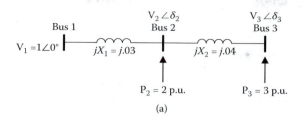

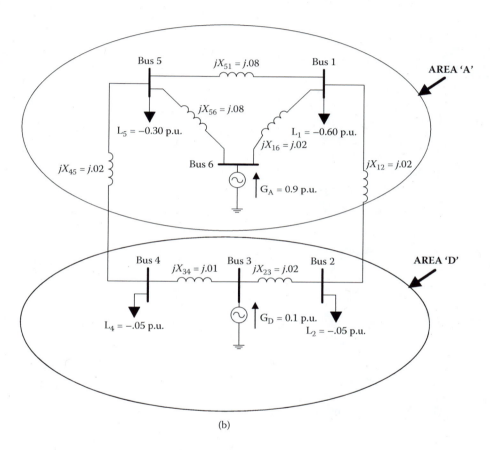

FIGURE E1.3
(a) Three-bus power system. (b) Interconnected power areas *A* and *D*. (c) Calculated flows on interconnected areas *A* and *D*.

obtained as

$$P_{21} = \frac{\delta_2 - \delta_1}{X_1} = \frac{.15}{.03} = 5 \quad \text{(p.u.)} \qquad P_{32} = \frac{\delta_3 - \delta_2}{X_2} = \frac{.27 - .15}{.04} = 3 \quad \text{(p.u.)}$$

The double subscript indicates the direction of power flow, from the first subscript (bus) toward the second subscript (bus). As all transmission lines are lossless, the sum of the network power demand (loads) equals the sum of power injections at

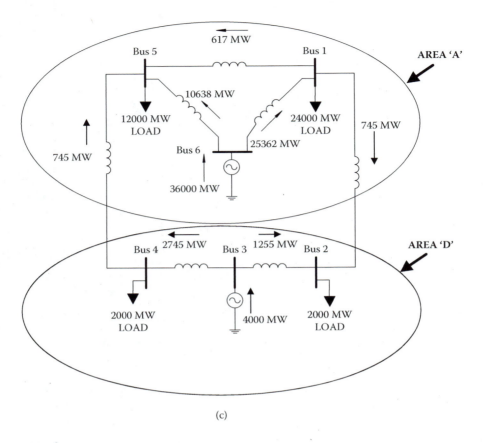

FIGURE E1.3
(Continued).

buses. Hence, bus 1 in the network receives 5 p.u. power, which is injected to the network at buses 2 and 3.

© This computation is performed as the base case in the software *dc_flow.exe* program, with the notation that 1.0 per-unit power is 100 MW. The data in the dc_flow program may be modified and extended to different network topologies to study more extensive, but idealized power flow cases. Results of the computation are saved in a file *dc_flow.sss*. It is suggested to execute this program to verify the above results, to understand identification of variables, and the method to input new cases. The data file for part (a) is *example13a.sss*, and for part (b) the data is in the file *example13.sss*.

(b) Part (b) of Figure E1.3b shows areas *A* and *D* connected by two tie lines of impedance *j*.02 between buses 1 and 2, as well as buses 4 and 5. Each area has its load balanced to its generation, with area *A* being a much larger power capacity area than area *D*.

Bus 6 is selected to be the slack bus, at 1.0 p.u. voltage and zero phase angle, even though any bus can be so designated. There are five buses for which to write the power injection equation and solve for the phase angles. The injections are defined in p.u. in part (b) of Figure E1.3. The simultaneous equations for buses 1 to 5, with bus 6 as the reference, are as follows:

$$
P = \begin{bmatrix} -.60 \\ -.05 \\ .10 \\ -.05 \\ -.30 \end{bmatrix} = \begin{bmatrix} \dfrac{1}{X_{12}} + \dfrac{1}{X_{51}} + \dfrac{1}{X_{16}} & \dfrac{-1}{X_{12}} & 0 & 0 & \dfrac{-1}{X_{51}} \\[2ex] \dfrac{-1}{X_{12}} & \dfrac{1}{X_{12}} + \dfrac{1}{X_{23}} & \dfrac{-1}{X_{23}} & 0 & 0 \\[2ex] 0 & \dfrac{-1}{X_{23}} & \dfrac{1}{X_{23}} + \dfrac{1}{X_{34}} & \dfrac{-1}{X_{34}} & 0 \\[2ex] 0 & 0 & \dfrac{-1}{X_{34}} & \dfrac{1}{X_{34}} + \dfrac{1}{X_{45}} & \dfrac{-1}{X_{45}} \\[2ex] \dfrac{-1}{X_{51}} & 0 & 0 & \dfrac{-1}{X_{45}} & \dfrac{1}{X_{12}} + \dfrac{1}{X_{51}} + \dfrac{1}{X_{16}} \end{bmatrix} \begin{bmatrix} \delta_1 \\ \delta_2 \\ \delta_3 \\ \delta_4 \\ \delta_5 \end{bmatrix} \quad (1)
$$

Numerical values are inserted for the impedances, and the 5×5 matrix is inverted to solve for phase angles of the bus voltages:

$$\delta_1 = -0.012681\,rad \qquad \delta_2 = -0.013053\,rad \qquad \delta_3 = -0.012426\,rad$$

$$\delta_4 = -0.013112\,rad \qquad \delta_5 = -0.013298\,rad$$

The tie-line power flows between areas A and D are

$$P_{12} = \frac{\delta_1 - \delta_2}{X_{12}} = .0186 \ \text{(p.u.)} \qquad P_{45} = \frac{\delta_4 - \delta_5}{X_{45}} = .0186 = -P_{54} \ \text{(p.u.)}$$

The net power flow from area A to area D, with flow directions taken into account, is $P_{AD} = P_{12} + P_{54} = 0$ because load is balanced with generation in both areas. The impedance on transmission lines in area A is such that in order to supply the demand at bus 5, power flows from area A through the tie lines, then through area D in order to supply the load at bus 5. The ACE signal for area D would therefore be zero. Power flow is determined by the entire interconnection.

Other flow results computed from the phase angles and impedances are as follows (in p.u.):

$$P_{23} = -0.0315 \qquad P_{34} = 0.0686 \qquad P_{51} = -0.0154 \qquad P_{56} = -0.2660 \qquad P_{61} = 0.6340$$

The power balance at bus 1 is such that 0.634 p.u. flows on the line from bus 6 to bus 1, the load absorbs 0.6, the flow 0.0154 goes from bus 1 to bus 5, and the balance of .0186 flows through area D to bus 5.

© These computed results may be verified by executing the *dc_flow* program using the file labeled *Example13.sss*. This file contains the injection and line impedance data for Figure E1.3b. Copy file *Example13.sss* into *dc_flow.sss* in order to reproduce the results of this example. The program uses 100 MVA as the base power for p.u.

(c) When areas A and D have capacities 36,000 and 4,000 MW, respectively, per-unit base power is the sum 40,000 MW. A new power base for *dc_flow* requires recomputation of the line impedances on the new base. Since the dc power flow is linear, the right-hand side of Equation (1) remains equal to the injected p.u. power. A change in per-unit impedances is countered by an opposite phase-angle change with the result that network power flows remain the same in per unit. Figure E1.3c shows the line power flows for areas A and D, with 745 MW circulating through D.

Example 1.3 demonstrated how power flows through an adjacent area in order to satisfy a local demand. In this example, the ACE for area D is zero because power flow in and out of area D sums to zero. Transmission lines of area D carry the extra burden of through power. This increases losses in the transmission lines and can lead to exceeding the current-carrying limit of the lines.

Through flows on a power area have an ACE = 0 for the area, but can lead to overloaded transmission lines when distant neighbors are exchanging power. In Example 1.3 the transmission line from bus 3 to bus 4 has an increase of 745 MW over its normal flow, while the line from bus 3 to bus 2 is reduced by this amount. When an area's transmission lines become overloaded due to power exchanges between neighboring systems, this is called congestion on the interconnection. The congestion is measured in the affected area, but has to be alleviated by adjustments in generation and flow control on the entire interconnection. The situation requires a super monitor that receives interchange data from all the areas of the interconnection, and can supply corrective actions to all members of the interconnection. Such super monitors exist in the United States and are called pool controllers or independent system operators (ISOs).

1.8   Area Lumped Dynamic Model

The power system models described thus far are for steady-state conditions, and do not address the dynamics necessary to move from one power flow condition to another. Often the network power flow transients are considered to be fast, or have very small time constants, compared to generators or major equipment time response. The generators and power demand (loads) due to rotating motors are described by macroscopic models. One can consider the macro models as having evolved from both power system operating experience and parallel operation of strongly coupled generators within the power system. These physically isolated generators could not share the area common load unless the turbine generator controllers have droop characteristics, or else a noninterruptible, differential power signal is available to every turbine generator of the system as well as the interconnection neighbors.

A power system macro model that neglects transmission line dynamics may be described by a block diagram as shown in Figure 1.23, where the following definitions are used:

H_A = effective inertia of the rotating machinery loads in the power system area as normalized by the generating capacity; the numerical value of $2H_A/10\beta_2$ is 2 to 8 sec for all systems [5]

β_2 = load-frequency characteristic, MW/0.1 Hz

$P_{i\,rate}$ = rated power output of generator i, MW

ΔP_i = electrical power increment for generator i, MW

$1/R_i$ = droop characteristic of generator i, Hz/MW (the algebraic sign is taken into account at the summing junction)

S = Laplace transform variable, 1/sec

This model represents an *isolated* power system without tie lines. The steady-state value of the frequency deviation, Δf, for a step load change $\Delta_L = \Delta_A/S$ is found using the final

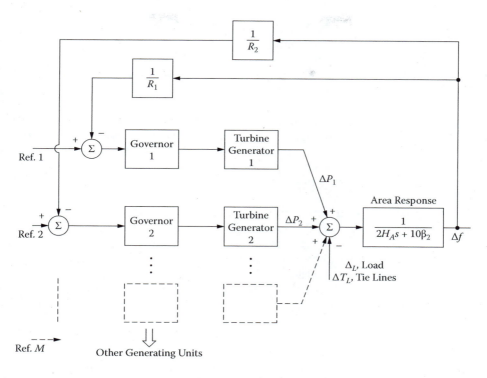

FIGURE 1.23
Equivalent model for generation and load within a power area.

value theorem of Laplace transforms as

$$\left(\frac{\Delta f}{\Delta_A}\right)_{s.s.} = \frac{1}{10\beta_1 - 10\beta_2} \tag{1.22}$$

where the droop characteristics of the M generating units combine according to Equation 1.19:

$$-10\beta_1 = \frac{P_{1\,rate}}{R_1} + \frac{P_{2\,rate}}{R_2} + \cdots + \frac{P_{M\,rate}}{R_M} \tag{1.23}$$

With this definition, the static response of the dynamic system of Figure 1.22 is consistent with Equation 1.5 for an area. The dynamic system is assumed to be a linearized, small-perturbation model valid for ±5% variations about the base generation.

The area dynamic response is approximated by a single time constant, $\tau = 2H_A/10\beta_2$. This response assumes that with no turbine generator action, if the load suddenly changes from base value by an increment Δ_A, the frequency deviation would exponentially reach its final value:

$$\Delta f = \frac{\Delta_A}{10\beta_2}(1 - e^{-t/\tau}) = \frac{\Delta_A}{10\beta_2}(1 - e^{-10\beta_2 t/2H_A}) \tag{1.24}$$

where t is measured in seconds. Numerical values for β_2 and H_A, which are measured by such a test, are difficult to predict for a power system area. The linear small-perturbation transfer functions for the governors and turbine generators are available from design data and unit experimental tests. As a result, the governor–turbine generator models are well known for fossil-fired units, hydro stations, combined cycle and nuclear, so that they may be modeled with any desired degree of accuracy ranging from a single time constant to multipath transfer functions, including deadbands and nonlinearities. The slower-response elements, such as boilers, dominate the governor–turbine generator transfer functions. For example, a drum-type steam unit with a turbine-leading control arrangement has its slowest time constant on the order of seconds, while a boiler-leading unit has pertinent time constants on the order of minutes. Integrated boiler–turbine control units and hydro stations range between these extremes for small disturbances.

Next, consider the isolated power area where the central computer has an AGC sensing only frequency in Equation 1.1. This is called flat-frequency control of the power system. Assume that the central computer command rate to each generator i is approximately a continuous signal $G_i'(s)$, as shown in Figure 1.24. In order for the AGC of the central computer to allocate the frequency error as a change in generation among units, the equivalent transfer function and gain in cascade with each turbine generator appear in the overall frequency response for a load change. A direct reduction of the block diagram by Mason's rule [6] for block diagrams yields the transfer function

$$\frac{\Delta f(s)}{\Delta L(s)} = \frac{-1}{(2H_A S + 10\beta_2) + \sum_{i=1}^{M}[G_i(s)/R_i] + 10B[\sum_{i=1}^{M}G_i'(s)G_i(s)]} \tag{1.25}$$

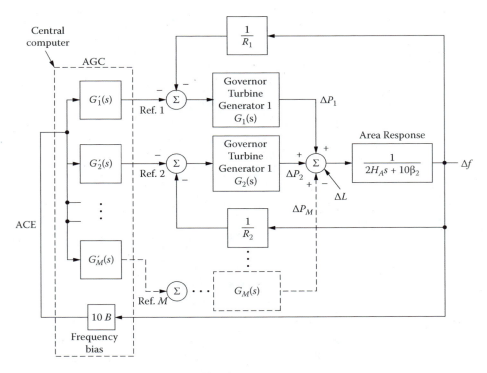

FIGURE 1.24
Isolated power area, AGC regulating frequency.

The general behavior of this transfer function is dependent on the complexity of the G_i and G_i', but for steady-state values assume the following:

$$G_i(s)\big|_{s=0} = 1 \equiv \text{per unit value for unit } i \tag{1.26}$$

$$G_i'(s)\big|_{s=0} = K_i \geq 0, \text{ gain allocation to unit } i \tag{1.27}$$

where the relative generating capacity of different units is included in R_i and K_i. With these assumptions G_i' is a type 0, or proportional AGC, so that the final value theorem of Laplace transforms yields the following for step changes in the load, $\Delta L(s) = \Delta_A/S$:

$$\lim_{S \to 0} \left\{ \frac{S\Delta f(s)}{\Delta L(s)} \right\} = \left(\frac{\Delta f}{\Delta_A} \right)_{s.s} = \frac{1}{10\beta_2 + \Sigma_{i=1}^M (1/R_i) + 10B\Sigma_{i=1}^M K_i} \tag{1.28}$$

In Equation 1.28 it is evident that an AGC sensing the frequency error additively contributes to the natural regulation of the area since the frequency change is less for a given disturbance. This contribution of the AGC is often called a supplemental control [7], and its effect on the system dynamic response time is dependent on $G_i'(s)$ and $G_i(s)$. For the purposes of illustration only, a sudden change in the load affects the area as shown schematically in Figure 1.25. The timescale is not defined in this figure, and the interval between sensing the error and corrective action by the AGC is expanded, but the effect of the bias, B, is evident. The entire AGC dynamic response, including ACE filtering and time lags associated with digital computer system scan rates, is included in $G_i'(s)$ and its digital equivalent.

If the AGC introduces an integrator into the loop, such that

$$G_i'(s) = \frac{K_i}{S} \text{ as } S \to 0$$

then the final value theorem of Laplace transforms yields

$$\lim_{S \to 0} \left\{ \frac{S\Delta f(s)}{\Delta L(s)} \right\} \sim \lim_{S \to 0} \left\{ \frac{1}{10\beta_2 + \Sigma_{i=1}^M 1/R_i + \frac{10B}{S} \Sigma_{i=1}^M K_i} \right\} = 0$$

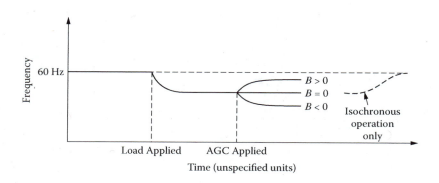

FIGURE 1.25
Representation of load and AGC effects.

which would make the power system isochronous, or force the system back to base frequency following a disturbance, as shown in Figure 1.25. An isolated system may have isochronous operation, but it is unacceptable for interconnected operation because the droop characteristic is *essential* for load sharing.

In the eastern U.S. power interconnection, the 24 h accumulated time error due to underfrequency operation (or overfrequency operation) is corrected at midnight EST by increasing (decreasing) the desired frequency f_0 in Equation 1.1 for a period of time. The western U.S. interconnection continuously introduces a time correction term into Equation 1.1. However, in both interconnections, the steady-state frequency remains very close to 60 Hz.

In subsequent chapters it is assumed the generator droop characteristics have forced the system into 60 Hz steady-state power flow condition. In the steady state, the individual line power flows and network state may be determined from voltages, loads, and line parameters. Initially, an isolated power area, or one in which the tie lines have been replaced by an equivalent load, is considered. This simplification limits the number of lines and elements to a reasonable size to begin the analysis of power systems.

Problems

1.1. A power system has a peak-load demand of 1,000 MW and is supplying this peak with the on-line units specified below. The area load is assumed to have a 1% change per 0.1 Hz frequency deviation. Determine the composite generation–load-frequency dependence for the area. Normalize the result to a 1,000 MW base. There is no automatic generation control (AGC).

Type of Unit	Nominal Rating (MW)	Regulation (%)
Coal fired	337	5
Coal fired	420	6
Oil burning	100	4
Oil burning	100	4
Oil burning	100	4
Natural gas (peaking)	50	1
Natural gas (peaking)	50	1

1.2. Areas *A*, *B*, and *C* (shown in Figure P1.2) comprising a power pool initially have zero tie-line flow before a 1,050 MW load decrease occurs in area *B*. Each area has AGC-assisted frequency regulation, which results in 1.5 times its combined load and generator natural characteristic. The frequency is 60.0 Hz before the load loss. Data for the areas are as follows:

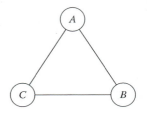

FIGURE P1.2

Area *A*:

Base generation = 10,000 MW

Natural regulation = 1% of base per 0.1 Hz

Area *B*:

Base generation = 7,000 MW

Natural regulation = 2% per 0.1 Hz

Area C:

Base generation = 11,500 MW

Natural regulation = 4% per 0.1 Hz

(a) What is the system frequency after the load loss?

(b) What net tie-line flow (MW) does each pool member have after the disturbance?

© Use the *dc_flow* program to compute the inter-area power flows for areas *A* and *D* of Example 1.3 after there has been a 10% load in area *D* and a frequency droop. The two-area base case is introduced by means copying the file ***example13.sss*** into *dc_flow.sss* to start the computations. For a 10% load disturbance in area *D*, increase the loads by 10% on buses 2 and 4 of area *D* from –5 MW to –5.5 MW. This corresponds to changes from –0.05 to –0.055 p.u. Generation in area *D* increases by .00069 p.u. due to the frequency droop, which corresponds to an increase to 10.069 MW for the generation at bus 3 in the *dc_flow* program. Execute *dc_flow* to obtain the results, and scale the results to a 40,000 MW base, as per Example 1.3c.

1.3. The purpose of this extended problem is to examine models for power areas without AGC, with proportional AGC regulating frequency, with isochronous AGC, and with a two-area interconnection. Time histories are to be calculated for 2 min, but the time increment should be on the order of 0.1 sec or $\frac{1}{50}$ of the smallest time constant of any dynamic element. Use any method available, including writing your own time-stepping program or using "canned" computer dynamic programs such as EMTP, Simplorer, Simulink, or MATLAB®. It is also possible, but not recommended, to use computer library programs to factor characteristic polynomials, perform a partial-fraction expansion, and compute time points from the time equivalents.

(a) Obtain the 2 min time history for the elementary generation-area model shown in Figure P1.3.1, for a 0.1 p.u. load increase. Plot Δ*P* and Δ*f* versus time.

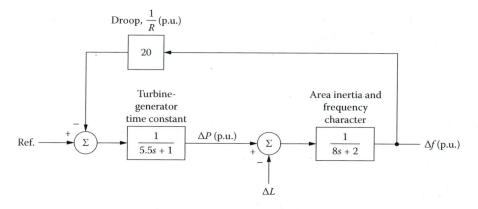

FIGURE P1.3.1

(b) Proportional frequency control, called bias, is added by means of an AGC as shown in Figure P1.3.2. Calculate the 2 min response for a 0.1 p.u. load increase. Plot ΔP and Δf versus time.

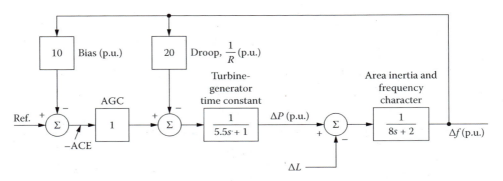

FIGURE P1.3.2

(c) Replace the AGC block by the Laplace term $0.05/S$, which performs an integration of frequency error. Calculate and plot ΔP, Δf versus time for a 0.1 p.u. load increase. Observe that this change makes the AGC an isochronous, or constant-frequency, type of control.

(d) The area is now ~99% of the interconnection shown in Figure P1.3.3, which can share disturbances. Plot a 2 min time history of ΔP_1, ΔP_2, interchange

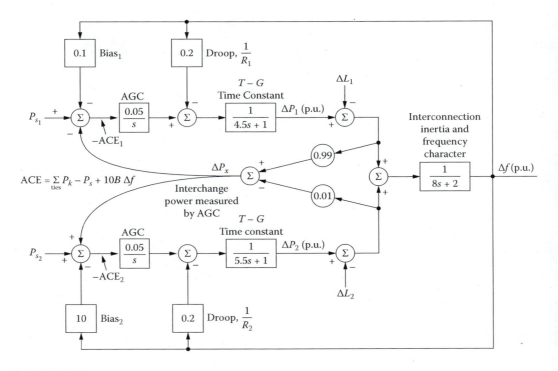

FIGURE P1.3.3

power ΔP_x, and both area control errors (ACE$_1$, ACE$_2$) when a 0.1 p.u. disturbance occurs in area 2. In other words, $\Delta L_2 = 0.1$ p.u.

© Use the MATLAB computational program to obtain transient responses for problem 1.3. The files are *prob1.3a.mdl*, *prob1.3b.mdl*, and *prob1.3d.mdl*. When MATLAB is called, each file presents the simulated block diagram. Start the simulation and double-click the mouse on a SCOPE element to observe the computed time history response at that point in the block diagram. When the time history is presented, it may be necessary to use the autoscale option in order to observe the entire response (right mouse-click to obtain the autoscale option).

1.4. Central digital computer control of the generators on a power system is called automatic generation control (AGC). The computer is employed in a periodic mode, measuring a generator's output every T seconds and commanding a new power level every T seconds. It is essentially a sampled-data control. Local servomechanisms on the turbine generator ensure that the unit has reasonable response to power demand changes. A typical computer-generating unit control loop is shown in Figure P1.4. Z^{-1} corresponds to a delay of one sampled-data point in terms of Z-transform theory. As part of the AGC design process, the properties of the basic control loop must be known. For $T = 8$ sec, do the following:

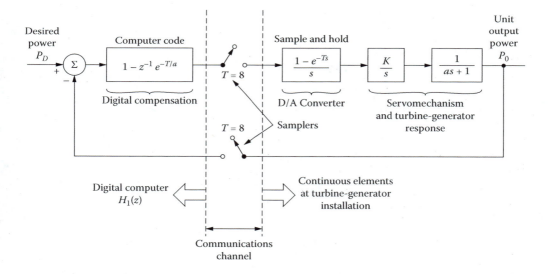

FIGURE P1.4

(a) Determine the largest value of K that is possible for a stable loop if the time constant of the unit is $a = 8$ sec and $a = 105$ sec. Convert continuous elements to Z-transforms and analyze as a numerical process.

(b) Compute the time history of the output power for a step change in power demand for both $a = 105$ sec and $a = 7$ sec. Use the value $K = 0.125$, which converts a 1 MW computer power demand into a 1 MW servomechanism change at the unit by means of integrating for 8 sec.

References

1. Cohn, N. 1966. *Control of generation and power flow on interconnected systems.* New York: John Wiley & Sons.
2. Kirchmayer, L. K. 1959. *Economic control of interconnected systems.* New York: John Wiley & Sons.
3. Jaleeli, N., and VanSlyck, L. S. 1999. NERC's new control performance standards. *IEEE Trans. Pow. Syst.* 14(3).
4. Yao, M., Shouts, R. R., and Kelm, R. 2000. AGC logic based on NERC's new control performance standard and disturbance control standard. *IEEE Trans. Pow. Syst.* 15(2).
5. DeMello, F. P., and Undrill, J. M. 1977. Automatic generation control. IEEE tutorial text. Summer power meeting.
6. Franklin, G. F., and Powell, J. D. 1980. *Digital control of dynamic systems.* Reading, MA: Addison-Wesley Publishing Co.
7. Cohn, N. 1957. Some aspects of tie-line bias control on interconnected power systems. *AIEE Transactions*, February.

2

Elements of Transmission Networks

Analysis of power systems is based on mathematical representations for the electrical components used to generate power, distribute it, and consume it. In general, linear representations are used to study the behavior of the circuits due to small disturbances, then when appropriate, nonlinear models are employed to investigate response due to large disturbances. The linear models permit theory and principles to be developed and often are sufficiently accurate to correlate physically measured data with theoretical results from the linear model.

To develop a vocabulary for circuit elements, examine Figure 2.1, which shows some elements of a transmission network connected in a *positive-sequence* or *one-line diagram*. The circuit breakers, switchgear, and protective relays are not shown on this diagram because emphasis is on the elements that affect the steady-state transmission of power. All physical elements represented in the diagram are assumed to be three-phase and *balanced*, in the sense that impedances are equal in all phases, even though it is possible to carry out the subsequent analysis for the more detailed unbalanced case. After preliminary definitions of phasors and symmetrical components, this chapter develops linear electrical circuit descriptions of common ac transmission circuit components. Some power system terminology for Figure 2.1 is as follows:

Bus. A common connection point for two or more three-phase elements of the power generation, transmission, distribution, and consumption network.

Bus charging capacitance. The capacitance that is present from phase conductors to earth ground in substations or switchyards. This is due to geometric factors in circuit breakers, switches, and proximity of the conductors to the earth plane.

FACTS. Flexible ac transmission system. Electronic devices on the power system to control the flow of real or reactive power. Two of these devices are:

1. *TCUL.* Tap-changing-under-load transformer. A transformer whose ratio of primary to secondary terminal voltage can be varied mechanically or by electronic switching to changing power demands. Phase-shift transformers are in this category to advance or retard the phase angle of input with respect to output voltage by means of selectively adding series voltages from other phases.

2. *SVC.* Static var compensator. Injects variable amounts of reactive power into the network from fixed capacitors and controlled inductors. The fundamental frequency component of current through the inductor is controlled by solid-state electronic thyristors or IGBT devices.

Shunt reactor. An inductive current element connected from line to neutral to compensate for capacitive current from transmission lines or underground cables.

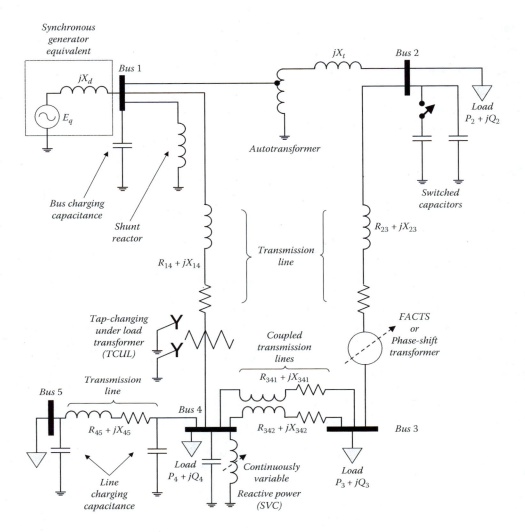

FIGURE 2.1
Positive-sequence diagram for a transmission network.

Switched capacitor. A capacitor bank whose segments can be connected to the network either automatically or manually, as needed, for power factor correction.

Load. The apparent power demand, in reactive megavolt-amperes and real power that is connected to this point on the power network. This is a convenient way to summarize the electrical effect of many circuits that may emanate from this point.

2.1 Phasor Notation

It is necessary to introduce phasor notation to simplify steady-state analysis and provide the necessary transforms that may approximate transient analysis. Let phase *a* of the three-phase voltages shown in Figure 2.2a be selected as a time reference, and the sinusoidal

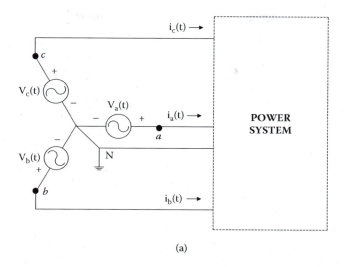

(a)

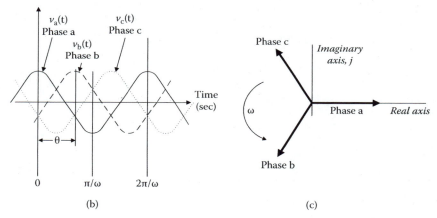

(b)   (c)

FIGURE 2.2
(a) Electrical schematic for three-phase voltage source. (b) Phasor representation of balanced, three-phase positive sequence.

line-to-neutral voltages for an *a, b, c* sequence, called the positive sequence, in the time domain are:

$$\text{Phase } a: \quad v_a(t) = \sqrt{2}VCos(\omega t) \tag{2.1a}$$

$$\text{Phase } b: \quad v_b(t) = \sqrt{2}VCos(\omega t - 2\pi/3) \tag{2.1b}$$

$$\text{Phase } v: \quad v_c(t) = \sqrt{2}VCos(\omega t + 2\pi/3) \tag{2.1c}$$

If the ±120° phase shift for phases *b* and *c* is interchanged, this is called a negative-sequence rotation: *a, c, b*. The units of *V* are root mean square (rms) volts, ω is in radians per second, and *t* is the time in seconds. When the voltage magnitudes are the same and the angle between phases is $2\pi/3$, as shown above, these comprise a set of *balanced* three-phase voltages. The time-dependent voltages of Figure 2.2b may be considered as projections onto the real axis of vectors that rotate at ω rad/sec, as shown in Figure 2.2c. These are *phasors*.

Projections on the j-axis, $j = \sqrt{-1}$, or imaginary axis are called the out-of-phase components with respect to the selected reference.

It is customary in power systems to express voltages as V (volts), the rms value, and currents as I (amperes), the rms value.

Let $\sqrt{2}V$ correspond to the length of the rotating vector. Euler's exponential $e^{jx} = \cos(x) + j\sin(x)$ is used to translate the rotating vector into the time domain by means of the real part according to the following definition:

DEFINITION 2.1

$$v(t) = \sqrt{2}V\text{Cos}(\omega t + \theta) = \Re\left\{v_{max}e^{j(\omega t + \theta)}\right\} = \Re\{\sqrt{2}Ve^{j\theta}e^{j\omega t}\}$$

The complex number $E = V\angle\theta = Ve^{j\theta} = (v_{max}/\sqrt{2})(\text{Cos}\theta + j\text{Sin}\theta)$ is called a *phasor*. At time $t = 0$, it is at the angle θ with respect to the positive real axis reference. For example, $\theta = -120°$ in Figure 2.2c for phase (b). Phasors are added, subtracted, multiplied, and divided according to rules established in basic textbooks of electrical engineering. A phasor is converted into a time-domain function according to definition 2.1. In terms of rotating vectors or phasors, the representation for the balanced, three-phase voltage is

$$v_a(t) \rightarrow E^a = V(1 + j0) = V\angle 0° \tag{2.2a}$$

$$v_b(t) \rightarrow E^b = V(-1/2 - j\sqrt{3}/2) = V\angle -120° \tag{2.2b}$$

$$v_c(t) \rightarrow E^c = V(-1/2 + j\sqrt{3}/2) = V\angle +120° \tag{2.2c}$$

where j is a positive phase operator, or projection on the imaginary axis. The arrow implies correspondence. Let **a** be defined as a phase-shift operator in terms of a Euler exponential:

$$\mathbf{a} = e^{j2\pi/3} = (-1/2 + j\sqrt{3}/2) = \text{Cos}(2\pi/3) + j\text{Sin}(2\pi/3) = 1\angle 120° \tag{2.3}$$

$$\mathbf{a}^2 = \mathbf{a}\,\mathbf{a} = e^{j4\pi/3} = (-1/2 - j\sqrt{3}/2) = 1\angle 240° \tag{2.4}$$

Then phasor representations for voltages $v_a(t)$, $v_b(t)$, and $v_c(t)$ may be written in vector form for a positive sequence as

$$E^{abc} = \begin{bmatrix} E^a \\ E^b \\ E^c \end{bmatrix} = \begin{bmatrix} 1 \\ \mathbf{a}^2 \\ \mathbf{a} \end{bmatrix}[E^a] \tag{2.5}$$

If the currents $i_a(t)$, $i_b(t)$, and $i_c(t)$ of Figure 2.2a have equal magnitudes and a $2\pi/3$ phase angle with respect to each other, they are called *balanced* three-phase currents. They may also be expressed as phasors similar to Equation 2.2a to c. Let ϕ be the phase angle between the currents and the voltage reference; then

$$i_a(t) = \sqrt{2}I\text{Cos}(\omega t + \phi) \rightarrow I^a = Ie^{j\phi} = I(\text{Cos}(\phi) + j\text{Sin}(\phi)) \tag{2.6a}$$

$$i_b(t) = \sqrt{2}I\text{Cos}(\omega t + \phi - 120°) \rightarrow I^b = Ie^{j(\phi-120°)} = I(\text{Cos}(\phi - 120°) + j\text{Sin}(\phi - 120°)) \tag{2.6b}$$

$$i_c(t) = \sqrt{2}I\text{Cos}(\omega t + \phi + 120°) \rightarrow I^c = Ie^{j(\phi +120°)} = I(\text{Cos}(\phi + 120°) + j\text{Sin}(\phi + 120°)) \tag{2.6c}$$

In terms of a vector, the rms phasor currents are expressed as

$$I^{abc} = \begin{bmatrix} 1 \\ \mathbf{a}^2 \\ \mathbf{a} \end{bmatrix} [Ie^{j\phi}] \qquad (2.7)$$

A useful property of balanced voltages and balanced currents is that their sum equals zero:

$$v_a(t) + v_b(t) + v_c(t) = 0 \rightarrow E^a + E^b + E^c = V + \mathbf{a}^2 V + \mathbf{a}V = 0 \qquad (2.8a)$$

$$i_a(t) + i_b(t) + i_c(t) = 0 \rightarrow I^a + I^b + I^c = Ie^{j\phi} + \mathbf{a}^2 Ie^{j\phi} + \mathbf{a}Ie^{j\phi} = 0 \qquad (2.8b)$$

A consequence of this property is that the neutral conductor shown in Figure 2.2a can be removed since it does not conduct current in balanced operating conditions. The neutral wire can be eliminated with a savings of ¼ the wire required. This is one of the main advantages of three-phase electrical systems. Many three-phase elements do not have a neutral connection, so the sum of the phase currents is always zero.

The line-to-line voltage is considered to be the **rated** voltage of a system. From Equations 2.2a,b,c, with the first subscript as the positive terminal, the line-to-line voltages are

$$E_{ab} = E^a - E^b = V - V(-1/2 - j\sqrt{3}/2) = \sqrt{3}V(\sqrt{3}/2 + j1/2) = \sqrt{3}V\angle+30° \qquad (2.9a)$$

$$E_{bc} = E^b - E^c = V(-1/2 - j\sqrt{3}/2) - V(-1/2 + j\sqrt{3}/2) = V(-j\sqrt{3}) = \sqrt{3}V\angle-90° \qquad (2.9b)$$

$$E_{ca} = E^c - E^a = V(-1/2 + j\sqrt{3}/2) - V = \sqrt{3}V(-\sqrt{3}/2 + j1/2) = \sqrt{3}V\angle+150° \qquad (2.9c)$$

Observe that the rms line voltages are $\sqrt{3}$ times the rms line-to-neutral voltages and comprise a balanced three-phase set. If the voltage E^a is used as the reference, the line-to-line voltages have a +30° phase shift with respect to the line-to-line neutral. Sometimes a device voltage is specified with both line-to-line and line-to-neutral voltages as 208/120 or 23kV/13.8 kV.

When sinusoidal, steady-state current flows through a linear resistor, inductor, or capacitor as shown in Figure 2.3, the integro-differential equations of Kirchhoff's voltage law for the elements may be written as phasor currents and impedances:

$$v_R = Ri(t) \rightarrow V_R = RIe^{j\phi} = RI^a \qquad (2.10a)$$

$$v_L = Ldi/dt = \sqrt{2}LI\frac{d}{dt}[Cos(\omega t + \phi)] = -\omega\sqrt{2}LISin(\omega t + \phi) \rightarrow V_L = j\omega LI^a = jX_L I^a \qquad (2.10b)$$

$$v_C = \frac{1}{C}\int i(t)\,dt = \frac{\sqrt{2}}{C}\int ICos(\omega t + \phi)dt = \sqrt{2}\frac{I}{\omega C}Sin(\omega t + \phi) \rightarrow V_C = \frac{-j}{\omega C}I^a = -jX_C I^a \qquad (2.10c)$$

where X_L and X_C are termed inductive and capacitive reactances, respectively. The *impedance* between points a and b of Figure 2.3 is the phasor relationship

$$Z = \frac{V_{ab}}{I^a} = \frac{V_R + V_L + V_C}{I^a} = R + jX_L - jX_C = R + jX \quad \text{ohms} \qquad (2.11a)$$

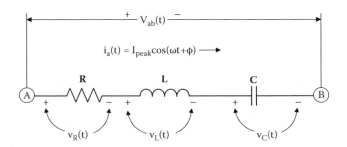

FIGURE 2.3
Circuit polarities for Kirchhoff's voltage law.

The inverse of the impedance, called an *admittance*, has real and imaginary parts, $G + jB$, termed *conductance* and *susceptance*, respectively:

$$Y = \frac{I^a}{V_{ab}} = \frac{1}{R + jX_L - jX_C} = G + jB \quad \text{siemens} \qquad (2.11b)$$

where a positive susceptance implies that capacitive reactance is numerically larger than the inductive reactance. Subsequent use of the symbols R, X, G, and B refers to phasor impedance quantities. Y is defined as an *admittance*.

2.2   Symmetrical Component Transformation

Consider a portion of a balanced three-phase circuit with a,b,c phase rotation as shown in Figure 2.4. The three-phase impedance has elements with mutual impedance equal to Z^m and series impedance Z^s. Currents and voltages of phases b and c may be expressed in terms of the phase-shift operator, $\mathbf{a}$. Let the vector representation of the internal source be e^{abc}, so a three-phase form of Kirchhoff's voltage law, $E^{abc} = e^{abc} + Z^{abc}I^{abc}$, applied to this circuit is

$$\begin{bmatrix} 1 \\ \mathbf{a}^2 \\ \mathbf{a} \end{bmatrix}[E^a] = \begin{bmatrix} 1 \\ \mathbf{a}^2 \\ \mathbf{a} \end{bmatrix}[e^a] + \begin{bmatrix} Z^s & Z^m & Z^m \\ Z^m & Z^s & Z^m \\ Z^m & Z^m & Z^s \end{bmatrix}\begin{bmatrix} 1 \\ \mathbf{a}^2 \\ \mathbf{a} \end{bmatrix}[I^a] \qquad (2.12)$$

Multiplying Equation 2.12 by the row vector

$$[\,1\,\mathbf{a}^2\,\mathbf{a}\,]^* = [1\,\mathbf{a}\,\mathbf{a}^2\,]$$

where the asterisk represents complex conjugate, yields

$$[1\,\mathbf{a}\,\mathbf{a}^2]\begin{bmatrix} 1 \\ \mathbf{a}^2 \\ \mathbf{a} \end{bmatrix}[E^a] = 3E^a = 3e^a + 3(Z^s - Z^m)I^a \qquad (2.13)$$

Therefore, current flow, potential distributions, and power flow in balanced conditions may be determined using only phase a. The impedance

$$Z_1 = (Z^s - Z^m) \qquad (2.14)$$

is called the *positive-sequence impedance*.

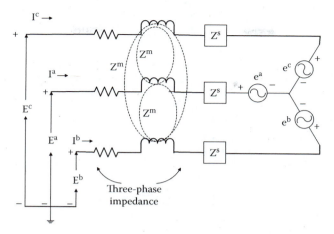

FIGURE 2.4
Three-phase balanced source and impedance.

Example 2.1

Find the single-phase equivalent for the balanced network shown in Figure E2.1a.

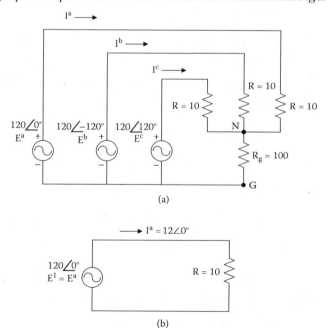

(a)

(b)

FIGURE E2.1
Balanced impedance with grounding resistor.

Solution

Kirchhoff's voltage law around the loop in each phase is:

$$
\begin{bmatrix} E^a \\ E^b \\ E^c \end{bmatrix} = \begin{bmatrix} R+R_g & R_g & R_g \\ R_g & R+R_g & R_g \\ R_g & R_g & R+R_g \end{bmatrix} \begin{bmatrix} I^a \\ I^b \\ I^c \end{bmatrix} = \begin{bmatrix} Z^s & Z^m & Z^m \\ Z^m & Z^s & Z^m \\ Z^m & Z^m & Z^s \end{bmatrix} \begin{bmatrix} I^a \\ I^b \\ I^c \end{bmatrix}
$$

Using Equation 2.13, the voltage and current in phase a are

$$E^a = (Z^s - Z^m)I^a = (R + R_g - R_g)I^a = RI^a$$

The equivalent circuit for the balanced network is shown in Figure E2.1b. The grounding resistor R_g has zero current and no voltage drop because $I^a + I^b + I^c = 0$, and it does not appear in the balanced equivalent.

For general three-phase voltages and unbalanced impedances, the symmetric component transformation (SCT) matrix is defined as

$$T_s = \begin{bmatrix} 1 & 1 & 1 \\ 1 & \mathbf{a}^2 & \mathbf{a} \\ 1 & \mathbf{a} & \mathbf{a}^2 \end{bmatrix} \tag{2.15}$$

The SCT has the following properties:

$$\frac{1}{3}T_s^* = T_s^{-1} \tag{2.16a}$$

$$\left[T_s^*\right]^t T_s = 3U \tag{2.16b}$$

where the t superscript denotes transpose and U is a real diagonal matrix of 1's.

The SCT is used to transform the three-phase phasor system from a, b, c components to a system of rotating vectors called zero sequence, positive sequence, and negative sequence. The a, b, c voltages are transformed as

$$E^{012} = \begin{bmatrix} E^0 \\ E^1 \\ E^2 \end{bmatrix} = T_s^{-1} \begin{bmatrix} E^a \\ E^b \\ E^c \end{bmatrix} = \frac{1}{3} \begin{bmatrix} E^a + E^b + E^c \\ E^a + \mathbf{a}E^b + \mathbf{a}^2 E^c \\ E^a + \mathbf{a}^2 E^b + \mathbf{a} E^c \end{bmatrix} \tag{2.17a}$$

Notice that the zero-sequence component, E^0, is the sum of phase voltages and hence indicates the unbalance in the voltages. The positive-sequence component, E^1, is the result of shifting phases b and c to align with the phase a voltage. For balanced voltages,

$$E^{012} = \begin{bmatrix} E^0 \\ E^1 \\ E^2 \end{bmatrix} = \begin{bmatrix} 0 \\ E^a \\ 0 \end{bmatrix} \quad \text{(balanced)} \tag{2.17b}$$

The negative-sequence component, E^2, is a negative or a, c, b rotation system. For balanced conditions the negative sequence is zero.

Currents in the a, b, c system are similarly transformed:

$$I^{012} = \begin{bmatrix} I^0 \\ I^1 \\ I^2 \end{bmatrix} = T_s^{-1} \begin{bmatrix} I^a \\ I^b \\ I^c \end{bmatrix} = \frac{1}{3} \begin{bmatrix} I^a + I^b + I^c \\ I^a + \mathbf{a}I^b + \mathbf{a}^2 I^c \\ I^a + \mathbf{a}^2 I^b + \mathbf{a} I^c \end{bmatrix} \tag{2.18a}$$

For balanced currents only the positive-sequence current is present:

$$I^{012} = \begin{bmatrix} I^0 \\ I^1 \\ I^2 \end{bmatrix} = \begin{bmatrix} 0 \\ I^a \\ 0 \end{bmatrix} \quad \text{(balanced)} \tag{2.18b}$$

If the SCT is applied to Kirchhoff's voltage law for the circuit of Figure 2.4, the result is

$$T_s^{-1}E^{abc} = E^{012} = T_s^{-1}[e^{abc} + Z^{abc}I^{abc}] = e^{012} + T_s^{-1}Z^{abc}T_s I^{012} = e^{012} + Z^{012}I^{012} \tag{2.19a}$$

In Equation 2.19a, new impedances, Z^{012}, have been defined in the symmetrical component reference frame. The SCT transforms *balanced three-phase elements* into

$$Z^{012} = T_s^{-1}Z^{abc}T_s = \frac{1}{3}\begin{bmatrix} 1 & 1 & 1 \\ 1 & a & a^2 \\ 1 & a^2 & a \end{bmatrix}\begin{bmatrix} Z^s & Z^m & Z^m \\ Z^m & Z^s & Z^m \\ Z^m & Z^m & Z^s \end{bmatrix}\begin{bmatrix} 1 & 1 & 1 \\ 1 & a^2 & a \\ 1 & a & a^2 \end{bmatrix}$$

$$= \begin{bmatrix} Z^s + 2Z^m & 0 & 0 \\ 0 & Z^s - Z^m & 0 \\ 0 & 0 & Z^s - Z^m \end{bmatrix} = \begin{bmatrix} Z_0 & 0 & 0 \\ 0 & Z_1 & 0 \\ 0 & 0 & Z_2 \end{bmatrix} \quad \text{(balanced three-phase impedances)}$$

$$\tag{2.19b}$$

Z_0, Z_1, Z_2 are respectively known as the zero-sequence, positive-sequence, and negative-sequence impedances. The sequence impedances are also *decoupled* from each other for balanced elements, i.e., the Z^{012} matrix is diagonal.

A fundamental feature of the SCT is that for balanced systems, only the phase a equivalent is needed. Complex mutually coupled a, b, c components are simpler to treat. For unbalanced elements, the SCT does not simplify calculations.

Because the positive-sequence voltage and current in balanced conditions represent only phase a, a conversion factor is required. The factor is 3.0, as may be seen by the following phasor power computation for any a, b, c system, whether balanced or unbalanced:

$$\boxed{\begin{aligned} S^{abc} &= P^{abc} + jQ^{abc} = [(I^{abc})^*]^t E^{abc} = E^a(I^a)^* + E^b(I^b)^* + E^c(I^c)^* \\ &= [(T_s I^{012})^*]^t T_s E^{012} = 3[(I^{012})^*]^t E^{012} = 3E^0(I^0)^* + 3E^1(I^1)^* + 3E^2(I^2)^* = 3S^{012} \end{aligned}} \tag{2.20}$$

There are other transforms used in power systems, but the symmetrical component transformation is the most common. It is subsequently used for the unbalanced fault-current calculations on networks with balanced supply voltages. An example will demonstrate its usefulness in balanced networks.

Example 2.2

Derive the symmetrical component impedances and power for balanced a, b, c phase voltages applied to the balanced impedance of Figure 2.5a.

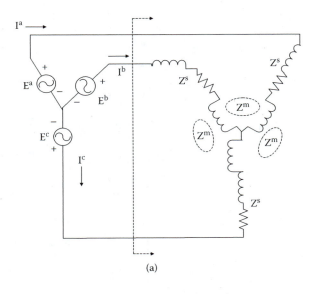

(a)

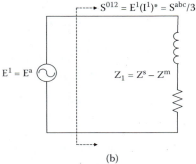

(b)

FIGURE 2.5
Positive-sequence equivalent for balanced excitation on balanced load.

Solution

The sequence voltages and the currents are given by Equation 2.17a and Equation 2.18a, respectively. For balanced *a*, *b*, *c* voltages, only the positive-sequence component is present and numerically equal to the line-to-neutral voltage magnitude:

$$E^{012} = \begin{bmatrix} 0 \\ E^1 \\ 0 \end{bmatrix} = T_s^{-1} E^{abc}$$

A similar transformation for balanced *a*, *b*, *c* currents yields

$$I^{012} = \begin{bmatrix} 0 \\ I^1 \\ 0 \end{bmatrix} = T_s^{-1} I^{abc}$$

The positive-sequence impedance for a balanced impedance is

$$Z_1 = Z^s - Z^m$$

Positive and negative impedances are equal for a balanced element because the element responds to an a, b, c phase rotation in the same way that it does for an a, c, b rotation.

Kirchhoff's voltage law for the element is

$$E^{012} = \begin{bmatrix} 0 \\ E^1 \\ 0 \end{bmatrix} = Z^{012} I^{012} = \begin{bmatrix} Z^s + 2Z^m & 0 & 0 \\ 0 & Z^s - Z^m & 0 \\ 0 & 0 & Z^s - Z^m \end{bmatrix} \begin{bmatrix} 0 \\ I^1 \\ 0 \end{bmatrix}$$

The positive-sequence current is found as

$$I^1 = \frac{E^1}{Z^1} = \frac{E^a}{Z^s - Z^m}$$

Only positive-sequence components are present, so the power in the sequence coordinates is

$$S^{012} = P^{012} + jQ^{012} = E^1 (I^1)^* = \frac{E^1 (E^1)^*}{(Z^s - Z^m)^*} = \frac{E^a (E^a)^*}{(Z^s - Z^m)^*}$$

According to Equation 2.20a, the power flow in a, b, c coordinates is

$$S^{abc} = 3S^{012} = 3E^1 (I^1)^* = \frac{3E^a (E^a)^*}{(Z^s - Z^m)^*}$$

Thus, for balanced sources and balanced circuit impedances, only phase a needs to be considered.

Figure 2.5b summarizes the positive-sequence (a, b, c) components of voltage and current for balanced three-phase voltages and currents. Except for rotating machinery, whose direction of rotation depends on the phase sequence, balanced electrical elements respond the same for a positive a, b, c sequence as they do for negative a, c, b phase sequences. Hence, the positive- and negative-sequence impedances are generally equal. For balanced networks, Z^{012} is diagonal and separate circuits can be used for each sequence component. The circuit configurations in the third row of Table 2.1 and the first row of Table 2.2 demonstrate this property. Observe the neutral impedances in these tables.

© In the enclosed computer software, the program *zabczpn* computes the symmetrical component impedances from the a, b, c elements. Use this program to calculate the 0, 1, 2 components for the values $Z^s = Z + Z_g = 2 + j2$, and $Z^m = Z_g = 0 + j1$ from the balanced a, b, c matrix:

$$Z^{abc} = \begin{bmatrix} 2 + j2 & j1 & j1 \\ j1 & 2 + j2 & j1 \\ j1 & j1 & 2 + j2 \end{bmatrix}$$

Observe the input to the program is $Z = Zan = Zbn = Zcn = 2 + j$ and $Zng = j$. The results are presented as Z^{012} components in the computer program to confirm the shunt elements in the third row of table 2.1. Open circuits are very large-valued impedances.

© The computer program *zpnzabc* computes Z^{abc} components from Z^{012} components. Use this program with $Z_0 = Z_1 = Z_2 = 1 + j$ to verify the first row of table 2.1

Observe that an infinite impedance may be approximated in *zabczpn* or *zpnzabc* by using large numerical values for the impedances, say $Z = j100000$, and interpreting the results accordingly.

TABLE 2.1

Z^{abc} to Z^{012} Conversions

Z^{abc}	Z^{abc} Schematic	Z^{012}	Z^{012} Schematic
Infinite Impedances to Neutral		Infinite Impedances to Neutral	
Infinite Impedances to Neutral		Infinite Impedances to Neutral	
$\begin{bmatrix} Z + Z_g & Z_g & Z_g \\ Z_g & Z + Z_g & Z_g \\ Z_g & Z_g & Z + Z_g \end{bmatrix}$		$\begin{bmatrix} Z + 3Z_g & 0 & 0 \\ 0 & Z & 0 \\ 0 & 0 & Z \end{bmatrix}$	
$\begin{bmatrix} Z_A + Z_g & Z_g & Z_g \\ Z_g & Z_B + Z_g & Z_g \\ Z_g & Z_g & Z_C + Z_g \end{bmatrix}$		$\begin{bmatrix} S + 3Z_g & M1 & M2 \\ M2 & S & M1 \\ M1 & M2 & S \end{bmatrix}$ $a = e^{j2\pi/3}$ $S = \dfrac{Z_A + Z_B + Z_C}{3}$ $M1 = \dfrac{Z_A + a^2 Z_B + a Z_C}{3}$ $M2 = \dfrac{Z_A + a Z_B + a^2 Z_C}{3}$	Coupled Circuits

TABLE 2.2

Y^{abc} to Y^{012} Conversions

Y^{abc}	abc Schematic	Y^{012}	Y^{012} Schematic
$\dfrac{1}{3Z}\begin{bmatrix} 2 & -1 & -1 \\ -1 & 2 & -1 \\ -1 & -1 & 2 \end{bmatrix}$	A, B, C each with Z; Open; ground	$\dfrac{1}{Z}\begin{bmatrix} 0 & 0 & 0 \\ 0 & 1 & 0 \\ 0 & 0 & 1 \end{bmatrix}$	0 Open; 1 with Z; 2 with Z
$\begin{bmatrix} \frac{1}{Z} & 0 & 0 \\ 0 & 0 & 0 \\ 0 & 0 & 0 \end{bmatrix}$	A with Z; B Open; C Open	$\dfrac{1}{3Z}\begin{bmatrix} 1 & 1 & 1 \\ 1 & 1 & 1 \\ 1 & 1 & 1 \end{bmatrix}$	0, 1, 2 with $-3Z$ couplings and Z each
$\dfrac{1}{Z^2+2ZZ_g}\begin{bmatrix} 0 & 0 & 0 \\ 0 & Z+Z_g & -Z_g \\ 0 & -Z_g & Z+Z_g \end{bmatrix}$	A Open; B, C with Z; Z_g	$\dfrac{1}{3(Z^2+2ZZ_g)}\begin{bmatrix} 2Z & -Z & -Z \\ -Z & 2Z+3Z_g & -Z-3Z_g \\ -Z & -Z-3Z_g & 2Z+3Z_g \end{bmatrix}$	Coupled Circuits
$\dfrac{1}{2Z}\begin{bmatrix} 0 & 0 & 0 \\ 0 & 1 & -1 \\ 0 & -1 & 1 \end{bmatrix}$	A Open; B, C with Z; Open	$\dfrac{1}{2Z}\begin{bmatrix} 0 & 0 & 0 \\ 0 & 1 & -1 \\ 0 & -1 & 1 \end{bmatrix}$	0 Open; 1 and 2 with $2Z$; Open

The symmetrical component transformation considerably simplifies balanced network calculations, as only a single voltage and current defined from phase *a* are necessary for power flow calculations. As Example 2.2 shows, the symmetrical component transformation for balanced impedances derives a single impedance in the positive-sequence circuit. Furthermore, by means of defining a base voltage as $V_{BASE} = \sqrt{3}\,|E^a|$ and a power base S_{BASE}, all electrical quantities for a balanced three-phase are calculated from one-line diagrams as shown in Figure 2.1. These are called *positive-sequence diagrams*.

For balanced networks it is interesting to note that the eigenvalues of a balanced impedance Z^{abc} correspond to the symmetrical component impedances, as may be calculated from

$$\det[\lambda - Z^{abc}] = \det\begin{bmatrix} \lambda - Z^s & -Z^m & -Z^m \\ -Z^m & \lambda - Z^s & -Z^m \\ -Z^m & -Z^m & \lambda - Z^s \end{bmatrix} = (\lambda - Z^s - 2Z^m)(\lambda - Z^s + Z^m)^2$$

Thus, the symmetrical component transformation essentially resolves the network along its natural modes (eigenvalues). Table 2.1 converts three-phase abc impedances into symmetrical component quantities. As the Z^{abc} impedances approach large values, the corresponding matrix values become infinite (i.e., open circuit). To avoid this problem, an admittance formulation of Table 2.2 has zeros for open-circuit conditions, and hence the admittance matrix values remain finite. This is one reason that some networks are expressed in admittance form rather than in impedance form.

A three-phase admittance form of Kirchhoff's voltage law for the circuit of Figure 2.4 is

$$I^{abc} = Y^{abc}(E^{abc} - e^{abc}) \tag{2.21a}$$

This equation is transformed into the symmetrical component reference frame as

$$I^{012} = T_s^{-1}I^{abc} = T_s^{-1}Y^{abc}(E^{abc} - e^{abc}) = T_s^{-1}Y^{abc}T_s(E^{012} - e^{012}) = Y^{012}(E^{012} - e^{012}) \tag{2.21b}$$

where symmetrical component admittances Y^{012} have been defined. Table 2.2 presents the results of transforming several admittances. The entries in Table 2.1 and Table 2.2 will be used later in short-circuit calculations. The reader is encouraged to verify a row of both Table 2.1 and Table 2.2 by means of the definitions of symmetrical component impedances and admittances:

$$\boxed{Z^{012} = T_s^{-1}Z^{abc}T_s \qquad\qquad Y^{012} = T_s^{-1}Y^{abc}T_s}$$

© The computer program **yabcypn.exe** can be used for admittance transformations that let open circuits be entered as zeros. The first row of Table 2.2, the Y^{abc} to Y^{012} conversion, can be verified by means of using only any balanced admittance phase-to-neutral entry. For this calculation, the results for 0, 1, 2 symmetrical admittance components show an open circuit for the zero-sequence circuit.

2.2.1   Floating Voltage Base Per-Unit Systems

Engineering units of kilovolts (kV), kiloamps (kA), ohms (Ω), megavolt-amperes reactive (Mvar), and so on, are descriptive of the electrical aspects of equipment, but very often it is necessary to compare the operating point of equipment with its rated value, or to compare one piece of equipment with another. For example, a power transmission network contains interconnected voltage levels, 220 kV, 110 kV, 69 kV, and others, that are coupled by transformers, so if each voltage were compared to nominal or percent of rated voltage, it is possible to determine quickly which are within standard acceptable limits of ±5%. The per-unit system is applied to power systems to aid in comparing magnitudes and accommodating various voltages.

To establish the per-unit system it is customary to select two of three basic values, the power base unit S_{BASE}, in megavolt-amperes, and a line-to-line base V_{BASE}, in kilovolts. The complex per-unit power and voltage on a per-unit base become

$$E_{p.u.} = \frac{E}{V_{BASE}} = \frac{V(\cos\theta + j\sin\theta)}{V_{BASE}} = \frac{V\angle\theta}{V_{BASE}} \quad \text{p.u.} \tag{2.22a}$$

$$X_{BASE} = \frac{V_{BASE}}{I_{BASE}} = \frac{V_{BASE}^2}{S_{BASE}} \tag{2.22b}$$

For three-phase elements, when the line-to-line voltage is selected as the base voltage, then factors of $\sqrt{3}$, or 3, are not required for power calculations. An example will clarify the use of per-unit values.

Example 2.3

For the transformer shown in Figure E2.3a, the excitation current is 12.5 A measured on the high-voltage terminals. It is wye-to-wye transformer rated at 50 MVA, 80:7.6 kV line to line. It is a three-phase transformer. The leakage reactance is 14.1 Ω referred to the high-voltage side.

Use the transformer MVA rating as a power base together with both high and low voltages as a voltage base to convert the excitation current and impedance into per-unit values.

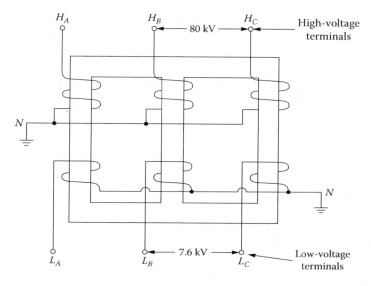

FIGURE E2.3a

Solution

The power base is selected to be the transformer rating, $S_{BASE} = 50$MVA. The high voltage, line to line, is selected as the base voltage:

$$V_{BASE} = 80kV = \sqrt{3}V_{line-neutral} = V_{HIGH} = \frac{N_1}{N_2}V_{LOW}$$

The turns ratio $N_1/N_2 = \frac{(80)}{(7.6)}$ is later used.

Using the MVA and voltage base, the current base is calculated as

$$I_{BASE} = \frac{S_{BASE}}{V_{BASE}} = \frac{50MVA}{80kV} = 625 \quad \text{amperes}$$

Currents are expressed in per unit using 0.625 kA as a normalization factor:

$$I_{p.u.} = \frac{I(Amperes)}{625}$$

The transformer-rated line current is calculated as

$$I_{rated} = \frac{50MVA/3}{V_{line-to-neutral}} = \frac{16.66MVA}{80kV/\sqrt{3}} = \frac{S_{BASE}/3}{V_{BASE}/\sqrt{3}} = \frac{I_{BASE}}{\sqrt{3}} = 360.7 \quad \text{amperes}$$

When rated voltage, or 1.0 p.u., is applied to the transformer, and 1.0 p.u. current flows into the transformer, its input is 1.0 p.u. (or 50 MVA) and the line current is 360.7 amps.

On the high-voltage base, the impedance and excitation current are converted to per-unit values as

$$X_{p.u.} = \frac{X(\Omega)}{X_{BASE}} = \frac{X(\Omega)}{V_{BASE}^2/S_{BASE}} = \frac{14.1}{80^2/50} = 0.110 \quad \text{p.u.}$$

$$I_{ex.p.u.} = \frac{12.5}{625} = 0.02 \quad \text{p.u.}$$

These are *positive-sequence* circuit values to be used to calculate the currents and voltages in the network. The transformer impedance and excitation current referred to the low-voltage side (using primes to indicate low-voltage quantities) are found using the turns ratio squared and inverse turns ratio, respectively, as

$$X' = X(\Omega)\left(\frac{V_{LOW}}{V_{HIGH}}\right)^2 = 14.1\left(\frac{7.6}{80}\right)^2 = 0.127 \quad \text{ohms}$$

$$I'_{ex} = I_{ex}\left(\frac{V_{HIGH}}{V_{LOW}}\right) = 12.5\left(\frac{80}{7.6}\right) = 131.579 \quad \text{amperes}$$

The power base for the low-voltage side is again the transformer rating, $S'_{BASE} = 50$ MVA $= S_{BASE}$, and $V'_{BASE} = V_{LOW} = 7.6$ kV is the low-voltage base. When X' and I'_{ex} are converted to per unit on the low-voltage base, the transformer voltage ratio enters the calculation to yield the same numerical per-unit values, as shown below:

$$X'_{p.u.} = X(\Omega)\left(\frac{V_{LOW}}{V_{HIGH}}\right)^2 \Big/ X'_{BASE} = X(\Omega)\left(\frac{V_{LOW}}{V_{HIGH}}\right)^2 \frac{S'_{BASE}}{(V_{LOW})^2} = X_{p.u.} = 0.110 \quad \text{p.u.}$$

$$I'_{p.u.} = I_{ex}\left(\frac{V_{HIGH}}{V_{LOW}}\right) \Big/ I'_{BASE} = I_{ex}\left(\frac{V_{HIGH}}{V_{LOW}}\right)\frac{S'_{BASE}}{(V_{LOW})^2} = X_{p.u.} = 0.02 \quad \text{p.u.}$$

Thus, the per-unit values are the same using either voltage base. If the excitation current is neglected, the positive-sequence impedance diagram in per unit for the transformer is as shown in Figure E2.3b.

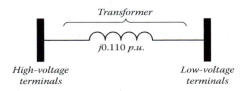

Transformer

$j0.110$ *p.u.*

High-voltage   *Low-voltage*
terminals   *terminals*

FIGURE E2.3b

For a general transmission network, as long as a common power base is used for all normalization, the transformer reflects impedances by the turns ratio squared and forces per-unit quantities to have the *same* numerical value regardless of the voltage level. This enables the analysis and power flow of mixed-voltage, interconnected transmission and distribution systems such as 69 kV, 138 kV, 230 kV, 345 kV, and so on, to be performed in per-unit values on a circuit diagram without concern as to the actual voltage magnitude. This is known as a *floating* voltage base per-unit system.

In the floating voltage base per-unit system, all impedances are *reflected through* the transformers (just as the leakage reactance of Example 2.3) from one voltage V_{B1} level to the new power system base voltage V_{B2}:

$$X_2 = \left(\frac{V_{B2}^2}{V_{B1}^2}\right) X_1 \quad \text{(ohms)}$$

Let $X_{p.u.B1}$ be a reactance on the previous power base S_{B1} at voltage base V_{B1}. Then the reflected reactance into the new power system base V_{B2}, S_{B2} in per unit is

$$X_{p.u.B2} = \frac{X_2}{X_{B2}} = \frac{S_{B2}}{V_{B2}^2} X_2 = \frac{S_{B2}}{V_{B2}^2}\left[\left(\frac{V_{B2}^2}{V_{B1}^2}\right) X_1\right] = \frac{S_{B2}}{V_{B2}^2}\left(\frac{V_{B2}^2}{V_{B1}^2}\right)\left(\frac{V_{B1}^2}{S_{B1}}\right) X_{p.u.B1} = \left(\frac{S_{B2}}{S_{B1}}\right) X_{p.u.B1}$$

All network elements must be converted to the same power system base. Stated as a general method to convert impedances or admittances from S_{B1} power base to another:

$$\boxed{Z_{p.u.B2} = Z_{p.u.B1}\left(\frac{S_{B2}}{S_{B1}}\right) \qquad Y_{p.u.B2} = Y_{p.u.B1}\left(\frac{S_{B1}}{S_{B2}}\right)}$$

where $Z_{p.u.B1}$ and $Z_{p.u.B2}$ are the per-unit values on power bases S_{B1} and S_{B2}, respectively. The voltage level does not directly enter the conversion.

In subsequent chapters of this book, for three-phase power systems, the base power base S_{BASE} is specified in megavolt-amperes (MVA) and the voltage base is selected to be the *nominal* (rated) line-to-line voltage at that point in the network; hence,

$$V_{BASE} = V_{line-to-line} = \sqrt{3} V_{line-to-neutral}$$
$$\text{(nominal)} \qquad \text{(nominal)}$$
$$\text{(rated)} \qquad \text{(rated)}$$

With the floating voltage base, the per-unit voltage calculated for power flow or short-circuit conditions directly indicates the operating condition with respect to the nominal voltage. In the same way, an element's current (e.g., the current flowing in a transmission line) expressed in per unit may be compared to a limit specified in per-unit value.

Another feature of this per-unit base is that it expresses three-phase power flow for balanced elements as the product of a single voltage and current:

$$S_{p.u.} = \frac{P + jQ}{S_{BASE}} = \frac{(\sqrt{3}E^1)(\sqrt{3}I^1)^*}{V_{BASE}I_{BASE}} = E_{p.u.}I_{p.u.}^*$$

In the last equation, three-phase power P is in megawatts and Q in megavolt-amperes, E^1 is the positive-sequence voltage, and I^1 is the positive-sequence current.

In a one-line, or positive-sequence, diagram, the circuit elements are linear resistances, inductances, and so on; thus, voltages and currents are linearly related. If voltages or magnitude $\sqrt{3} E^1$ instead of E^1 are impressed on the circuit, resulting currents are $\sqrt{3} I^1$ instead of I^1. Therefore, when per-unit voltages are used, per-unit currents result and power flow is calculated in per unit. The floating voltage base per-unit system and the positive-sequence diagram (e.g., Figure 2.1) allow currents, voltages, and power to be calculated in per-unit values for *balanced* networks.

2.3   Overhead Transmission Line Representation

Three-phase transmission lines (Figure 2.6) are used to transfer power from a point of generation to a point of consumption. The voltage levels or potentials are usually selected to be as high as possible to minimize conductor I^2R losses. However, the height above ground level, tower construction costs, insulation costs, right-of-way clearance, and equipment cost also increase with voltage levels. Therefore, the voltage used for transmission lines depends on the amount of power to be transmitted and the distance the line must traverse. Various utilities consider voltages as low as 23 kV line to line to be a transmission level as opposed to a distribution, but most utilities consider 69 kV line to line and above to be transmission levels. The subject of transmission line design is extensive, as is the detailed modeling. Our purpose here is to derive the lumped impedances describing balanced, transposed lines for several simple configurations and aspects of unbalanced lines.

Figure 2.7 shows the dimensions and cross-section geometry of a double-circuit three-phase transmission line. The conductors are shown as dots for a bundle. There exists self-inductance

FIGURE 2.6
A three-phase 138 kV transmission line under construction.

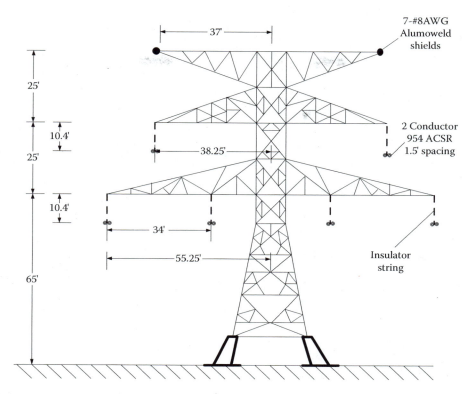

FIGURE 2.7
Dimensions of a double-circuit, 345 kV, three-phase transmission line.

for each conductor, mutual inductance with other phases, and phase-to-phase capacitance. These parameters also exist between the conductors and their *reflections* in the earth plane. The resistivity and permeability (and hence image electrical parameters) of the earth vary widely with moisture content and type of terrain. Proximity to the ground plane or electrically grounded lightning shield wires and tower also affect inductances and capacitances. By means of rotating the *a*, *b*, *c* phase with respect to the tower, or in other words, transposing the line every one-third of its traversed distance, it is possible to balance the geometric factors such as line-to-ground capacitance. In actual practice, very few transmission lines are truly transposed because of the additional equipment cost.

Overhead transmission lines in the United States are usually standardized at 69, 115, 138, 230, 345, 500, or 765 kV line to line to avoid purchasing specialized equipment and protection devices. The analytical methods that follow may be used to determine ideal electrical parameters of an overhead line of any voltage rating. Underground lines, usually in the form of coaxial cables, are not considered here. The following methods are also applicable to calculate the parameters of high-voltage direct-current (HVDC) lines, which usually have a simpler geometry than three-phase ac lines.

The electrical description of transmission lines relies on linearization and superposition to derive models suitable for specific applications. Consider two very long parallel, distributed parameter conductors, oriented in the *x*-direction and isolated in space as shown in Figure 2.8a. For a potential difference between conductors, $v(x, t)$, and current flow, $i(x, t)$, the intrinsic electrical properties of the conductors are represented by the incremental equivalent circuit shown in Figure 2.8b.

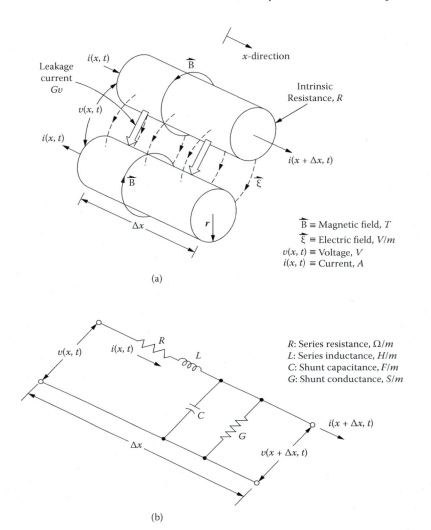

FIGURE 2.8
Two-conductor, distributed parameter transmission line. (a) Currents and electromagnetic fields. (b) Incremental lossy equivalent circuit.

As the incremental length becomes very small, $\Delta x \to 0$, the two-conductor voltage and current are described by partial differential equations:

$$-\frac{\partial v}{\partial x} = Ri + L\frac{\partial i}{\partial t} \tag{2.23a}$$

$$-\frac{\partial i}{\partial x} = Gv + C\frac{\partial v}{\partial t} \tag{2.23b}$$

R, L, G, C are, respectively, ohms/meter, Henrys per meter, Siemens/meter and Farads/meter. By appropriate differentiations and substitutions, Equation 2.23a and Equation 2.23b are separable into a voltage equation:

$$\frac{\partial^2 v}{\partial x^2} = RGv + (RC + LG)\frac{\partial v}{\partial t} + LC\frac{\partial^2 v}{\partial t^2} \tag{2.23c}$$

and a similar partial differential equation with i replacing v. These partial differential equations in v and i are known as the *equations of telegraphy* and describe traveling waves on transmission lines.

As a solution to the traveling-wave equation for the voltage $v(x, t)$, let the $x = 0$ terminals of the distributed line be excited by a steady-state input sinusoidal voltage,

$$v(0,t) = V\cos(\omega t) \rightarrow V \angle 0°$$

and the output terminals be open circuited at $x = l$. After initial transients have decayed, it may be shown [1] that the steady-state voltage at any point along the two-conductor line is given by

$$V(x, j\omega) = V \cosh[\gamma(l - x)]/\cosh(\gamma l)$$

$$= V[\cosh\{\alpha(l - x)\}\cos\{\beta(l - x)\} + j\sinh\{\alpha(l - x)\}\sin\{\beta(l - x)\}]/\cosh(\gamma l) \quad (2.24a)$$

where α and β are, respectively, the attenuation and phase constants defined in terms of ω and the line electrical properties:

$$\gamma = \alpha + j\beta = \sqrt{(R + j\omega L)(G + j\omega C)} \quad (\Omega/m) \quad (2.24b)$$

For the special case of $R \sim 0$, $G \sim 0$, with the conductors widely spaced in Figure 2.8, and for ideal L and C, the voltage wave travels down the parallel lines at a speed

$$c = 3 \times 10^8 = \frac{1}{\sqrt{\mu_0 \varepsilon_0}} = \lambda f = \frac{\omega}{\beta} = \frac{1}{\sqrt{LC}} \quad (m/s) \quad (2.24c)$$

The symbol λ is the wavelength in meters. For $f = 60$ Hertz excitation, the wavelength is calculated as

$$\lambda = \left(\frac{1}{f}\right) \times 3 \times 10^8/1609 = \left(\frac{1}{60}\right) \times 3 \times 10^8/1609 = 3108 \quad (miles) \quad (2.24d)$$

Almost every power transmission line is much less than $\lambda/4$, such that the distributed parameter effect can be ignored and simple lumped circuits used to describe transmission lines. The two-conductor distributed parameter solution is later used as a comparison to a lumped-circuit, single-phase π-equivalent.

2.3.1 Inductance of Long Parallel Conductors

To calculate the R,L,C parameters of long parallel conductors, consider first the geometry of two lines shown in Figure 2.9. The admittance G between the two wires is neglected.

Let the current, I, be uniformly distributed on a solid cross section of the line normal to its length. The magnetic flux density at any radius $\rho < r$ inside the wire due only to the current in the wire is found using Ampere's law [2] applied to the geometry of Figure 2.9. Let $\mathbf{B} = \mu \mathbf{H}$ and Θ be, respectively, the vector representation of the circumferential flux density in the wire B_i and the angular direction θ; then the line integral for Ampere's law yields

$$\oint \mathbf{H} \circ \partial l = I_{enclosed} = \frac{\pi \rho^2 I}{\pi r^2} = \int_0^{2\pi} \frac{B_i \rho \partial \theta}{\mu} = \frac{2\pi \rho B_i}{\mu} \quad (2.25)$$

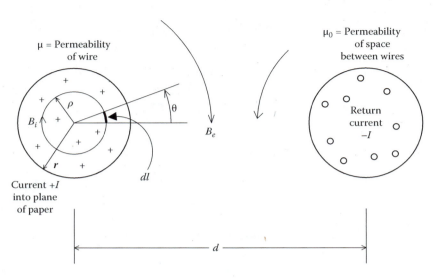

FIGURE 2.9
Magnetic field due to uniform current in a conductor.

where the dot represents an inner product of vectors. Solving this last equation for the internal radial flux density, we have

$$B_i = \frac{\mu I \rho}{2\pi \mathbf{r}^2} \quad \text{(Tesla)} \quad (\rho < \mathbf{r}) \tag{2.26}$$

where ρ and $\mathbf{r}$ have the dimensions of meters and μ is the permeability of the wire material. Observe that each radius within the wire encloses a significant part of the current within the wire. The magnetic flux also encloses a varying amount of current, such that the wire internal inductance per length is given as an integral:

$$L_i = \frac{\int I_{enclosed} \partial \phi}{I^2} = \frac{\int (\rho^2 I / \mathbf{r}^2) \partial \phi}{I^2} \quad \text{(H/m)} \tag{2.27}$$

where ϕ is the flux per length of the wire. The flux density vector is circumferential, so the differential element of magnetic flux per length of wire is given by

$$\partial \phi = B_i \partial \rho = \frac{\mu I \rho \partial \rho}{2\pi \mathbf{r}^2} \quad \text{(webers/m)} \tag{2.28}$$

Introducing Equation 2.28 into Equation 2.27 and performing the integration yields the wire internal inductance:

$$L_i = \frac{\mu}{8\pi} = \frac{\phi_i}{I} \quad \text{(H/m)} \tag{2.29}$$

This is the internal inductance for any diameter wire with uniform current density.

At radii $r > \mathbf{r}$, the magnetic field and incremental flux per length due to the current, I, are, respectively,

$$B = \frac{\mu_0 I}{2\pi r} \quad \text{(Tesla)} \quad (r > \mathbf{r}) \tag{2.30}$$

$$\partial \phi_{external} = B \partial r = \frac{\mu_0 I \partial r}{2\pi r} \quad \text{(webers/m)} \tag{2.31}$$

where μ_0 is the permeability of air or free space. Equation 2.31 is integrated to calculate the magnetic flux in the space $\mathbf{r} \le r \le d$ to obtain the wire external inductance up to distance d:

$$L_e = L_{external} = \frac{\phi_{external}}{I} = \frac{\mu_0}{2\pi} \ln\left(\frac{d}{\mathbf{r}}\right) \quad (\text{H/m}) \tag{2.32}$$

Combining the internal and external inductance of a single wire, a long single conductor has the self-inductance to a distance d as

$$L_l = L_i + L_e = \frac{\mu}{8\pi} + \frac{\mu_0}{2\pi} \ln\left(\frac{d}{\mathbf{r}}\right) \quad (\text{H/m}) \tag{2.33}$$

Often the permeability of the conductors is approximated as $\mu \approx \mu_0$, so the self-inductance is written in parts as

$$L_l = \frac{\mu_0}{2\pi} \ln\left(\frac{1}{e^{-4}}\right) + \frac{\mu_0}{2\pi} \ln\left(\frac{1}{\mathbf{r}}\right) + \frac{\mu_0}{2\pi} \ln(d) = \frac{\mu_0}{2\pi} \ln\left(\frac{1}{\mathbf{r}e^{-4}}\right) + \frac{\mu_0}{2\pi} \ln(d)$$

$$= \frac{\mu_0}{2\pi} \ln\left(\frac{1}{GMR}\right) + \frac{\mu_0}{2\pi} \ln(d) \quad (\text{H/m}) \tag{2.34}$$

The geometric mean radius (GMR) is here defined for a single homogeneous conductor. Later, the GMR is derived for composite conductors of aluminum and steel.

Instead of a solid conductor, bundles of smaller-radii wires, two adjacent wires, combinations of copper and steel, or aluminum and steel, may comprise the unidirectional current path. Superposition methods are used to calculate the magnetic fields internally and around the conductors. The term *geometric mean radius* (GMR) is defined as the radius of a tubular conductor with an infinitesimal thin wall that has the same magnetic flux out to a radius of 1 ft as does the original configuration internal and external fluxes. This definition reduces bundled and nonuniform conductors to a GMR at 1 ft. Beyond 1 ft, the equivalent magnetic fields due to all conductors are dependent upon the d term in Equation 2.34. Subsequent calculations involve three-phase currents with a common return path, so a *general* form of Equation 2.34 is written in terms of a conductor-dependent term, GMR, and distance d. For a *single* wire at 60 Hz excitation, this expression is converted into convenient English units as

$$X_l = \omega L_l = 0.27942\left(\frac{f}{60}\right)\log_{10}\left(\frac{1}{GMR}\right) + 0.27942\left(\frac{f}{60}\right)\log_{10}(d)$$

$$= X_a + X_b \quad \Omega/\text{conductor-mile} \tag{2.35}$$

where $X_a \equiv$ internal reactance of all subconductors for internal flux and external flux to a distance of 1 ft, $X_b \equiv$ inductive reactance due to the magnetic field surrounding the conductors from a radius of 1 ft to a distance of d feet, $f \equiv$ frequency (Hz), and $d \equiv$ distance (ft).

English units rather than the RMKS system are commonly used in the United States for equations such as 2.34. In order to convert from English to RMKS:

$$X_{RMKS} = (X_a + X_b) \times (1/1.609) \quad \Omega/\text{conductor-km} \tag{2.35a}$$

A current return, $-I$, located at distance d establishes an additive magnetic field between the wires. The region outside the wires does not contribute to the net magnetic flux linkages in the circuit. Therefore, a parallel pair of wires separated by a distance d has twice the inductance per length:

$$L_{pair} = 2L_l \quad (\text{H/m}) \quad X_{pair} = 2\omega L_l = 2X_l \quad (\text{ohms}) \quad (\text{2 parallel wires}) \tag{2.36}$$

The following example shows the properties of two parallel solid conductors.

Example 2.3.1

Neglect resistance and leakage conductance between two 4/0 AWG parallel solid copper conductors (diameter 0.46 in.) located 12 ft apart on centers. Calculate the 60 Hz reactances per mile for the line and the characteristic wavelength, λ. Assume air as the dielectric material. The permeability of copper is approximately the same as air. The geometry of the line is shown in Figure E2.3.1. Express the results in physical units and per unit for a 100 MVA, 100 kV base.

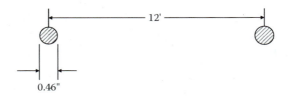

FIGURE E2.3.1

Solution

For a 1-mile length of line with a small radius $r \ll d$, a pair of conductors has the capacitance (1609 m/mile)

$$C_{pair} = \frac{1609\pi\varepsilon_0}{\ln\left[\frac{d}{2r}+\sqrt{\left(\frac{d}{2r}\right)^2-1}\right]} \approx \frac{1609\pi\varepsilon_0}{\ln\left(\frac{d}{r}\right)} = \frac{1609\pi\times 8.85\times 10^{-12}}{\ln[12\times 12/0.23]} = 6.947\times 10^{-9} \quad \text{F/mile}$$

The capacitive reactance for a pair of conductors of 1-mile length in per unit is

$$X'_{pair} = \frac{1}{\omega C_{pair}}\left(\frac{S_{BASE}}{V_{BASE}^2}\right) = 3818 \quad \text{p.u. mile}$$

The GMR of the solid conductor is

$$re^{-1/4} = \frac{0.23\times e^{-1/4}}{12} = 0.013842 \quad \text{ft}$$

The pair of conductors (Equation 2.35) has the reactance

$$X_{pair} = 2\left[0.27942\,\log_{10}\left(\frac{1}{GMR}\right)+0.27942\,\log_{10}(d)\right] = 1.02018+0.60282 = 1.623 \quad (\Omega/\text{mile})$$

Converted into per unit, the inductive reactance is

$$X_{p.u.} = 1.623\left(\frac{S_{BASE}}{V_{BASE}^2}\right) = 0.01623 \quad \text{p.u./mile} \quad \text{pair of conductors}$$

The characteristic wavelength for this conductor pair is

$$\lambda = \frac{2\pi}{\beta} = \frac{1}{\omega\sqrt{L_{pair}C_{pair}}} = 2\pi\sqrt{\frac{X'_{pair}}{X_{pair}}} = 3048 \quad \text{miles}$$

The difference for the shorter wavelength calculated here, compared to Equation 2.24d, is that Equation 2.24d does not take into account the internal flux of the conductor. The internal flux increases the inductance of the conductor.

Most transmission lines are much shorter than $\lambda/4$, or 777 miles, so that standing waves and reflections at terminations or discontinuities are not significant for 60 Hz power flow. For lines less than 150 miles in length, the lumped-circuit R, L, C parameters are considered adequate and the line is often represented by a π-equivalent circuit, as shown in Figure 2.10, where $\omega \equiv 2\pi f$, where f is the line frequency (Hz); $R_1 \equiv$ conductor intrinsic and skin effect resistance (Ω or p.u.); $jX_1 = j\omega L_1 \equiv$ positive sequence, mutual and self-inductive reactance (Ω or p.u.); and $Y_1/2 = j\omega C_1/2 \equiv$ half the positive-sequence line charging susceptance due to other conductors and ground (Siemens or p.u.).

For long lines, those greater than 150 miles, a positive-sequence equivalent for balanced lines is often calculated using the distributed parameters as derived from Equation 2.23 and Equation 2.24, or else several "short" π sections are cascaded to approximate the long line.

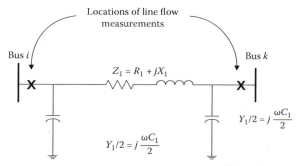

FIGURE 2.10
Single-phase equivalent for a short- or medium-length line.

Example 2.4

Excite one end of the π-equivalent circuit and calculate the voltage at the open terminals at the other end for Example 2.3.1. Compare the receiving end voltage with the distributed parameter equivalent, Equation 2.24, for lines up to 350 miles in length. For both lines, assume a line resistance of 0.606 Ω/mile (two conductors at 50°C, 60 Hz frequency). Express results as a ratio of $|V_{OUT}/V_{IN}|$.

Solution

The open-circuit π-equivalent line excited by the 60 Hz phasor voltage V_{IN} is shown in Figure E2.4.1 for a 1-mile length.

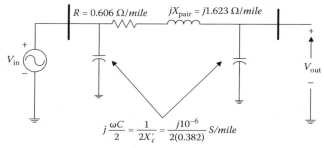

FIGURE E2.4.1
π-equivalent for two-conductor 4/0 line with 12 ft spacing.

Let l be the length of the line in miles. The ratio of receiving-end to sending-end voltage for the π-equivalent is computed from lumped impedances:

$$\left|\frac{V_{out}}{V_{in}}\right| = \left|\frac{\frac{2X_c'}{jl}}{Rl + jX_{pair}l - j2\frac{X_c'}{l}}\right| = \left|\frac{2X_c'}{jRl^2 - X_{pair}l^2 + 2X_c'}\right| \tag{1}$$

Usually the resistance is comparatively small, so the ratio is always greater than unity for any line length.

For the distributed parameter line, the attenuation and phase constants are calculated from the incremental values for impedances:

$$\gamma = \alpha + j\beta = \sqrt{(R + jX_{pair})(j\omega C)} = \sqrt{(R + jX_{pair})(j/X_c')} = 0.00039 + j0.002096 \tag{2}$$

These numerical values for α and β, at progressive lengths $l = 50, 100, \ldots, 350$ miles, are inserted into Equation 2.24a to compute the output/input ratio for the distributed parameter line. A tabulation of computed values for Equation 1 and the distributed parameter line is given in Table E2.4. Observe the close agreement in voltage gain between the two circuits.

TABLE E2.4

Comparison of π-Equivalent and Distributed Parameter Voltage for Two Lines (Gain for Open-Circuit Conditions, Line of Example 2.4)

Line Length (miles)	$\left\|\dfrac{V_{out}}{V_{in}}\right\|$ π-Equivalent	Distributed Parameter
50	1.0051	1.0053
100	1.0217	1.0216
150	1.0500	1.0497
200	1.0922	1.0910
250	1.1512	1.1481
300	1.2316	1.2244
350	1.3403	1.3252

For very short lines, less than 50 miles, the line charging is often negligible. In the next sections, equivalent line impedances X_1 (positive-sequence reactance) and $2X_c'$ are calculated from geometry and intrinsic electrical parameters for general, balanced, three-phase lines.

The π model is a good fit to the distributed parameter line for medium- and short-length lines, such that the voltage rise for a line energized at one end and open circuited at the other is

$$\left|\frac{V_{out}}{V_{in}}\right| = \left|\frac{2X_c'}{jRl^2 - X_{pair}l^2 + 2X_c'}\right| \approx \left|\frac{2X_c'}{-X_{pair}l^2 + 2X_c'}\right| = \left|\frac{1}{1 - X_{pair}l^2/2X_c'}\right| = \left|\frac{1}{1 - \Gamma \times 10^{-6}l^2}\right|$$

Here the parameter $\Gamma = 10^6 X_1/2X_c'$ is a convenient way to compare lines for the voltage rise when a line is excited at only one end. X_1 is the positive-sequence reactance and $2X_c'$ is the capacitance reactance at the ends of the π-equivalent. The graph of Figure E2.4.2 shows the voltage rise for various Γ values.

The two-conductor line of Example E2.3.1 has a calculated value for the Γ parameter:

$$\Gamma_{2\text{-}conductor} = \frac{10^6(1.623)}{0.764 \times 10^6} = 2.12$$

Such that the voltage rise is found by interpolation between two curves.

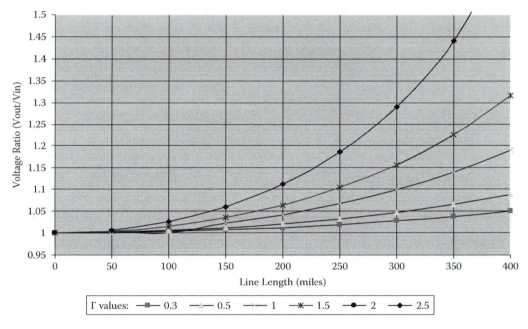

FIGURE E2.4.2
Voltage rise at open-circuit end of a transmission rise as dependent upon the ratio $\Gamma = 10^6 X_1/(2X_c')$.

2.3.2 Balanced Three-Phase Lines

Consider the case where balanced three-phase currents I^a, I^b, and I^c flow in three circular conductors triangularly spaced, as shown in Figure 2.11. The line is isolated from all other conductors such as lightning shield wires. Temporarily neglect the resistance to consider only the self- and mutual reactances of the lines.

Assume further that balanced three-phase line-to-neutral voltages E^a, E^b, and E^c are applied at terminal p, and that the voltages at terminal q are balanced. Either positive-sequence a, b, c or negative-sequence a, c, b can be used. Because the currents are balanced, $I^a + I^b + I^c = 0$ and no current flows in the return point k. Considering the flux linkages around conductor a, and operation at electrical frequency ω, the reactive voltage drop

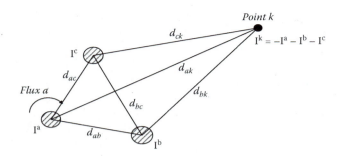

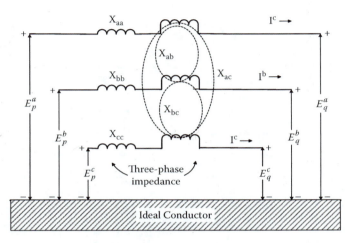

FIGURE 2.11
Triangularly spaced three-phase transmission lines.

across the length of line for phase *a* is

$$E_p^a - E_q^a = j(I^a X_{aa} + I^b X_{ab} + I^c X_{ac}) \quad \text{(V/mi)} \tag{2.37}$$

where the following reactances are defined for conductor *a*:

$$X_{aa} = X_a + X_b = 0.27942\left(\frac{f}{60}\right)\left[\log_{10}\left(\frac{1}{GMR}\right) + \log_{10}(d_{ak})\right]$$

X_{aa} = self-inductive reactance of phase *a* to distance d_{ak}
X_{ab} = mutual reactance between conductors *a* and *b*
X_{ac} = mutual reactance between conductors *a* and *c*

Superposition of currents is next used to calculate the mutual reactances. Define positive flux around conductor *a* to be in the clockwise direction. When the only current flowing is I^b in conductor b and returning at point k at distance d_{bk}, the mutual inductive reactance X_{ab} due to the currents I^b and $-I^b$ is expressed as

$$X_{ab} = 0.27942\left(\frac{f}{60}\right)[\log_{10}(d_{ak}) - \log_{10}(d_{ab})] \quad (\Omega/\text{mi}) \tag{2.38}$$

Next, when only current I^c is flowing in conductor c and returning at point k, the mutual reactance with phase a is given by

$$X_{ac} = 0.27942\left(\frac{f}{60}\right)[\log_{10}(d_{ak}) - \log_{10}(d_{ac})] \quad (\Omega/\text{mi}) \tag{2.39}$$

When all three currents are flowing simultaneously, including the current-I^a at distance d_{ak}, the voltage drop from terminal p to terminal q is found using superposition:

$$E_p^a - E_q^a = jI^a X_a + j0.27942\left(\frac{f}{60}\right)[-I^b \log_{10}(d_{ab}) - I^c \log_{10}(d_{ac})]$$

$$+ j0.27942\left(\frac{f}{60}\right)[I^a \log_{10}(d_{ak}) + I^b \log_{10}(d_{ak}) + I^c \log_{10}(d_{ak})] \tag{2.40}$$

In this last equation, $I^a = -I^b - I^c$, so the d_{ak} terms sum to zero. When the three conductors are equally spaced, $d_{ab} = d_{ac} = d$ (the three conductors are equilateral in space), then the voltage difference in phase a becomes

$$E_p^a - E_q^a = jI^a X_a + j0.27942\left(\frac{f}{60}\right)[I^a \log_{10}(d)] = jI^a(X_a + X_b) \quad (\text{V/mi}) \tag{2.41}$$

Therefore, the positive or negative reactance per phase for a three-phase circuit with equilateral triangular spacing, d, is the *same* as a *single* conductor reactance to a distance d. In Ω/mile, the reactance for a balanced line (without shields or ground) is

$$\boxed{\begin{array}{c} \text{Balanced, isolated line} \\[6pt] X_1 = X_2 = X_a + X_b = 0.27942\left(\frac{f}{60}\right)\log_{10}\left(\frac{1}{GMR}\right) + 0.27942\left(\frac{f}{60}\right)\log_{10}(d) \\[6pt] (\Omega/\text{mi}) \end{array}} \tag{2.42}$$

The GMR term in Equation 2.42 has been previously defined for a solid conductor. For many types of composite steel-aluminum, steel-copper stranded, and multiple-conductor configurations, extensive tabulations of the GMR and X_a are available [3]. Table 2.3 has columns for the GMR and single conductor X_a of aluminum-conductor-steel-reinforced (ACSR) cable. All ACSR items are given code names of birds. By means of the table and the center-to-center phase spacing d, the positive- and negative-sequence reactances of the line may be calculated.

When two or more conductors are symmetrically bundled together in order to reduce resistive losses or reduce grain effects, an equivalent geometric mean radius, GMR_B, is calculated. Figure 2.12 shows two conductors sharing phase current, $I = I_1 + I_2$.

TABLE 2.3

Inductive Reactance (X_a) of Aluminum-Conductor-Steel-Reinforced (ACSR) Bundled Conductors at 60 Hz in Ohms per Mile for 1 Ft Spacing

Code	kcmil A1	Sq. mm Tot.	Strand	Dia. (in.)	GMR (ft)	Single Cond.	2-Cond. Spacing			3-Cond. Spacing			4-Cond. Spacing		
							12	18	24	12	18	24	12	18	24
Joree	2515.	1344.	76/19	1.880	0.0621	0.337	0.169	0.144	0.127	0.112	0.080	0.056	0.074	0.037	0.011
Thrasher	2312.	1235.	76/19	1.802	0.0595	0.342	0.171	0.147	0.129	0.114	0.081	0.058	0.075	0.038	0.012
Kiwi	2167.	1146.	72/7	1.735	0.0570	0.348	0.174	0.149	0.132	0.116	0.083	0.060	0.076	0.039	0.013
Bluebird	2156.	1181.	84/19	1.762	0.0588	0.344	0.172	0.147	0.130	0.115	0.082	0.059	0.075	0.039	0.012
Chuckar	1781.	976.	84/19	1.602	0.0534	0.355	0.178	0.153	0.136	0.118	0.086	0.062	0.078	0.041	0.015
Falcon	1590.	908.	54/19	1.545	0.0521	0.358	0.179	0.155	0.137	0.119	0.087	0.063	0.079	0.042	0.016
Plover	1431.	817.	54/19	1.465	0.0494	0.365	0.182	0.158	0.140	0.122	0.089	0.066	0.081	0.044	0.018
Bobolink	1431.	775.	45/7	1.426	0.0472	0.371	0.185	0.161	0.143	0.124	0.091	0.067	0.082	0.045	0.019
Pheasant	1272.	726.	54/19	1.382	0.0466	0.372	0.186	0.161	0.144	0.124	0.091	0.068	0.082	0.046	0.019
Bittern	1272.	689.	45/7	1.345	0.0445	0.378	0.189	0.164	0.147	0.126	0.093	0.070	0.084	0.047	0.021
Finch	1114.	636.	54/19	1.293	0.0436	0.380	0.190	0.165	0.148	0.127	0.094	0.071	0.085	0.048	0.021
Bluejay	1113.	603.	45/7	1.258	0.0416	0.386	0.193	0.168	0.151	0.129	0.096	0.073	0.086	0.049	0.023
Ortolan	1034.	560.	45/7	1.213	0.0401	0.390	0.195	0.171	0.153	0.130	0.097	0.074	0.087	0.050	0.024
Cardinal	954.	546.	54/7	1.196	0.0404	0.389	0.195	0.170	0.153	0.130	0.097	0.074	0.087	0.050	0.024
Rail	954.	517.	45/7	1.165	0.0385	0.395	0.198	0.173	0.156	0.132	0.099	0.076	0.088	0.051	0.025
Drake	795.	469.	26/7	1.108	0.0375	0.399	0.199	0.175	0.157	0.133	0.100	0.077	0.089	0.052	0.026
Condor	796.	456.	54/7	1.093	0.0369	0.400	0.200	0.176	0.158	0.133	0.101	0.077	0.090	0.053	0.027
Cuckoo	795.	455.	24/7	1.092	0.0366	0.402	0.201	0.176	0.159	0.134	0.101	0.078	0.090	0.053	0.027
Tern	795.	431.	45/7	1.063	0.0352	0.406	0.203	0.179	0.161	0.135	0.103	0.079	0.091	0.054	0.028
Rook	636.	364.	24/7	0.977	0.0327	0.415	0.208	0.183	0.165	0.138	0.106	0.082	0.093	0.056	0.030
Grosbeak	636.	375.	26/7	0.990	0.0335	0.412	0.206	0.181	0.164	0.137	0.105	0.081	0.093	0.056	0.029
Parakeet	557.	319.	24/7	0.914	0.0306	0.423	0.212	0.187	0.170	0.141	0.108	0.085	0.095	0.058	0.032
Dove	556.	328.	26/7	0.927	0.0313	0.420	0.210	0.186	0.168	0.140	0.107	0.084	0.095	0.058	0.031
Osprey	556.	298.	18/1	0.879	0.0284	0.432	0.216	0.191	0.174	0.144	0.111	0.088	0.097	0.061	0.034
Hawk	477.	281.	26/7	0.858	0.0290	0.430	0.215	0.190	0.173	0.143	0.110	0.087	0.097	0.060	0.034

Source: Reprinted with permission from Electric Power Research Institute, *Transmission Line Reference Book 345 kV and Above*. Copyright © 1975.

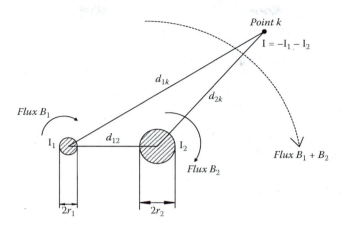

FIGURE 2.12
Conductors sharing phase current *I*.

The flux linkage in Figure 2.12 due only to current I_1 from its center to point k is

$$\phi'_{1k} = \frac{\mu I_1}{8\pi} + \frac{\mu_0 I_1}{2\pi} \ln\left(\frac{d_{1k}}{r_1}\right) \approx \frac{\mu_0 I_1}{2\pi} \ln\left(\frac{d_{1k}}{GMR_1}\right) \quad \text{(W/m)} \tag{2.43a}$$

The contribution of I_2 to the flux ϕ_{1k} has both positive and negative parts:

$$\phi_{1k} = \frac{\mu_0 I_1}{2\pi} \ln\left(\frac{d_{1k}}{GMR_1}\right) - \frac{\mu_0 I_2}{2\pi} \ln\left(\frac{d_{12}}{GMR_2}\right) + \frac{\mu_0 I_2}{2\pi} \ln\left(\frac{d_{2k}}{GMR_2}\right)$$

$$= \frac{\mu_0 I_1}{2\pi} \ln\left(\frac{d_{1k}}{GMR_1}\right) + \frac{\mu_0 I_2}{2\pi} \ln\left(\frac{d_{2k}}{d_{12}}\right) \quad \text{(W/m)} \tag{2.43b}$$

The flux ϕ_{2k} is similarly calculated by interchanging subscripts. Since currents are twice taken into account, the flux due to two conductors that share current I is

$$\phi = \frac{\phi_{1k} + \phi_{2k}}{2} = \frac{1}{2}\left[\frac{\mu_0 I_1}{2\pi} \ln\left(\frac{d_{1k}}{GMR_1}\right) + \frac{\mu_0 I_2}{2\pi} \ln\left(\frac{d_{2k}}{d_{12}}\right) + \frac{\mu_0 I_2}{2\pi} \ln\left(\frac{d_{2k}}{GMR_2}\right) + \frac{\mu_0 I_1}{2\pi} \ln\left(\frac{d_{1k}}{d_{12}}\right)\right]$$

$$= \frac{\mu_0 I_1}{2\pi} \ln\left(\frac{d_{1k}}{\sqrt{GMR_1 d_{12}}}\right) + \frac{\mu_0 I_2}{2\pi} \ln\left(\frac{d_{2k}}{\sqrt{GMR_2 d_{12}}}\right) \quad \text{(W/m)} \tag{2.43c}$$

As the point k becomes far away from the spacing and radii of the conductors, the distances are about the same, $d \approx d_{1k} \approx d_{2k} \approx d_{3k} \approx \cdots d_{Nk}$ for any number of conductors.

The case of two conductors sharing current generalizes for a current I shared by N unequal strands conducting unequal currents to the following expression:

$$\Phi = \left(\frac{1}{N}\right)\frac{\mu_0}{2\pi}\left[\begin{array}{l} I_1 \ln\left(\dfrac{d}{GMR_1}\right) + \displaystyle\sum_{n\neq 1}^{N} I_n \ln\left(\dfrac{d}{d_{1n}}\right) + I_2 \ln\left(\dfrac{d}{GMR_2}\right) + \displaystyle\sum_{n\neq 2}^{N} I_n \ln\left(\dfrac{d}{d_{2n}}\right) \\[4mm] + I_3 \ln\left(\dfrac{d}{GMR_3}\right) + \displaystyle\sum_{n\neq 3}^{N} I_n \ln\left(\dfrac{d}{d_{3n}}\right) + I_4\left(\dfrac{d}{GMR_4}\right) + \displaystyle\sum \cdots + \cdots + \cdots N^{th} term \end{array}\right] \tag{2.43d}$$
(W/m)

where $I = I_1 + I_2 + \cdots I_N$. This expression is used to calculate the equivalent GMR for composites such as ACSR cables where the steel and aluminum strands do not conduct equal current. Figure 2.12a shows a seven-strand ACSR conductor. When a conductor is a composite of steel and aluminum strands, such as ACSR, the current-carrying capacity of the steel is often neglected. Figure A.1 in appendix A shows the geometry of several such multistrand conductors.

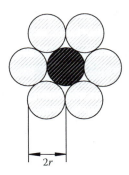

FIGURE 2.12a Composite ACSR conductor of one steel strand (center) and six aluminum strands, equal diameters.

Another application of Equation 2.43d is to employ a bundle of widely separated conductors in order to dissipate heat more easily, and to reduce the electric field near the conductors. A reduced electric field at the surface of a conductor results in less ionization of the local air. Cyclic ionization of the air is a power loss (called corona loss) and is a source of audio and radio frequency noise. Example 2.5 demonstrates how the GMR of a single conductor is used to obtain the equivalent reactance of a bundle. The bundle separation is maintained by spacer-dampers. Table 2.3 lists the reactance of bundles of various ACSR conductors. The GMR given in Table 2.3 is that of a single conductor.

Example 2.4a

Three bundled Joree phase conductors of a transmission line are separated by $D = d_{ab} = d_{bc} = d_{ca} = 1.5\,ft$ as shown in Figure E2.4a. Each conductor of the bundle carries current $I/3$.

Find the phase reactance to a distance d for the equilateral triangle bundle that is used for the line.

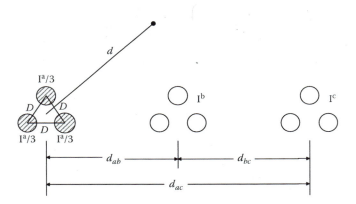

FIGURE E2.4a
Bundled conductors of a three-phase transmission line.

Solution

Let the GMR be the same for each conductor. A direct substitution into Equation 2.43d yields the flux to a distance d due to a phase bundle:

$$\Phi = \left(\frac{1}{3}\right)\frac{\mu_0}{2\pi}\left[\left(\frac{I}{3}\right)\ln\left(\frac{d^3}{(GMR_1)^3}\right) + 6\left(\frac{I}{3}\right)\ln\left(\frac{d}{D}\right)\right] \quad (W/m) \qquad (2.44a)$$

The reactance of the phase bundle to the distance d is

$$X = \omega L = \omega \Phi / I = \frac{\omega \mu_0}{2\pi} \ln\left(\frac{d}{(GMR_1)^{1/3} D^{2/3}}\right) \quad (\Omega/\text{m}) \tag{2.44b}$$

This last equation for a three-bundle of conductors may be rewritten similar to Equation 2.35 as

$$X = X_a + X_b = 0.27942\left(\frac{f}{60}\right)\log_{10}\left(\frac{1}{GMR_1^{1/3}D^{2/3}}\right) + 0.27942\left(\frac{f}{60}\right)\log_{10}(d) \quad \Omega/\text{mi}$$

From Table 2.3, a single Joree conductor has a $GMR_1 = 0.337\,ft$. When this numerical value and $D = 1.5\,ft$ are substituted in the above equation,

$$X_a = .079609 \quad \Omega/\text{mi for a 3-conductor Joree bundle separated by 1.5 ft:}$$

This computed value matches (within round-off) Table 2.3 for a bundle of three Joree conductors separated by 1.5 ft.

For the case of N subconductors in a symmetrical circular bundle, each carrying the current I/N, let γ be the bundle radius in feet and GMR_1 the geometric mean radius of each subconductor. For a distance D between adjacent subconductors, such as specified in Table 2.3, the bundled GMR_B [3] is calculated from:

$$GMR_B = \sqrt[N]{N\gamma^{N-1}GMR_1} \quad \gamma = \frac{D}{2\sin(\pi/N)} \quad (N > 1 \text{ bundle inductive})$$

For capacitance calculations, the surface of the subconductor determines charge distribution so $GMR = \rho$, the subconductor radius. The capacitive reactance geometric mean radius for a bundle is:

$$GMR_B' = \sqrt[N]{N\gamma^{N-1}\rho} \quad (\text{bundle capacitive})$$

Two typical transmission line configurations that employ bundled subconductors are shown in Figure 2.13. The spacer-dampers are shown in the Figure 2.13a and Figure 213b.

Most practical transmission lines cannot maintain equilateral triangular spacing of conductors because of construction considerations, so the voltage drop per length of line is different for each phase. Rotating the phase configuration every one-third the length, or in other words, transposing the line, reduces these differences. Transposing also balances the effect of the conducting ground plane and reduces the mutual coupling of lines parallel to the three-phase system.

Consider next the transposed three-conductor, three-phase system shown in Figure 2.14. Let the distance from the phases to the ideal conductor ground plane be very large, so the reactive voltage drop in the first section, from p to q, for phase a is derived from Equation 2.40 as

$$E_p^a - E_q^a = jI^a X_a - j0.27942\left(\frac{f}{60}\right)[I^b \log_{10}(d_1) + I^c \log_{10}(d_3)] \quad (\text{V/mi}) \tag{2.45a}$$

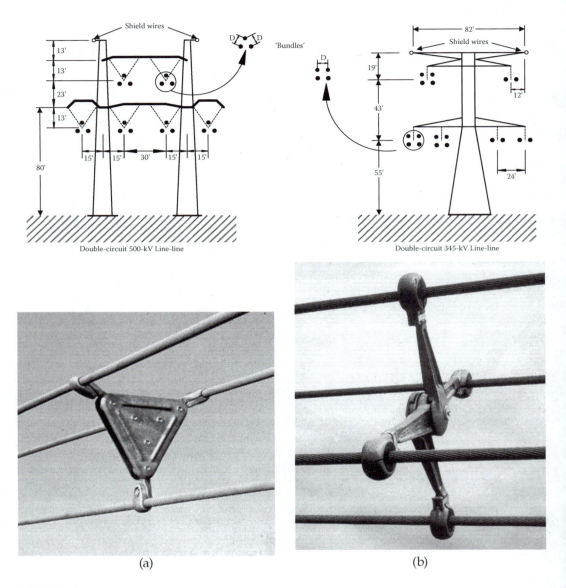

FIGURE 2.13
Typical bundled conductor transmission line configurations. (a,b) Two spacer-dampers for bundled subconductors. The spacers prevent clashing (mutual attraction due to electromagnetic forces) and provide damping of Aeolian vibration. (Photograph courtesy of Aluminum Company of America, Pittsburgh, PA)

In the second section phase *a* the voltage drop is

$$E_q^a - E_r^a = jI^a X_a - j0.27942\left(\frac{f}{60}\right)[I^b \log_{10}(d_{22}) + I^c \log_{10}(d_{23})] \quad (\text{V/mi}) \qquad (2.45b)$$

A similar equation describes the third section. Each conductor is in each of three positions, where $d_1 = d_{12} = d_{23} = d_{31}$, etc. The average voltage drop in phase *a* for the entire length

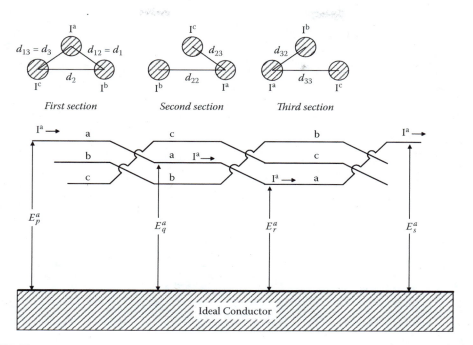

FIGURE 2.14
Schematic representation of a transposed three-phase line.

is the sum of the three sections divided by 3, and is written as

$$\Delta E^a_{average} = jI^a X_a - \left(\frac{1}{3}\right) j0.27942\left(\frac{f}{60}\right)[I^b \log_{10}(d_1 d_2 d_3) + I^c \log_{10}(d_1 d_2 d_3)] \quad \text{(V/mi)} \qquad (2.46a)$$

For balanced currents, $I^a = -I^b - I^c$, so the average voltage drop for the transposed line is

$$\Delta E^a_{average} = jI^a X_a + j0.27942\left(\frac{f}{60}\right) I^a \log_{10}(d_1 d_2 d_3)^{\frac{1}{3}}$$

$$= jI^a \left[X_a + 0.27942\left(\frac{f}{60}\right)\log_{10}(GMD)\right] \quad \text{(V/mi)}$$

$$= jI^a (X_a + X_b) = jI^a X_1 = jI^a X_2 \qquad (2.46b)$$

The quantity GMD is called the *geometric mean distance*. It is defined as

$$GMD = (d_1 d_2 d_3)^{1/3}$$

such that the reactance from Equation 2.46b defines

> Isolated three-phase, transposed line:
>
> $$X_1 = X_2 = X_a + 0.27942\left(\frac{f}{60}\right)\log_{10}(GMD) \quad (\Omega/\text{mi})$$

(2.47)

For the line isolated in space, there is no neutral or ground, so the zero-sequence impedance Z_o is an open circuit. If the line is not transposed, the GMD is often used as an approximation in computing X_1, or X_2 for power flow. Equation 2.42 is a special case for $GMD = d$.

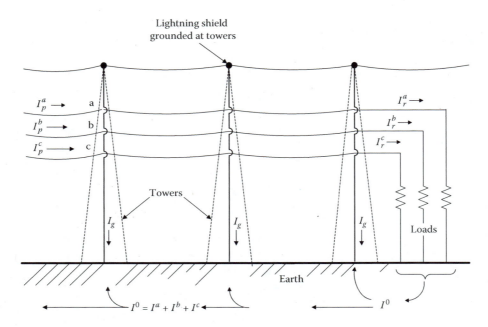

FIGURE 2.15
Grounded lightning shield and earth return.

2.3.3 Unbalanced Lines

The positive- and negative-sequence reactances derived to this point have neglected the earth, or ground plane, near the transmission line. Zero-sequence currents are in phase and flow through the a, b, c conductors to return through the neutral or earth plane. The earth and any grounded aerial lines used for lightning protection are effectively in the path of zero-sequence currents, as shown in Figure 2.15.

In order to derive the zero-sequence impedance, the earth plane and neutral effects are next examined. The earth is assumed to be an infinite plane of uniform resistivity ρ (Ω-m). Two parallel conductors at constant heights h_m and h_n over the earth employ the earth as a current return, as shown in Figure 2.16. The *effective* height of the conductor over a flat plane is assumed to be its height at the tower minus one-third of the line sag.

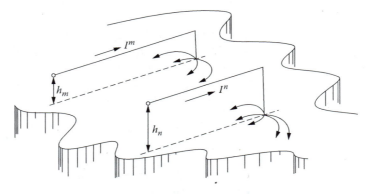

FIGURE 2.16
Two parallel wires with earth return.

The two grounded conductors have self- and mutual inductive impedances due to the earth return as derived from the fundamental work of Carson as referenced by Hesse [4]. The first two terms of an infinite series for the resistive and reactive Carson's corrections components due *only to the earth* in Ω/mile are:

$$\Delta r_{ii} = 0.095304\left(\frac{f}{60}\right) - 1.633 \times 10^{-6} h_i \sqrt{\frac{f}{\rho}} + + + \cdots \tag{2.48}$$

$$\Delta r_{mn} = 0.095304\left(\frac{f}{60}\right) - 0.8165 \times 10^{-6} (h_m + h_n) \sqrt{\frac{f}{\rho}} + + + \cdots \tag{2.49}$$

$$\Delta X_{ii} = 0.27942\left(\frac{f}{60}\right) \log_{10}\left(2162.5361 \sqrt{\frac{f}{\rho}}\right) + 1.633 \times 10^{-6} h_i \left(\frac{f^{\frac{3}{2}}}{\sqrt{\rho}}\right) + + + \cdots \tag{2.50}$$

$$\Delta X_{mn} = 0.27942\left(\frac{f}{60}\right) \log_{10}\left(2162.5361 \sqrt{\frac{f}{\rho}}\right) + 1.633 \times 10^{-6} (h_m + h_n) \sqrt{\frac{f}{\rho}} + + + \cdots \tag{2.51}$$

where the *ii* and *mn* subscripts refer to self- and mutual corrections, respectively.

Most investigators employ only the first term of the series and neglect the terms involving conductor heights above ground. The self- and mutual terms due to the earth are *added* to line parameters to obtain the three-phase zero-sequence impedance. In order to calculate the zero-sequence impedance of the transposed three-phase line, consider an unbalanced current *I* to flow in each phase line, so that unit zero-sequence current is in the earth return, as shown in Figure 2.17a.

The voltage difference across a line length $p - q$ with unbalanced current *I* flowing in all three phases, as shown in Figure 2.14, is

$$\begin{bmatrix} E_p^a - E_q^a \\ E_p^b - E_q^b \\ E_p^c - E_q^c \end{bmatrix} = \begin{bmatrix} Z_{aa} & Z_{ab} & Z_{ac} \\ Z_{ab} & Z_{bb} & Z_{bc} \\ Z_{ac} & Z_{bc} & Z_{cc} \end{bmatrix} \begin{bmatrix} I \\ I \\ I \end{bmatrix} \tag{2.52}$$

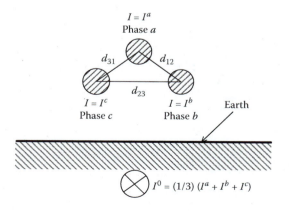

FIGURE 2.17a
Unbalanced current flow in phases with earth return.

Let the q terminal of the line be connected to the earth, and the p terminal of each phase be excited with E^0. Adding the rows of Equation 2.52 to obtain an average for the three-phase transposed line, as done in Equation 2.45, yields the zero-sequence impedance:

$$Z_0 = \left(\frac{E^0}{I}\right) = \left(\frac{1}{3}\right)(Z_{aa} + Z_{bb} + Z_{cc} + 2Z_{ab} + 2Z_{ac} + 2Z_{bc}) \tag{2.53}$$

The self- and mutual impedances of Equation 2.53 for the earth return are defined as:

$$
\boxed{
\begin{array}{c}
\text{Earth-corrected self and mutual impedances} \\
\text{(identical conductors)} \\
Z_{aa} = Z_{bb} = Z_{cc} = r_c + \Delta r_{ii} + j(X_a + \Delta X_{ii}) \quad (\Omega/\text{mi}) \\
Z_{ab} = Z_{bc} = Z_{ac} = \Delta r_{mn} + j\left[\Delta X_{mn} - 0.27942\left(\frac{f}{60}\right)\log_{10}(d_{mn})\right] \quad (\Omega/\text{mi})
\end{array}
}
$$

$$\tag{2.54}$$
$$\tag{2.55}$$

where r_c is the intrinsic resistance of the phase conductor (Ω/mile), and X_a is given in Equation 2.35. The Δ terms are Carson's earth corrections. Using only the first-order terms from Carson's earth corrections (those without heights h), the zero-sequence impedance is found by substituting Equation 2.54 and Equation 2.55 into Equation 2.53:

$$Z_0 = r_c + 0.285912\left(\frac{f}{60}\right) + 0.27942\left(\frac{f}{60}\right)\log_{10}\left[\frac{(2162.5361\sqrt{\rho/f})^3}{(d_1 d_2 d_3)^{\frac{2}{3}}\,GMR}\right] \quad (\Omega/\text{mi}) \tag{2.56}$$

It is convenient to define an equivalent depth of a fictitious return conductor as

$$D_{earth} = 2162.5361\sqrt{\rho/f}$$

such that the zero-sequence impedance for a three-phase line over resistive ground is

$$
\boxed{
\begin{array}{c}
\textbf{Zero sequence for transposed three-phase with earth return} \\
Z_0 = r_c + 0.285912\left(\frac{f}{60}\right) + 0.27942\left(\frac{f}{60}\right)\log_{10}\left[\frac{(D_{earth})^3}{(GMD)^2 GMR}\right] \quad (\Omega/\text{mi})
\end{array}
}
\tag{2.57}
$$

where the equivalent geometric mean distance between conductors, GMD, has been introduced to use consistent notation. Observe that Carson's first-order corrections do not affect $Z_1 = Z_2$ because they are expressed as a *difference* of a, b, c series and mutual impedances, $Z^s - Z^m$ (see Equation 2.19b). Hence, Equation 2.42 and Equation 2.47 are valid in the presence of earth for transposed lines without lighting shields. See appendix A for conductor resistance calculations, r_c, as dependent upon temperature.

2.3.4   Capacitance of Transmission Lines

The positive- and negative-sequence reactances due to line capacitances are next derived. For transmission lines, any charge on a conductor is assumed to be at the center of the

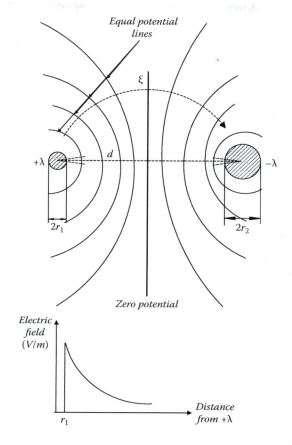

FIGURE 2.17b
Potential lines and electric fields due to two charged, long lines.

conductor radius because the distance between lines or between line and earth is great compared to the line radius. Figure 2.17b shows two long, parallel, charged lines isolated in space. The charges on the lines are $+\lambda$ and $-\lambda$ Coulombs per meter for a length l, considering conservation of charge. With the small radius assumption, Gauss's law for electric flux leaving the surface of a conductor r_1 is written as

Gauss's Law

$$\varepsilon_0 \oint \vec{\xi} \bullet d\vec{A} = \varepsilon_0 \xi 2\pi r_1 l = Q_{enclosed} = \lambda l \quad \text{(Colombs)}$$

In this last equation, ξ is the electric field in V/m, ε_0 is the permittivity of free space or air, and l is the length of line in meters. The voltage difference between the wires is a superposition due to both line charges. For the left-side conductor from radius r_1 to point d, the potential due to $+\lambda$ is

$$V_{r_1 d} = \int_{r_1}^{d} \vec{\xi} \bullet d\vec{r} = \int_{r_1}^{d} \frac{\lambda}{2\pi \varepsilon_0 r} dr = \frac{\lambda}{2\pi \varepsilon_0} \ln\left(\frac{d}{r_1}\right) \quad \text{(V)} \qquad (2.58a)$$

A similar equation holds for the $-\lambda$ charge. The voltage difference between r_1 and r_2 is

$$V_{r_1 r_2} = \frac{\lambda}{2\pi\varepsilon_0} \ln\left(\frac{d}{r_1}\right) - \frac{-\lambda}{2\pi\varepsilon_0} \ln\left(\frac{d}{r_2}\right) = \frac{\lambda}{2\pi\varepsilon_0} \ln\left(\frac{d^2}{r_1 r_2}\right) = \frac{\lambda}{\pi\varepsilon_0} \ln\left(\frac{d}{\sqrt{r_1 r_2}}\right) \qquad (2.58b)$$

For two wires separated by a distance d, with charges $+\lambda$ and $-\lambda$ Coulombs/m on the conductors, the capacitance of a pair of wires is the ratio of charge to voltage:

$$C_p = \frac{\lambda}{V_{r_1 r_2}} = \frac{\pi\varepsilon_0}{\ln\left(\frac{d}{\sqrt{r_1 r_2}}\right)} \quad \text{(F/m)} \qquad (2.58c)$$

For transmission lines, it is more significant to consider the capacitance of a single line. When the line is one of two wires of equal radii, the voltage at their mid-plane between the wires is zero, which is true only if each conductor has capacitance $2C_p$ to the distance d. Therefore, the capacitance for a single line of radius r to a distance d is found from Equation 2.58c to be

capacitance of a single conductor to distance d:

$$C_{single} = \frac{\lambda}{V_{rd}} = \frac{2\pi\varepsilon_0}{\ln\left(\frac{d}{r}\right)} \quad \text{(F/m)} \qquad (2.58d)$$

Initially consider the equilateral triangle configuration of the three-phase lines with the apparent line-to-line capacitances shown in Figure 2.18. There is no neutral wire and the capacitance C_p is the conductor to conductor as given by Equation 2.58c.

The positive- or negative-sequence impedance is also the phase a impedance to neutral:

$$-jX_1' = -jX_2' = \left.\frac{E^a}{I^a}\right|_{E^b=E^c=0}$$

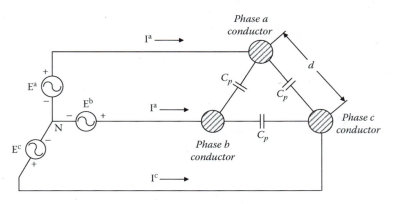

FIGURE 2.18
Equidistant conductors without neutral wire.

This is the Thévenin impedance presented to source E^a. The impedance into the circuit is found by short-circuiting voltage sources E^b and E^c, which effectively shorts the capacitance between conductors b and c. Hence, $2C_p$ is across the terminals of E^a and the neutral terminal. This is the positive-sequence impedance presented to phase a. In other words, the positive- or negative-sequence shunt reactance for a three-phase circuit is the *same* as a *single* conductor. Let the radius of the conductor be r and K be a unit conversion factor, then the capacitive reactance is

$$-jX_1' = -jX_2' = \frac{E^a}{I^a}\bigg|_{E^b=E^c=0} = K\left(\frac{-j}{\omega 2C_p}\right)$$

$$= -j0.0683\left(\frac{60}{f}\right)\left[\log_{10}\left(\frac{1}{r}\right) + \log_{10}(d)\right] = -j(X_a' + X_b') \quad \text{(M}\Omega\text{-miles)} \quad (2.59)$$

To extend this single conductor result to a line transposed every 1/3 its length, with spacing d_1, d_2, and d_3 between conductors, the capacitive shunt reactance is

$$\boxed{\begin{array}{l}
\text{Shunt capacitive reactance for transposed, isolated lines:} \\[4pt]
-jX_1' = -jX_2' = -j(X_a' + X_b') \\[4pt]
= -jX_a' - j0.0683\left(\frac{60}{f}\right)[\log_{10}(d_1) + \log_{10}(d_2) + \log_{10}(d_3)] \quad \text{(M}\Omega\text{-miles)} \\[4pt]
= -j0.0683\left(\frac{60}{f}\right)\left[\log_{10}\left(\frac{1}{GMR_B'}\right) + \log_{10}(GMD)\right]
\end{array}} \quad (2.60)$$

The bundle equivalent radius GMR_B' reduces to r for a single conductor. The zero-sequence network for shunt reactances is open circuited for the equilateral triangle or transposed when there is no neutral and ground-plane effects are negligible.

For high voltages, a bundle of conductors is used to reduce the electric field at the surface of the wires and thereby decrease the ionization of surrounding air, a phenomenon called corona. Ionization of the air causes power loss, audible noise, and radio frequency noise, and increases the probability of flashover from the conductor to other conductors or nearby structures. The electric field as seen from Gauss's law is a maximum V/m at the surface of the conductor. The electric field near r_1 can be expressed [3] in terms of the voltage between conductors, $V_{r_1r_2} = V_{LL}$, Equation 2.58b and Gauss's law as:

$$\xi = \frac{\lambda}{2\pi\varepsilon_0 r_1} = \frac{V_{LL}}{2r_1 \ln\left(\frac{d}{\sqrt{r_1 r_2}}\right)} \quad \text{(line-to-ground electric field near } r_1, \text{ V/m)}$$

When the transposed line is at a height $h = d/2$ above the ground plane, this is the midplane of two lines. In this case, the electric field for a voltage V_{LG} with respect to ground, for a conductor of radius r, may be expressed at any distance r as

$$\xi = \frac{\lambda}{2\pi\varepsilon_0 r} = \frac{V_{LG}}{r \ln\left(\frac{2h}{r}\right)} \quad \text{(line-to-ground electric field near } r, \text{ V/m)}$$

Extensive descriptions of electric field effects and transmission line losses in fair and rainy weather, called corona losses, are available in reference [3].

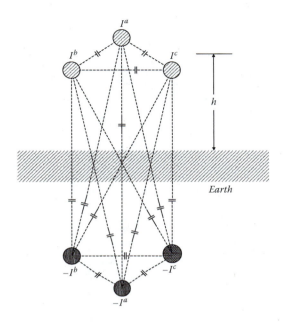

FIGURE 2.19
Equivalent shunt capacitances in the presence of earth.

When transmission lines are above electrically conductive ground, $\rho < \infty$, electric images are induced in the ground plane. The method of images may be used to determine the equivalent shunt capacitances as shown in Figure 2.19. The effective height of the conductor over a plane is its height at the tower minus one-third of the line sag between towers. Over rough terrain an average value of h is used.

For the transmission lines and images of Figure 2.19, it may be shown that the zero-sequence capacitive reactance is given by

$$
\begin{array}{l}
\textbf{Shunt zero-sequence reactance, transposed, above earth:} \\[6pt]
-jX_0' = -j\left(X_a' - 2X_b' + X_{earth}'\right) \\[6pt]
\quad = -j0.0683\left(\dfrac{60}{f}\right)\left[\log_{10}\left(\dfrac{1}{GMR_B'}\right) - 2\log_{10}(GMD) + 3\log_{10}(2h)\right] \quad (\text{M}\Omega\text{-miles}) \quad (2.61)
\end{array}
$$

where GMR_B' reduces to r for the case of a single conductor.

In general, the conductor-dependent term X_a' is computed for various materials and bundle configurations, then the line geometry terms are added to compute X_b' as in Equation 2.61. Table 2.4 is a summary of X_a' for various sizes of aluminum-conductor-steel-reinforced (ACSR) cables and bundle configurations. Table 2.5 gives single-conductor resistance and other properties of ACSR. *Linear interpolation* is used to calculate conductor resistance for operating temperatures between values specified in Table 2.5. The geometry of ACSR cable is shown in appendix A. An example will demonstrate use of the tables.

TABLE 2.4

Capacitive Reactance (X'_a) of Aluminum-Conductor-Steel-Reinforced (ACSR) Bundled Conductors at 60 Hz in Megaohm-Miles for 1 Ft Spacing

Code	kcmil Al	Sq. mm Tot.	Strand	Dia. (in.)	Single Cond.	2-Cond. Spacing			3-Cond. Spacing			4-Cond. Spacing		
						12	18	24	12	18	24	12	18	24
Joree	2515.	1344.	76/19	1.880	0.0755	0.0377	0.0317	0.0275	0.0252	0.0171	0.0115	0.0163	0.0073	0.0009
Thrasher	2312.	1235.	76/19	1.802	0.0767	0.0384	0.0323	0.0281	0.0256	0.0176	0.0119	0.0166	0.0076	0.0012
Kiwi	2167.	1146.	72/7	1.735	0.0778	0.0389	0.0329	0.0286	0.0259	0.0179	0.0123	0.0169	0.0079	0.0015
Bluebird	2156.	1181.	84/19	1.762	0.0774	0.0387	0.0327	0.0284	0.0258	0.0178	0.0121	0.0168	0.0078	0.0014
Chuckar	1781.	976.	84/19	1.602	0.0802	0.0401	0.0341	0.0298	0.0267	0.0187	0.0130	0.0175	0.0085	0.0021
Falcon	1590.	908.	54/19	1.545	0.0813	0.0406	0.0346	0.0304	0.0271	0.0191	0.0134	0.0178	0.0087	0.0023
Plover	1431.	817.	54/19	1.465	0.0828	0.0414	0.0354	0.0312	0.0276	0.0196	0.0139	0.0181	0.0091	0.0027
Bobolink	1431.	775.	45/7	1.426	0.0836	0.0418	0.0358	0.0315	0.0279	0.0199	0.0142	0.0183	0.0093	0.0029
Pheasant	1272.	726.	54/19	1.382	0.0846	0.0423	0.0363	0.0320	0.0282	0.0202	0.0145	0.0186	0.0096	0.0032
Bittern	1272.	689.	45/7	1.345	0.0854	0.0427	0.0367	0.0324	0.0285	0.0205	0.0148	0.0188	0.0098	0.0034
Finch	1114.	636.	54/19	1.293	0.0866	0.0433	0.0373	0.0330	0.0289	0.0208	0.0152	0.0191	0.0101	0.0037
Bluejay	1113.	603.	45/7	1.258	0.0873	0.0437	0.0377	0.0334	0.0291	0.0211	0.0154	0.0193	0.0103	0.0039
Ortolan	1034.	560.	45/7	1.213	0.0884	0.0442	0.0382	0.0340	0.0295	0.0215	0.0158	0.0195	0.0105	0.0041
Cardinal	954.	546.	54/7	1.196	0.0889	0.0444	0.0384	0.0342	0.0296	0.0216	0.0159	0.0196	0.0106	0.0042
Rail	954.	517.	45/7	1.165	0.0896	0.0448	0.0388	0.0345	0.0299	0.0219	0.0162	0.0198	0.0108	0.0044
Drake	795.	469.	26/7	1.108	0.0911	0.0456	0.0396	0.0353	0.0304	0.0224	0.0167	0.0202	0.0112	0.0048
Condor	796.	456.	54/7	1.093	0.0915	0.0458	0.0398	0.0355	0.0305	0.0225	0.0168	0.0203	0.0113	0.0049
Cuckoo	795.	455.	24/7	1.092	0.0916	0.0458	0.0398	0.0355	0.0305	0.0225	0.0168	0.0203	0.0113	0.0049
Tern	795.	431.	45/7	1.063	0.0923	0.0462	0.0402	0.0359	0.0308	0.0228	0.0171	0.0205	0.0115	0.0051
Rook	636.	364.	24/7	0.977	0.0949	0.0474	0.0414	0.0372	0.0316	0.0236	0.0179	0.0211	0.0121	0.0057
Grosbeak	636.	375.	26/7	0.990	0.0944	0.0472	0.0412	0.0370	0.0315	0.0235	0.0178	0.0210	0.0120	0.0056
Parakeet	557.	319.	24/7	0.914	0.0968	0.0484	0.0424	0.0381	0.0323	0.0243	0.0186	0.0216	0.0126	0.0062
Dove	556.	328.	26/7	0.927	0.0964	0.0482	0.0422	0.0379	0.0321	0.0241	0.0184	0.0215	0.0125	0.0061
Osprey	556.	298.	18/1	0.879	0.0980	0.0490	0.0430	0.0387	0.0327	0.0247	0.0190	0.0219	0.0129	0.0065
Hawk	477.	281.	26/7	0.858	0.0987	0.0493	0.0433	0.0391	0.0329	0.0249	0.0192	0.0221	0.0131	0.0067

Source: Reprinted with permission from Electric Power Research Institute, *Transmission Line Reference Book 345 kV and Above.* Copyright © 1975.

TABLE 2.5

Characteristics of Multilayer Aluminum Conductor Steel Reinforced (ACSR)

Code[a]	Cross Section kcmil Al	Cross Section Sq. mm Al	Cross Section Sq. mm Tot.	Stranding Aluminum	Stranding Steel	Diameter Cond. (in.)	Diameter Core (in.)	Layers	Wt (lb/1,000 ft)	STRG (kips)	DC 25°C	Resistance (Ω/mile) AC at 60 Hz 25°C	50°C	75°C	100°C	GMR (ft)	X_a (Ω/mile)	X'_a (MΩ/mile)
Joree	2515.	1274.	1344.	76 × .1819	19 × .0849	1.880	.425	4	2749	61.7	.0365	.0418	.0450	.0482	.0516	.0621	.337	.0755
Thrasher	2312.	1171.	1235.	76 × .1744	19 × .0814	1.802	.407	4	2526	57.3	.0397	.0446	.0482	.0518	.0554	.0595	.342	.0767
Kiwi	2167.	1098.	1146.	72 × .1735	7 × .1157	1.735	.347	4	2303	49.8	.0424	.0473	.0511	.0550	.0589	.0570	.348	.0778
Bluebird	2156.	1092.	1181.	84 × .1602	19 × .0961	1.762	.480	4	2511	60.3	.0426	.0466	.0505	.0544	.0584	.0588	.344	.0774
Chukar	1781.	902.	976.	84 × .1456	19 × .0874	1.602	.437	4	2074	51.0	.0516	.0549	.0598	.0646	.0695	.0534	.355	.0802
Falcon	1590.	806.	908.	54 × .1716	19 × .1030	1.545	.515	3	2044	54.5	.0578	.0602	.0657	.0712	.0767	.0521	.358	.0813
Lapwing	1590.	806.	862.	45 × .1880	7 × .1253	1.504	.376	3	1792	42.2	.0590	.0622	.0678	.0734	.0790	.0497	.364	.0821
Parrot	1510.	765.	862.	54 × .1672	19 × .1003	1.505	.502	3	1942	51.7	.0608	.0631	.0689	.0748	.0806	.0508	.362	.0821
Nuthatch	1510.	765.	818.	45 × .1832	7 × .1221	1.465	.366	3	1702	40.1	.0622	.0652	.0711	.0770	.0830	.0485	.367	.0828
Plover	1431.	725.	817.	54 × .1628	19 × .0977	1.465	.489	3	1840	49.1	.0642	.0663	.0725	.0787	.0849	.0494	.365	.0828
Bobolink	1431.	725.	775.	45 × .1783	7 × .1189	1.427	.357	3	1613	38.3	.0656	.0685	.0747	.0810	.0873	.0472	.371	.0836
Martin	1351.	685.	772.	54 × .1582	19 × .0949	1.424	.475	3	1737	46.3	.0680	.0700	.0765	.0831	.0897	.0480	.368	.0837
Dipper	1351.	685.	732.	45 × .1733	7 × .1155	1.386	.347	3	1522	36.2	.0695	.0722	.0788	.0855	.0922	.0459	.374	.0845
Pheasant	1272.	645.	726.	54 × .1535	19 × .0921	1.382	.461	3	1635	43.6	.0722	.0741	.0811	.0881	.0951	.0466	.372	.0846
Bittern	1272.	644.	689.	45 × .1681	7 × .1121	1.345	.336	3	1434	34.1	.0738	.0764	.0835	.0906	.0977	.0445	.378	.0854
Grackle	1192.	604.	681.	54 × .1486	19 × .0892	1.338	.446	3	1533	41.9	.0770	.0788	.0863	.0938	.1013	.0451	.376	.0855
Bunting	1193.	604.	646.	45 × .1628	7 × .1085	1.302	.326	3	1344	32.0	.0787	.0811	.0887	.0963	.1039	.0431	.382	.0863
Finch	1114.	564.	636.	54 × .1436	19 × .0862	1.293	.431	3	1431	39.1	.0825	.0842	.0922	.1002	.1082	.0436	.380	.0866
Bluejay	1113.	564.	603.	45 × .1573	7 × .1049	1.258	.315	3	1255	29.8	.0843	.0866	.0947	.1029	.1111	.0416	.386	.0873
Curlew	1033.	523.	591.	54 × .1383	7 × .1383	1.245	.415	3	1331	36.6	.0909	.0924	.1013	.1101	.1190	.0420	.385	.0877
Ortolan	1033.	523.	560.	45 × .1515	7 × .1010	1.212	.303	3	1165	27.7	.0909	.0930	.1018	.1106	.1195	.0401	.390	.0885
Merganser	954.	483.	596.	30 × .1785	7 × .1783	1.248	.535	2	1493	46.0	.0987	.0995	.1092	.1189	.1286	.0430	.382	.0876
Cardinal	954.	483.	546.	54 × .1329	7 × .1329	1.196	.399	3	1229	33.8	.0984	.0998	.1094	.1191	.1287	.0404	.389	.0889
Rail	954.	483.	517.	45 × .1456	7 × .0971	1.165	.291	3	1075	25.9	.0984	.1004	.1099	.1195	.1291	.0385	.395	.0896
Baldpate	900.	456.	562.	30 × .1732	7 × .1732	1.212	.520	2	1410	43.3	.1046	.1054	.1156	.1259	.1362	.0417	.385	.0885
Canary	900.	456.	515.	54 × .1291	7 × .1291	1.162	.387	3	1159	31.9	.1043	.1056	.1158	.1260	.1362	.0392	.393	.0897
Ruddy	900.	456.	487.	45 × .1414	7 × .0943	1.131	.283	3	1015	25.4	.1043	.1062	.1163	.1265	.1367	.0374	.399	.0905
Crane	875.	443.	501.	54 × .1273	7 × .1273	1.146	.382	3	1126	31.4	.1073	.1086	.1191	.1296	.1401	.0387	.395	.0901
Willet	874.	443.	474.	45 × .1394	7 × .0929	1.115	.279	3	987	25.0	.1073	.1092	.1196	.1301	.1406	.0369	.400	.0909

Skimmer	795.	403.	497.	30 × .1628	7 × .1628	1.140	.488	2	1246	38.3	.1183	.1191	.1307	.1423	.1540	.0392	.393	.0903
Mallard	795.	403.	495.	30 × .1628	19 × .0977	1.140	.489	2	1235	38.4	.1183	.1191	.1307	.1423	.1540	.0392	.393	.0903
Drake	795.	403.	469.	26 × .1749	7 × .1360	1.108	.408	2	1094	31.5	.1180	.1190	.1306	.1422	.1538	.0375	.399	.0911
Condor	795.	403.	455.	54 × .1213	7 × .1213	1.092	.364	3	1024	28.2	.1181	.1193	.1309	.1425	.1541	.0368	.401	.0916
Cuckoo	795.	403.	455.	24 × .1820	7 × .1213	1.092	.364	2	1024	27.9	.1181	.1193	.1308	.1424	.1540	.0366	.402	.0916
Tern	795.	403.	431.	45 × .1329	7 × .0886	1.063	.266	3	896	22.1	.1181	.1197	.1313	.1428	.1544	.0352	.406	.0923
Coot	795.	403.	414.	36 × .1486	1 × .1486	1.040	.149	3	805	16.5	.1175	.1197	.1311	.1426	.1540	.0337	.411	.0930
Buteo	715.	362.	447.	30 × .1544	7 × .1544	1.081	.463	2	1119	34.4	.1316	.1322	.1452	.1581	.1711	.0372	.399	.0919
Redwing	715.	362.	445.	30 × .1544	19 × .0926	1.081	.463	2	1111	34.6	.1316	.1322	.1452	.1581	.1711	.0372	.399	.0919
Starling	716.	363.	422.	26 × .1659	7 × .1290	1.051	.387	2	985	28.4	.1312	.1321	.1450	.1579	.1707	.0355	.405	.0927
Crow	715.	362.	409.	54 × .1151	7 × .1151	1.036	.345	3	921	26.3	.1312	.1323	.1452	.1580	.1709	.0350	.407	.0931
Stilt	716.	363.	410	24 × .1727	7 × .1151	1.036	.345	2	922	25.5	.1311	.1323	.1451	.1579	.1708	.0347	.408	.0931
Grebe	716.	363.	388.	45 × .1261	7 × .0841	1.009	.252	3	807	20.6	.1312	.1327	.1455	.1583	.1712	.0334	.413	.0939
Gannet	666.	338.	393.	26 × .1601	7 × .1245	1.014	.374	2	917	26.6	.1409	.1417	.1555	.1694	.1832	.0343	.409	.0937
Gull	667.	338.	382.	54 × .1111	7 × .1111	1.000	.333	3	858	24.5	.1408	.1418	.1557	.1695	.1833	.0337	.411	.0942
Flamingo	667.	338.	382	24 × .1667	7 × .1111	1.000	.333	2	859	23.7	.1407	.1418	.1556	.1694	.1832	.0335	.412	.0942
Scoter	636.	322.	397.	30 × .1456	7 × .1456	1.019	.437	2	993	30.8	.1480	.1486	.1631	.1777	.1923	.0351	.406	.0936
Egret	636.	322.	396.	30 × .1456	19 × .0874	1.019	.437	2	988	31.5	.1480	.1485	.1631	.1777	.1923	.0351	.406	.0936
Grosbeak	636.	322.	375.	26 × .1564	7 × .1216	0.990	.365	2	875	25.2	.1476	.1484	.1629	.1774	.1920	.0335	.412	.0944
Goose	636.	322.	364.	54 × .1085	7 × .1085	.0977	.326	3	819	23.6	.1477	.1486	.1631	.1776	.1922	.0330	.414	.0949
Rook	636.	322.	364.	24 × .1628	7 × .1085	.977	.326	2	819	22.0	.1476	.1485	.1630	.1775	.1920	.0327	.415	.0949
Kingbird	636.	322.	340.	18 × .1880	1 × .1880	.940	.188	2	691	15.7	.1468	.1484	.1627	.1771	.1915	.0304	.424	.0960
Swift	636.	322.	331.	36 × .1329	1 × .1329	.930	.133	3	644	13.4	.1469	.1487	.1630	.1774	.1918	.0302	.425	.0963
Wood Duck	605.	307.	378.	30 × .1420	7 × .1420	.994	.426	2	947	29.4	.1556	.1561	.1714	.1868	.2021	.0342	.410	.0943
Teal	605.	307.	376.	30 × .1420	19 × .0852	.994	.426	2	940	30.0	.1556	.1561	.1714	.1868	.2021	.0342	.410	.0943
Squab	605.	306.	356.	26 × .1525	7 × .1186	.966	.356	2	833	23.6	.1552	.1560	.1713	.1866	.2018	.0327	.415	.0952
Peacock	605.	307.	346.	24 × .1588	7 × .1588	.953	.318	2	780	21.6	.1551	.1560	.1712	.1865	.2018	.0319	.418	.0956
Duck	606.	307.	347.	54 × .1059	7 × .1059	.953	.318	3	779	22.5	.1550	.1559	.1711	.1864	.2016	.0322	.417	.0956
Eagle	557.	282.	348.	30 × .1362	7 × .1362	.953	.409	2	872	27.2	.1691	.1696	.1863	.2029	.2196	.0328	.415	.0956
Dove	556.	282.	328.	26 × .1463	7 × .1138	.927	.341	2	766	22.4	.1687	.1694	.1860	.2026	.2192	.0313	.420	.0964
Parakeet	557.	282.	319.	24 × .1523	7 × .1015	.914	.305	2	717	19.8	.1686	.1695	.1860	.2026	.2192	.0306	.423	.0968
Osprey	556.	282.	298.	18 × .1758	1 × .1758	.879	.176	2	604	13.7	.1679	.1693	.1857	.2022	.2187	.0284	.432	.0980
Hen	477.	242.	298.	30 × .1261	7 × .1261	.883	.378	2	747	23.8	.1973	.1977	.2171	.2366	.2560	.0304	.424	.0979
Hawk	477.	242.	281.	26 × .1354	7 × .1053	.858	.316	2	657	19.5	.1969	.1975	.2169	.2363	.2557	.0290	.430	.0987
Flicker	477.	242.	273.	24 × .1410	7 × .0940	.846	.282	2	615	17.2	.1967	.1975	.2168	.2362	.2556	.0283	.432	.0991
Pelican	477.	242.	255.	18 × .1628	1 × .1628	.814	.163	2	518	11.8	.1958	.1970	.2162	.2355	.2547	.0263	.441	.1003

(continued)

TABLE 2.5

Characteristics of Multilayer Aluminum Conductor Steel Reinforced (ACSR) (Continued)

Code[a]	Cross Section			Stranding		Diameter		Layers	Wt (lb/1,000 ft)	STRG (kips)	DC 25°C	Resistance (Ω/mile) AC at 60 Hz				GMR (ft)	Reactance: 1 ft rad., 60 Hz	
	kcmil Al	Sq. mm Al	Sq. mm Tot.	Aluminum	Steel	Cond. (in.)	Core (in.)					25°C	50°C	75°C	100°C		X_a (Ω/mile)	X'_a (MΩ/mile)
Ibis	397.	201.	234.	26 × .1236	7 × .0961	0.783	.288	2	547	16.3	.2363	.2368	.2601	.2834	.3067	.0265	.441	.1014
Brant	398.	201.	228.	24 × .1287	7 × .0858	0.772	.257	2	512	14.7	.2361	.2367	.2600	.2833	.3066	.0259	.444	.1018
Chickadee	397.	201.	213.	18 × .1486	1 × .1486	0.743	.149	2	432	9.9	.2350	.2360	.2591	.2822	.3054	.0240	.452	.1030
Oriole	336.	170.	210.	30 × .1059	7 × .1059	0.741	.318	2	527	17.0	.2797	.2800	.3076	.3352	.3628	.0255	.445	.1030
Linnet	336.	170.	198.	26 × .1137	7 × .0884	0.720	.265	2	463	14.0	.2793	.2797	.3072	.3348	.3623	0.243	.451	.1039
Widgeon	336.	170.	193.	24 × .1184	7 × .0789	0.710	.237	2	433	12.5	.2790	.2795	.3070	.3345	.3621	.0238	.454	.1043
Merlin	336.	170.	180.	18 × .1367	1 × .1367	0.684	.137	2	366	8.6	.2777	.2785	.3059	.3332	.3606	.0221	.463	.1054
Piper	300.	152.	187.	30 × .1000	7 × .1000	0.700	.300	2	470	15.5	.3137	.3139	.3449	.3758	.4068	.0241	.452	.1047
Ostrich	300.	152.	177.	26 × .1074	7 × .0835	0.680	.251	2	413	12.7	.3130	.3134	.3443	.3751	.4060	.0230	.458	.1056
Gadwall	300.	152.	172.	24 × .1118	7 × .0745	0.671	.224	2	386	11.2	.3129	.3134	.3442	.3751	.4060	.0225	.461	.1060
Phoebe	300.	152.	160.	18 × .1281	1 × .1291	0.646	.129	2	326	7.7	.3114	.3121	.3428	.3735	.4042	.0209	.469	.1071
Junco	267.	135.	167.	30 × .0943	7 × .0943	0.660	.283	2	418	13.7	.3527	.3530	.3878	.4226	.4574	.0227	.459	.1065
Partridge	267.	135.	157.	26 × .1013	7 × .0788	0.642	.236	2	367	11.3	.3518	.3522	.3869	.4216	.4563	.0217	.465	.1073
Waxwing	267.	135.	143.	18 × .1217	1 × .1217	0.609	.122	2	289	6.9	.3504	.3510	.3856	.4201	.4547	.0197	.477	.1089

Source: Reprinted with permission from Electric Power Research Institute, *Transmission Line Reference Book 345 kV and Above.* Copyright © 1975.

[a] The bird code names in the first column that are *not* indented are sizes preferred by the American Society for Testing and Materials (ASTM).

Example 2.5

Calculate the symmetrical component reactances $Z_0, Z_1 = Z_2, X_0', X_1' = X_2'$ for a horizontal, three-phase, transposed transmission line, above 100 Ω-m earth, three conductors per phase of 18-in. spacing. This is a code name Rail-type conductor, 1.165 in. diameter line, with 38-ft phase spacing as shown in Figure E2.5a. There are no lightning shield wires. Show the π-equivalent circuit for a 100-mile line in a positive-sequence diagram in per unit for a 100 MVA, 345 kV base.

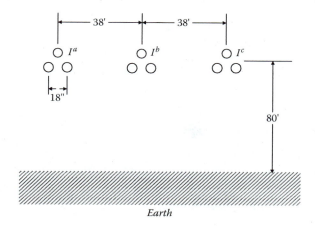

FIGURE E2.5a
Three conductor bundles of a 345 kV transmission line.

Solution

First the inductive reactance up to 1 ft for the bundled conductors from Table 2.3 is:

$$X_a = 0.27942 \log_{10} \left(\frac{1}{GMR_B} \right) = 0.099 \quad (\Omega/\text{mile})$$

which is directly read from the twelfth column. In more detailed steps, GMR = 0.0385 ft is calculated for unequal current distribution in the composite steel and aluminum conductor of 1.165 in. diameter, then GMR_B is calculated for conductor spacing $D = 1.5$ ft. The GMR_B of the bundled conductor configuration is found using the X_a reactance

$$GMR_B = \frac{1}{anti \log_{10}(X_a/0.2794)} = 0.4423 \quad (\text{ft})$$

The GMD between phases is calculated as

$$GMD = \sqrt[3]{38 \times 38 \times 76} = 47.9 \quad (\text{ft})$$

and the inductive reactance of the line is calculated using Equation 2.47 as

$$X_1 = X_2 = X_a + 0.27942 \log_{10}(GMD) = 0.099 + 0.4695 = 0.5685 \quad (\Omega/\text{mile})$$

The resistance of a conductor varies with wind, ambient temperature, and current, but a typical value at 50°C is 0.1099 Ω/mile for a Rail ACSR conductor from Table 2.5. There are three conductors per phase; hence the positive-sequence reactance is

$$Z_1 = Z_2 = 0.0366 + j0.5685 \quad (\Omega/\text{mile})$$

The multiple-conductor configuration is taken into account by GMR_B, in calculating the zero-sequence impedance due to earth return. The depth of a fictitious return conductor within the earth is

$$D_e = 2162.5361\sqrt{\frac{\rho}{f}} = 2792 \quad \text{(ft)}$$

The zero-sequence impedance is given by substituting values into Equation 2.57:

$$Z_0 = r_c + 0.285912 + j0.27942\log_{10}\left(\frac{D_e^3}{GMR_B(GMD)^2}\right)$$

$$= 0.0366 + 0.285912 + j0.27942\log_{10}\left(\frac{(2792)^3}{0.4423(47.9)^2}\right)$$

$$= 0.3225 + j2.049 \quad (\Omega/\text{mile})$$

The shunt positive-sequence capacitive reactance is calculated using Equation 2.60 as

$$X_1' = X_2' = 0.0623\left(\frac{60}{f}\right)\left[\log_{10}\left(\frac{1}{GMR_B'}\right) + \log_{10}(GMD)\right] \quad (\text{M}\Omega\text{-mi})$$

where the GMR_B' is calculated using the conductor radius, as follows:

$$GMR_B' = \sqrt[N]{N\gamma^{N-1}r} = \sqrt[3]{3\left[\frac{1.5}{2\sin(\pi/3)}\right]^2\left(\frac{1.165}{24}\right)} = 0.4780 \quad \text{(ft)}$$

Substituting numerical values into the X_1' expression, we obtain

$$X_1' = X_2' = 0.0623\left(\frac{60}{f}\right)\left[\log_{10}\left(\frac{1}{0.4780}\right) + \log_{10}(47.9)\right] = 0.1366 \quad (\text{M}\Omega - \text{mi})$$

The zero-sequence shunt reactance is calculated from Equation 2.61 for $h = 80$ ft:

$$X_0' = 0.0683\left[\log_{10}\left(\frac{1}{GMR_0'}\right) - 2\log_{10}(GMD) + 3\log_{10}(2h)\right]$$

$$= 0.02189 - 2(0.1148) + 0.4516 = 0.2440 \quad (\text{M}\Omega - \text{mi})$$

The impedance base to normalize the calculated parameters is

$$X_{BASE} = \frac{V_{BASE}^2}{S_{BASE}} = \frac{(345)^2}{100} = 1190.25 \quad \Omega$$

A 100-mile length of line has the following per-unit values:

$$Z_0 = 0.027 + j0.1721 \qquad Z_1 = Z_2 = 0.0031 + j0.04776$$

$$X_0' = 2.050 \quad X_1' = X_2' = 1.148$$

These values are shown on the π-equivalents on Figure E2.5b and Figure E2.5c. The shunt capacitance is equally divided at both ends of the transmission line equivalent circuit. A line-charging capacitive reactance of 1.148 p.u. is usually considered excessive and would be compensated with reactors in a practical transmission line.

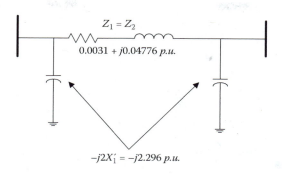

FIGURE E2.5b
Positive- and negative-sequence π-equivalent circuit in p.u. for a 100-mile line.

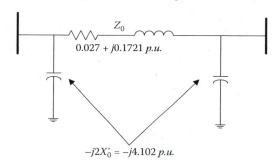

FIGURE E2.5c
Zero-sequence π-equivalent circuit in p.u. for a 100-mile line.

Observe that series sequence reactances are essentially *decoupled* from the shunt capacitive parameters of the π-equivalent circuit assumed to model the short- and medium-length lines. For more accurate representation with regard to frequency characteristics, it is possible to cascade several π-sections.

© The computer programs *l_param.exe* and *c_param.exe* may be used to respectively calculate the symmetrical component series impedances and shunt capacitive reactances for nontransposed lines. Copy the data file *example25.sss* and paste it into *l_param.sss*, then execute both programs. The series impedances match the above results. However, because of images in the ground plane (developed in the next section), the shunt capacitive reactances differ from the estimate for ideally transposed, isolated lines in space.

As the number of conductors increases, such as double-hung circuits, or when there are multiple lightning shields, and ground effects are present, it is clear that a general numerical approach is necessary. The general approach can also be used to calculate mutual coupling effects between adjacent transmission lines. A general method is presented in the next session.

2.3.5 General Method to Determine Aerial Transmission Line Parameters

Methods of the preceding section to compute self-impedance of a conductor or mutual impedance between conductors, with or without the presence of earth and neutral wires, may be extended, in a general formulation. Consider a double-hung, parallel circuit, with three-phase lines for circuit A, circuit B, lightning shields S, plus conducting earth, as

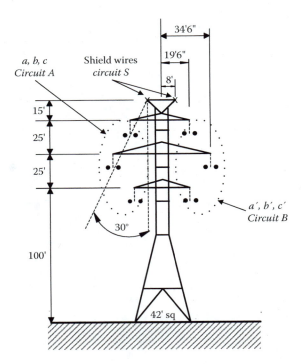

FIGURE 2.20
Typical double-hung, offset vertical configuration with lightning shields.

shown in Figure 2.20. A 30° "shielding angle" between the lightning shield wire and the nearest phase conductor is shown in this figure. The angle is a significant factor in effective protection from lightning strokes [3].

Circuits *A* and *B* of Figure 2.20 are often operated with *a-a'*, *b-b'*, and *c-c'* in parallel.

Because of geometrical differences between conductors and nearby grounds, unbalanced three-phase electrical conditions exist. Unless symmetry is present between both circuits with respect to earth and the shields, balanced three-phase terminal currents induce unbalanced sequence potentials in both circuits. The unbalanced potentials cause a *circulating* current to flow through the phases. Since line impedances are low, the circulating current can be appreciable. The circulating currents due to unbalanced geometry in a double-hung circuit can be calculated in terms of unbalanced factors [3], but a general formulation calculates the unbalances directly.

Each conductor (or bundle) of both circuits is to be approximated by a π-equivalent. As the first step, the line series impedances are determined. Consider a length of line between terminals *p* and *q* with fixed center-to-center distances d_{mn} between conductors (or bundles). Each conductor (or bundle) in the *A*, *B*, and *S* group has the following self- and mutual impedances (in the presence of earth) to all other conductors (Ω/mile):

$$Z_{ii} = r_c + \Delta r_{ii} + j(X_a + \Delta i X_{ii}) \tag{2.62}$$

$$Z_{mn} = \Delta r_{mn} + j\left(\Delta X_{mn} - 0.27942\left(\frac{f}{60}\right)\log_{10}(d_{mn})\right) \tag{2.63}$$

where the Δ terms are Carson's corrections specified in Equation 2.48 to Equation 2.51, and r_c is the conductor (or bundle) resistance. The self-impedance to a distance of 1 ft, X_a, is the

same as previously defined. Writing a matrix form of Kirchhoff's voltage law for groups
A, *B*, and *S*, the voltage difference across a length of line using the series impedances of
groups of conductors is

$$
\mathbf{E}_p - \mathbf{E}_q = \begin{bmatrix} \mathbf{E}_A \\ \mathbf{E}_B \\ \mathbf{E}_S \end{bmatrix} = \begin{bmatrix} \mathbf{Z}^x_{AA} & \mathbf{Z}^x_{AB} & \mathbf{Z}^x_{AS} \\ \mathbf{Z}^x_{BA} & \mathbf{Z}^x_{BB} & \mathbf{Z}^x_{BS} \\ \mathbf{Z}^x_{SA} & \mathbf{Z}^x_{SB} & \mathbf{Z}^x_{SS} \end{bmatrix} \begin{bmatrix} \mathbf{I}_A \\ \mathbf{I}_B \\ \mathbf{I}_S \end{bmatrix} = \mathbf{Z}^x\mathbf{I} \tag{2.64}
$$

where $\mathbf{E}_A$ is the vector of circuit *A* phase voltage differences and $\mathbf{E}_B$ corresponds to circuit
B and $\mathbf{E}_S$ to the shields. The submatrices of Equation 2.64 are defined as:

$\mathbf{Z}^x_{AA} \equiv$ self-and mutual impedances for phases *a*, *b*, and *c* of circuit *A* of the double-
    hung line

$\mathbf{Z}^x_{AB} = (\mathbf{Z}^x_{BA})^t \equiv$ mutual coupling between three-phase circuits *A* and *B*

$\mathbf{Z}^x_{AS} \equiv (\mathbf{Z}^x_{SA})^t \equiv$ mutual coupling between circuit *A* and the shield circuit *S*

$\mathbf{Z}^x_{BB} \equiv$ circuit *B*, impedances for phases *a′*, *b*, and *c′*

$\mathbf{Z}^x_{BS} = (\mathbf{Z}^x_{SB})^t \equiv$ mutual coupling between circuit *B* and the shield wires

$\mathbf{Z}^x_{SS} \equiv$ self- and mutual impedances of shield wires

The conductors are grouped in the *A*, *B*, and *S* arrangements to eliminate the shield
wires by grounding, $\mathbf{E}_S = 0$, at every other tower or at points along the length of line.
Equation 2.64 is also suitable to derive the mutual coupling between the *A* and *B* three-
phase circuits. Because there are two systems of conductors, to achieve balance each
phase conductor must be transposed within its group and with respect to the parallel
three-phase line. For a double circuit, there are nine possible arrangements, as shown in
Figure 2.21.

If the circuits are always operated with *a-a′*, *b-b′*, *c-c′* connected electrically, then configurations (b) and (c), (d) and (e), and (g) and (i) in Figure 2.21 are equivalent, reducing the number
of combinations to six. For the phase *a* conductor in any of nine positions shown in Figure 2.21,
the submatrices of Equation 2.64 take on different values. Various combinations of the nine
configurations are possible to transpose the parallel system of conductors. Let each of *k* =
1, 2, ..., *K* successive transpositions be defined as $\mathbf{Z}(k)$. The average impedance for the transposed lines is

$$
\bar{\mathbf{Z}} = \frac{1}{K}\sum_{k=1}^{K} \mathbf{Z}(k) = \begin{bmatrix} \bar{\mathbf{Z}}_{AA} & \bar{\mathbf{Z}}_{AB} & \bar{\mathbf{Z}}_{AS} \\ \bar{\mathbf{Z}}_{BA} & \bar{\mathbf{Z}}_{BB} & \bar{\mathbf{Z}}_{BS} \\ \bar{\mathbf{Z}}_{SA} & \bar{\mathbf{Z}}_{SB} & \bar{\mathbf{Z}}_{SS} \end{bmatrix} \tag{2.65}
$$

The lightning shield wires are eliminated using $\mathbf{E}_S = 0$ and substituting the shield currents into the phase equations:

$$
\begin{bmatrix} \mathbf{E}_A \\ \mathbf{E}_B \end{bmatrix} = \left\{ \begin{bmatrix} \bar{\mathbf{Z}}_{AA} & \bar{\mathbf{Z}}_{AB} \\ \bar{\mathbf{Z}}_{BA} & \bar{\mathbf{Z}}_{BB} \end{bmatrix} - \begin{bmatrix} \bar{\mathbf{Z}}_{AS} \\ \bar{\mathbf{Z}}_{BS} \end{bmatrix} \bar{\mathbf{Z}}^{-1}_{SS} [\bar{\mathbf{Z}}_{SA} \quad \bar{\mathbf{Z}}_{SB}] \right\} \begin{bmatrix} \mathbf{I}_A \\ \mathbf{I}_B \end{bmatrix}
$$

$$
= \begin{bmatrix} \bar{\mathbf{Z}}_{AA} & \bar{\mathbf{Z}}_{AB} \\ \bar{\mathbf{Z}}_{BA} & \bar{\mathbf{Z}}_{BB} \end{bmatrix} \begin{bmatrix} \mathbf{I}_A \\ \mathbf{I}_B \end{bmatrix} = \mathbf{Z}_{av}\mathbf{I} \tag{2.66}
$$

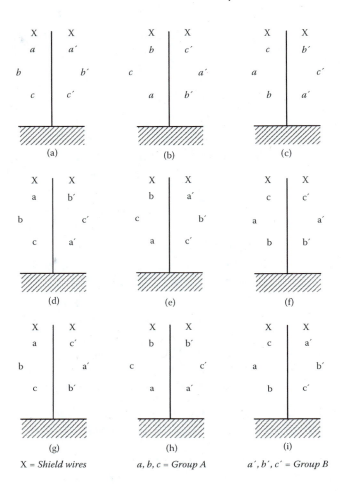

FIGURE 2.21
Transposing a double-hung vertical circuit for balance.

The symmetrical component transformation is applied to both the A and B conductor systems and yields the impedance matrix:

$$\mathbf{Z}_S^{012} = \begin{bmatrix} \mathbf{T}_s^{-1} & 0 \\ 0 & \mathbf{T}_s^{-1} \end{bmatrix} \mathbf{Z}_{av} \begin{bmatrix} \mathbf{T}_s & 0 \\ 0 & \mathbf{T}_s \end{bmatrix}$$

$$= \begin{bmatrix} \mathbf{T}_s^{-1}\mathbf{Z}_{AA}\mathbf{T}_s & \mathbf{T}_s^{-1}\mathbf{Z}_{AB}\mathbf{T}_s \\ \mathbf{T}_s^{-1}\mathbf{Z}_{BA}\mathbf{T}_s & \mathbf{T}_s^{-1}\mathbf{Z}_{BB}\mathbf{T}_s \end{bmatrix} \tag{2.67}$$

where $\mathbf{Z}_{AB} = \mathbf{Z}'_{BA}$. The submatrix $\mathbf{T}_s^{-1}\mathbf{Z}_{AB}\mathbf{T}_s$ in Equation 2.67 indicates symmetrical component coupling between the A and B systems. Mutual coupling of the zero-sequence component is *always* present between the circuits, and only in the case of ideal transposing of phases from one circuit with respect to phases of the other does the positive-sequence coupling go to zero. For example, all nine transpositions of Figure 2.21 are required to eliminate positive-sequence coupling between circuits A and B.

For a double-hung circuit, if A and B have identical conductors and are symmetric with respect to each other as well as grounds (e.g., (a), (f), and (h) of Figure 2.21), then $\mathbf{Z}_{AA} = \mathbf{Z}_{BB}$.

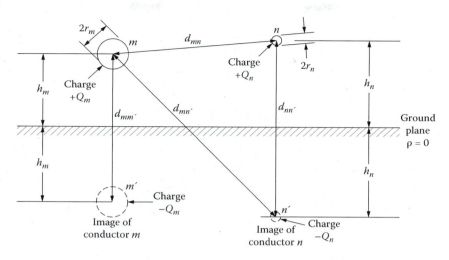

FIGURE 2.22
Two-wire conductors above infinitely conductive earth (cross-sectional view).

If these two circuits are operated in parallel, connected at their terminals, it is easily shown that the paralleled circuit impedance, $\mathbf{Z}_p^{012}$, is

> Parallel operations, symmetrical double-hung circuit:
>
> $$Z_p^{012} = \frac{1}{2}\mathbf{T}_s^{-1}(\mathbf{Z}_{AA} + \mathbf{Z}_{AB})\mathbf{T}_s = \frac{1}{2}\mathbf{T}_s^{-1}(\mathbf{Z}_{BB} + \mathbf{Z}_{AB})\mathbf{T}_s$$

(2.68)

The capacitances between lines of a multiconductor system are determined next. The electrostatic potential in a linear dielectric space (or a conductor) is superposition of the electric fields due to all local electrostatic charges. For an ac system in steady state, the ac charges and currents are related by $I = j\omega Q$. Therefore, the ratio of charge to applied voltage for the geometry of an electrostatic field problem is used to calculate the interconductor capacitances. For line conductors with charges $+Q_m$ and $+Q_n$ coulombs/mile, of radii r_m and r_n respectively, parallel to a zero-resistivity earth plane, the geometry is as shown in Figure 2.22. Equivalent image charges $-Q_m$ and $-Q_n$ replace charges induced in the earth surface.

For distance $d_{ik} \gg r_i$, the $\pm Q_m$ and $\pm Q_n$ charges on conductors can be considered concentrated at the centers. The relative potential at conductor m (charge Q_m) with respect to all other charges is calculated by integrating the electric field gradient [2] due to charges $-Q_m$, $+Q_n$, and $-Q_n$. Assign the potential V_m to conductor m:

$$
\begin{aligned}
V_m &= -\left(\frac{-Q_m}{2\pi\varepsilon_0}\right)\int_{r_m}^{d_{mm'}}\frac{dr}{r} - \left(\frac{Q_n}{2\pi\varepsilon_0}\right)\int_{r_n}^{d_{mn}}\frac{dr}{r} - \left(\frac{-Q_n}{2\pi\varepsilon_0}\right)\int_{r_n}^{d_{mn'}}\frac{dr}{r} \\
&= \frac{Q_m}{2\pi\varepsilon_0}\ln\left(\frac{d_{mm'}}{r_m}\right) - \frac{Q_n}{2\pi\varepsilon_0}\ln\left(\frac{d_{mn}}{r_n}\right) + \frac{Q_n}{2\pi\varepsilon_0}\ln\left(\frac{d_{mn'}}{r_n}\right) \\
&= \frac{Q_m}{2\pi\varepsilon_0}\ln\left(\frac{d_{mm'}}{r_m}\right) + \frac{Q_n}{2\pi\varepsilon_0}\ln\left(\frac{d_{mn'}}{d_{mn}}\right) \\
&= p_{11}Q_m + p_{12}Q_n
\end{aligned}
$$

(2.69)

In Equation 2.69, p_{11} and p_{12} are called potential coefficients and are conveniently expressed in units of miles per farad. A simultaneous equation may be written for the potential of conductor n and solved in conjunction with Equation 2.69 for the ratio of charge in terms of potentials.

For the general case of N conductors, including phase conductors and lightning shield wires, the potential at the conductors is written as

$$\mathbf{E} = \mathbf{PQ} = \frac{1}{j\omega}\mathbf{PI'} = \mathbf{Z'I'} \tag{2.70}$$

In Equation 2.70, $\mathbf{E}$ is an N-dimensional vector of the potential of each conductor with respect to the ground plane, $\mathbf{Q}$ is an N vector of charge, $\mathbf{I'}$ is a vector of shunt currents (A/mile) due to the capacitances, $\mathbf{Z'}$ is an $N \times N$ matrix of impedances (Ω-mile) due to shunt capacitances, and $\mathbf{P}$ is an $N \times N$ matrix of potential coefficients, defined as follows:

$$p_{mm'} = 2.5718 \times 10^7 \log_{10}\left(\frac{d_{mm'}}{r_m}\right) \quad \text{miles/F} \tag{2.71}$$

$$p_{mn} = 2.5718 \times 10^7 \log_{10}\left(\frac{d_{mn'}}{d_{mn}}\right) \quad m \neq n \quad \text{miles/F} \tag{2.72}$$

where $d_{mn} \equiv$ center-to-center distance (ft) from conductor m (or bundle m) to conductor n (or bundle n), $d_{mn'} \equiv$ center-to-center distance (ft) from conductor m to *image n'*, and $r_m \equiv$ radius (ft) of conductor m (or geometric mean radius for a bundle, GMR$'_B$).

If the N conductors include shield wires that are grounded along the length of line, their potentials are set to zero. For a double-hung circuit, let the N conductors be separated into a three-phase A circuit, three-phase B circuit, and shield, so Equation 2.70 is partitioned as

$$\begin{bmatrix} \mathbf{E}_A \\ \mathbf{E}_B \\ 0 \end{bmatrix} = \begin{bmatrix} \mathbf{Z}'_{AA} & \mathbf{Z}'_{AB} & \mathbf{Z}'_{AS} \\ \mathbf{Z}'_{BA} & \mathbf{Z}'_{BB} & \mathbf{Z}'_{BS} \\ \mathbf{Z}'_{SA} & \mathbf{Z}'_{SB} & \mathbf{Z}'_{SS} \end{bmatrix} \begin{bmatrix} \mathbf{I}'_A \\ \mathbf{I}'_B \\ \mathbf{I}'_S \end{bmatrix} \tag{2.73}$$

The shield wires are eliminated to yield

$$\begin{bmatrix} \mathbf{E}_A \\ \mathbf{E}_B \end{bmatrix} = \left\{ \begin{bmatrix} \mathbf{Z}'_{AA} & \mathbf{Z}'_{AB} \\ \mathbf{Z}'_{BA} & \mathbf{Z}'_{BB} \end{bmatrix} - \begin{bmatrix} \mathbf{Z}'_{AS} \\ \mathbf{Z}'_{BS} \end{bmatrix} [\mathbf{Z}'_{SS}]^{-1} [\mathbf{Z}'_{SA} \quad \mathbf{Z}'_{SB}] \right\} \begin{bmatrix} \mathbf{I}'_A \\ \mathbf{I}'_B \end{bmatrix}$$

$$= \frac{1}{j\omega}\mathbf{P'} \begin{bmatrix} \mathbf{I}'_A \\ \mathbf{I}'_B \end{bmatrix} \tag{2.74}$$

where $\mathbf{P'}$ is the reduced matrix of potential coefficients.

As a last step, the shunt admittance matrix $\mathbf{Y}$ is calculated by inverting $\mathbf{P'}/j\omega$:

$$\mathbf{Y} = j\omega[\mathbf{P'}]^{-1} \quad \text{S/mile} \tag{2.75}$$

If the lines are transposed, P' is an average over the length of lines, P'_{av}. The symmetrical components in admittance form are given as

$$\mathbf{Y}_S^{012} = j\omega \left\{ \begin{bmatrix} \mathbf{T}_s^{-1} & 0 \\ 0 & \mathbf{T}_s^{-1} \end{bmatrix} \mathbf{P}'_{av} \begin{bmatrix} \mathbf{T}_s & 0 \\ 0 & \mathbf{T}_s \end{bmatrix} \right\}^{-1} \quad \text{S/mile} \tag{2.76}$$

Half the admittance $\mathbf{Y}_S^{012}$ is allocated to each terminal of the π-equivalent for the line. A numerical example is used to demonstrate the application of the general approach to calculate line parameters.

Example 2.5.1

A double-hung vertical-offset 60 Hz transmission line has the dimensions shown in Figure E2.5.1 for conductors in position (a) of Figure 2.21.

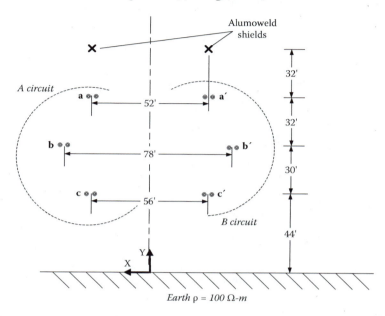

FIGURE E2.5.1

The phase wires are two ACSR conductors, code name Kiwi, 1.735 in. diameter, at 18 in. spacing. The two lightning shield wires are 0.548-in.-diameter aluminum-clad steel (Alumoweld) with the following properties:

$r_c = 1.2021\ \Omega/\text{mile}$ at 20°C
$\text{GMR} = 0.00296\ \text{ft}$
$X_a = 0.707\ \Omega/\text{mile}$
$X'_a = 0.1121\ \text{M}\Omega\text{-mile}$

Assume that the phase conductor temperature is 75°C with 60 Hz resistance, 0.055 Ω/mile. The resistivity of the earth is 100 Ω-m. Use Carson's first-order term for earth corrections and do the following:

(a) Let circuits A and B be energized by different sources. Calculate self- and mutual symmetrical component series impedances per mile for a single section of untransposed line.
(b) Repeat part (a) for a transposed line, employing transpositions (a), (f), and (h) of Figure 2.21. Calculate the line-charging capacitances for the transposed lines.
(c) Calculate the symmetrical component series impedances and shunt capacitances for an ideally transposed line employing all nine conductor positions of Figure 2.21.

Solution

(a) The dual transmission line conductors have $X_a = 0.149$ as read from Table 2.3 and an equivalent $X'_a = 0.0329$ as read from Table 2.4. This value of X'_a yields an equivalent $\text{GMR}'_B = 0.330\ \text{ft}$. Using the first term from Carson's earth corrections, numerical values are introduced in Equation 2.62 and Equation 2.63 in order to calculate the impedance matrix for circuits A and B and shields. The line conductors are positioned as shown in Figure E2.5.1. The Z^x matrices are

$$
\mathbf{Z}^x = \begin{bmatrix} \mathbf{Z}^x_{AA} & \mathbf{Z}^x_{AB} & \mathbf{Z}^x_{AS} \\ & \mathbf{Z}^x_{BB} & \mathbf{Z}^x_{BS} \\ \diagup & \text{symmetry} & \mathbf{Z}^x_{SS} \end{bmatrix}
$$

$$
\mathbf{Z}^x_{AA} = \mathbf{Z}^x_{BB} = \begin{bmatrix} 0.12280 + j1.1118 & 0.095304 + j0.53301 & 0.095304 + j0.46196 \\ \diagup & 0.12280 + j1.1118 & 0.095304 + j0.54246 \\ \text{Symmetry} & & 0.12280 + j1.1118 \end{bmatrix}
$$

$$
\mathbf{Z}^x_{AB} = \mathbf{Z}^x_{BA} = \begin{bmatrix} 0.095304 + j0.48336 & 0.095304 + j0.44312 & 0.095304 + j0.42777 \\ 0.095304 + j0.44312 & 0.095304 + j0.43416 & 0.095304 + j0.44152 \\ 0.095304 + j0.42777 & 0.095304 + j0.44152 & 0.095304 + j0.47437 \end{bmatrix}
$$

$$
\mathbf{Z}^x_{AS} = \begin{bmatrix} 0.095304 + j0.54228 & 0.095304 + j0.46388 \\ 0.095304 + j0.45571 & 0.095304 + j0.41516 \\ 0.095304 + j0.41149 & 0.095304 + j0.39421 \end{bmatrix} \quad \Omega/\text{mile}
$$

$$
\mathbf{Z}^x_{SS} = \begin{bmatrix} 1.2974 + j1.6698 & 0.095304 + j0.48336 \\ 0.095304 + j0.44336 & 1.2974 + j1.6698 \end{bmatrix} \quad \Omega/\text{mile}
$$

The lightning shield wires are eliminated using Equation 2.66; then the phase impedances are transformed into symmetrical components according to Equation 2.67. The result for an untransposed, symmetrical line is

$$
\mathbf{T}_s^{-1}\mathbf{Z}_{AA}\mathbf{T}_s = \begin{bmatrix} 0.38978 + j1.65230 & 0.02205 - j0.04720 & 0.00021 - j0.03965 \\ 0.0021 + j0.03965 & 0.02881 + j0.59746 & -0.03883 + j0.02929 \\ 0.02205 - j0.04720 & 0.04088 + j0.02767 & 0.02881 + j0.59746 \end{bmatrix} \quad \Omega/\text{mile}
$$

$$
\mathbf{T}_s^{-1}\mathbf{Z}_{AB}\mathbf{T}_s = \begin{bmatrix} 0.35967 + j0.85697 & -0.00620 - j0.02321 & 0.02657 - j0.01643 \\ 0.02657 + j0.01643 & 0.00091 + j0.02498 & -0.01884 + j0.01306 \\ 0.00620 - j0.02321 & 0.02031 + j0.01200 & 0.00091 + j0.02498 \end{bmatrix} \quad \Omega/\text{mile}
$$

When the *untransposed* lines are operated in parallel,

$$
\mathbf{Z}_p^{012} = \frac{1}{2}\mathbf{T}_s^{-1}[\mathbf{Z}_{AA} + \mathbf{Z}_{AB}]\mathbf{T}_s =
$$

	0 – 0′	1 – 1′	2 – 2′	
	0.37473 + j1.25463	0.00792 − j0.03521	0.01339 − j0.02804	Ω/mile
	0.01339 + j0.02804	0.01486 + j0.31122	−0.02883 + j0.02117	
	0.00792 − j0.03521	0.03060 + j0.01983	0.01486 + j0.3112	

(b) The transmission line is transposed through symmetrical positions (a), (f), and (h) of Figure 2.21 and the line impedances are averaged. The averaging can be performed using off-diagonal terms of matrices $\mathbf{Z}^x_{AA}$ and $\mathbf{Z}^x_{AB}$ above. Alternately, the *x-y* positions of the conductors can be rotated to recompute these matrices for each configuration. The resulting symmetrical component matrices are

$\mathbf{Z}_S^{012} =$

	0	1	2	0'	1'	2'
	0.38978 + j1.6523	0	0	0.35967 + j0.85697	0	0
	0	0.02750 + j0.59937	0	0	j0.02650	0
	0	0	0.02750 + j0.59937	0	0	j0.02650
	0.35967 + j0.85697	0	0	0.38978 + j1.6523	0	0
	0	j0.0265	0	0	0.02750 + j0.59937	0
	0	0	j0.0265	0	0	0.02750 + j0.59937

Ω/mile

Notice that the mutual coupling between the parallel lines has a large-magnitude zero-sequence component, but a comparatively small positive-sequence component. If nonsymmetrical transpositions are used, the positive-sequence coupling has both resistance and reactance. When the *symmetrically transposed lines* are operated in parallel, the result is

$\mathbf{Z}_p^{012} =$

0–0'	1–1'	2–2'
0.37473 + j1.25463	0	0
0	0.01375 + j0.31294	0
0	0	0.01375 + j0.31294

Ω/mile

The diagonal terms of the transposed line impedance matrices differ by several percent with the results in part (a) for an untransposed line. Often, line power flow calculations assume that the lines are transposed even if they are not.

The corresponding symmetrical component shunt admittances due to line charging are calculated using Equation 2.70 to Equation 2.76. The results are given as a capacitance matrix for the line transposed through three positions:

$C_S^{012} = \dfrac{1}{j\omega}\mathbf{Y}_S =$

	0	1	2	0'	1'	2'
	0.011915	0	0	−0.0026803	0	0
	0	0.018800	0	0	−000075112	0
	0	0	0.018800	0	0	−0.00075112
	−0.0026803	0	0	0.011915	0	0
	0	−0.00075112	0	0	0.018800	0
	0	0	−0.0007512	0	0	0.01800

μF/mile

(c) The parallel lines are *ideally transposed*, and the resulting series symmetrical component impedances are

$$\mathbf{Z}_S^{012} =$$

0	1	2	0'	1'	2'
0.38978 + j1.65230	0	0	0.35967 + j0.85697	0	0
0	0.02750 + j0.59937	0	0	0	0
0	0	0.02750 + j0.59937	0	0	0
0.35967 + j0.85697	0	0	0.38978 + j1.65230	0	0
0	0	0	0	0.02750 + j0.59937	0
0	0	0	0	0	0.02750 + j0.59937

Ω/mile

where only zero-sequence coupling is present between the parallel circuits. The shunt capacitance matrix for the *ideally transposed* lines is

$$\mathbf{C}_S^{012} =$$

0	1	2	0'	1'	2'
0.011915	0	0	−0.0026803	0	0
0	0.018770	0	0	0	0
0	0	0.018770	0	0	0
−0.0026803	0	0	0.011915	0	0
0	0	0	0	0.018770	0
0	0	0	0	0	0.018770

μF/mile

The shunt capacitance matrix has only zero-sequence coupling between the ideally transposed lines.

©The results calculated for nontransposed Z^{012} series impedance (Ω/mi) and shunt susceptance Y^{012} ($\Omega S/m$) matrices may be calculated by means of the computer programs. To do this, copy the file *example251.sss* and paste it into *l_param.sss*, then execute *l_param.exe* and *c_param.exe*.

In the following chapters, the transmission lines are assumed to be transposed for purposes of short-circuit calculations, contingencies, and power flow. For short circuit calculations, the mutual coupling impedance of parallel lines is retained, but shunt capacitor coupling between lines is neglected. These approximations introduce errors in the numerical calculations, but the errors are small in comparison with idealizing the geometry of the line. Parameter corrections due to line sag, coupling to grounded towers, variations in clearance to earth, variations in resistivity, switchyard capacitances, and so on, are usually larger than the errors introduced by theoretical approximations.

2.4 Transformer Representation

The preceding sections have demonstrated the advantages of symmetrical components, so the analytical description of transformers such as shown in Figure 2.23 will be treated from this point of view. Also, a π-equivalent for transformers is desirable in order to be compatible with transmission line models. There are a great variety of transformers used in transmission networks, such as two-winding wye to wye, single-winding autotransformers, wye-to-delta combinations, three-winding, phase-shifting transformers, and so on. Only several of the most common types are treated here. A suitable analytical description can be derived from the two-winding linearized transformer equivalent circuit shown in Figure 2.24. There are N_1 turns on the primary winding and N_2 turns on the secondary of the ideal transformer in the equivalent circuit. Because the core losses and magnetization current for power transformers are on the order of 1% of maximum ratings, the core of the transformer is often neglected for power flow and fault calculations. The saturation effects and harmonic distortion are also beyond the detail required here. As a result, only a

FIGURE 2.23
345/118 kV autotransformer, oil-immersed, external oil coolers, rated at 400 MVA. (Photograph courtesy of Westinghouse Electric Corporation Transformer Division, Muncie, IN)

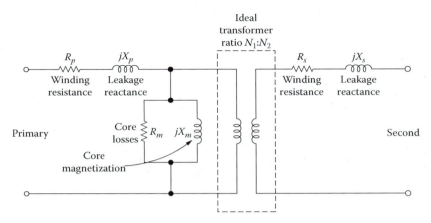

FIGURE 2.24
Single-phase linearized circuit equivalent for an iron-core transformer.

leakage reactance plus a resistance representing winding and core loss (as reflected to one winding) are used to approximate the transformer electrical characteristics.

Consider the single-phase equivalent of a three-phase transformer with negligible excitation currents as shown in Figure 2.25. The off-nominal turns ratio, α, is defined as:

$$\alpha = \frac{\text{actual secondary voltage}}{\text{rated secondary voltage}} \times \frac{\text{rated primary voltage}}{\text{actual primary voltage}} \tag{2.77}$$

This normalizes <u>open circuit</u> primary and secondary voltages by the nameplate rating. An $\alpha = 1$ means the secondary is at rated voltage, while $\alpha < 1$ means the secondary is below rated voltage, and $\alpha > 1$ is higher than rated. This definition converts both windings of the transformer to the same voltage base, which is required for network calculations, but the per-unit impedance is different from both sides of the transformer. A π-equivalent with different shunt elements to ground, as shown in Figure 2.26, is used to describe the impedance differences because of off-nominal voltage levels. The unknown impedances Z_A, Z_B, and Z_C are found from open- and short-circuit tests performed on the transformer of Figure 2.25a. These tests and the combination of impedances from Figure 2.26 are:

Excite q, short p:

$$Z_{psc} = Z_{pq} = \frac{Z_A Z_B}{Z_A + Z_B} \tag{2.78a}$$

Excite q, open p:

$$\infty = \frac{Z_B(Z_A + Z_C)}{Z_B + Z_A + Z_C} \rightarrow Z_C = -Z_A - Z_B \tag{2.78b}$$

Excite p, short q:

$$Z_{qsc} = \alpha^2 Z_{pq} = \frac{Z_A Z_C}{Z_A + Z_C} \tag{2.78c}$$

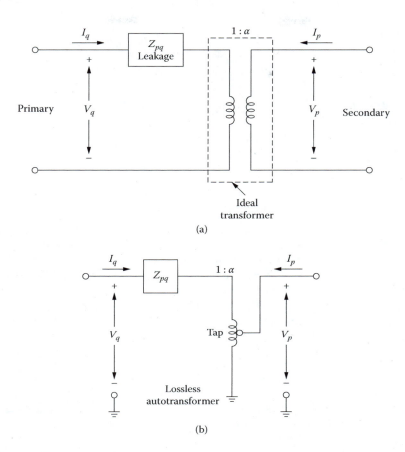

(a)

(b)

FIGURE 2.25
Two-winding transformers (one- or three-phase): (a) separate windings; (b) autotransformer.

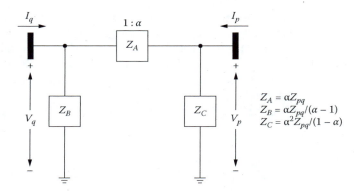

$$Z_A = \alpha Z_{pq}$$
$$Z_B = \alpha Z_{pq}/(\alpha - 1)$$
$$Z_C = \alpha^2 Z_{pq}/(1 - \alpha)$$

FIGURE 2.26
π-equivalent for a two-winding transformer.

Excite p, open q:

$$\infty = \frac{Z_C(Z_A + Z_B)}{Z_C + Z_Z + Z_B} \rightarrow Z_C = -Z_A - Z_B \tag{2.78d}$$

Using Equation 2.78d and simultaneously solving Equations 2.78a and 2.78c, we obtain

$$Z_A = \alpha Z_{pq} \tag{2.79a}$$

$$Z_B = \frac{\alpha}{\alpha - 1} Z_{pq} \tag{2.79b}$$

$$Z_C = \frac{\alpha^2}{1 - \alpha} Z_{pq} \tag{2.79c}$$

where Z_{pq} is the per-unit leakage or equivalent impedance of the transformers. Figure 2.26 shows the resulting off-nominal π-equivalent.

If the autotransformer shown in Figure 2.25b has a variable tap on the winding, the secondary voltage may be regulated by moving the tap setting either manually, with an automatic voltage sensor to adjust the tap (tap changing under load (TCUL)), or in response to a command from a remote dispatch computer. The incremental change in tap settings of TCUL transformers is on the order of $\frac{5}{8}$ of 1% per step. If the voltage at bus q is regulated, the tap is moved whenever

$$\left[|E_q| - |E_q|_{desired} \right] > 0.01 + \epsilon \tag{2.80}$$

where ϵ is a fraction of 1% and selected to avoid unnecessary movement, and hence wear on the mechanism. There are high and low limits of travel for the tap-adjusting mechanism, and often time delays in the voltage-sensing circuits to prevent tracking of transients or short time disturbances. The tap may be moved to control a voltage at some point remote to the transformer, or to regulate Mvar from the transformer. Whenever the tap is disturbed, the π-equivalent derived in Equation 2.79 also changes.

2.4.1 Wye–Delta and Phase-Shift Transformers

For protecting a system due to ground faults and zero-sequence currents, many power systems employ wye-to-delta transformers that introduce 30° phase shifts in primary with respect to secondary voltages. Multiwinding transformers may employ wyes on primary-to-secondary autotransformers and also have a floating delta tertiary. To describe analytically the wye–delta phase shift or special transformers that incrementally shift the input-output voltages, consider the leakage reactance or equivalent impedance in series with an ideal shift element as shown in Figure 2.27. The phase-shift element has the property

$$E_t = (a + jb)E_q \quad a^2 + b^2 = 1 \tag{2.81}$$

such that complex conjugate voltages are obtained from q to t versus t to q. The power into terminal q equals that into t because the phase shifter is lossless:

$$S_q = S_t = E_q I_q^* = E_t I_{tp}^* \tag{2.82}$$

and the currents are related by

$$I_q = \frac{E_t^*}{E_q^*} I_{tp} = (a - jb)I_{tp} \tag{2.83}$$

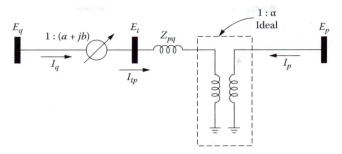

FIGURE 2.27
Phase-shifting transformer.

The current I_{tp}, described by the π-equivalent circuit of Figure 2.26, is

$$I_{tp} = \frac{E_t}{Z_B} + \frac{E_t - E_p}{Z_A} = \left(\frac{1}{Z_B} + \frac{1}{Z_A} \right)(a + jb)E_q - \frac{E_p}{Z_A} \qquad (2.84)$$

Substituting the current from Equation 2.84 into Equation 2.83 gives us

$$I_q = \left(\frac{1}{Z_B} + \frac{1}{Z_A} \right) E_q - \frac{a - jb}{Z_A} E_p \qquad (2.85)$$

Considering the p terminal of the transformer, the current I_p may be derived as

$$I_p = \left(\frac{1}{Z_C} + \frac{1}{Z_A} \right) E_p - \frac{a + jb}{Z_A} E_q \qquad (2.86)$$

Comparing Equation 2.85 and Equation 2.86, the mutual impedance between input and output of the transformer is Hermitian (complex conjugate), and hence it makes the power network nonsymmetric. The power flow program of Chapter 5 takes this into account.

Phase-shifting transformers operate incrementally over a limited range of angles. The movement of the mechanical actuator can again be automatic to regulate real or reactive power, or may be commanded from a remote signal or dispatching source.

Example 2.6

Neglect magnetization for an 80:7.6 kV, 50 MVA, 0.11 p.u. leakage, wye–delta transformer. The transformer is delivering rated or 1.0 p.u. power at 0.95 p.u. voltage to a balanced load at 0.80 lagging power factor. For a load on the low-voltage side, find the transformer input voltage and power. Express the results in per unit and in engineering units.

Solution

Let the voltage at the load be used as the phase-angle reference. From Equation 2.20a and the definition of per-unit power, the power to the balanced load is expressed as

$$S_{p.u.} = E_{p.u.} I_{p.u.}^* = \frac{S^{abc}}{S_{BASE}} = \frac{3S^{012}}{S_{BASE}} = 0.8 + j0.6$$

This expression is "looking out" from the low-voltage terminals into the load. Substituting numerical values for the load power and voltage, the per-unit current is

$$I_{p.u.} = \left(\frac{S_{p.u.}}{E_{p.u.}} \right)^* = \left(\frac{0.8 + j0.6}{0.95 + j0} \right)^* = 0.842 - j0.632$$

which is a lagging power factor load.

Loads or generation are usually expressed "looking into" the networks for power systems. In this case the network is a transformer, so the algebraic signs of the load power and current are reversed for injections into the network. The equivalent circuit is shown in Figure E2.6 using the subscripts L for the load, S for an internal point of the transformer (without the phase shift), and p for the high-voltage primary. The equivalent voltage at the S terminal of the transformer is calculated from Kirchhoff's voltage law as

$$E_s = I_s Z_t + E_L = (-I_L)Z_t + E_L$$

$$= (0.842 - j0.632)(j0.110) + (0.95 + j0)$$

$$= 1.0195 + j0.093 = 1.024\angle 5.21° \quad \text{p.u.}$$

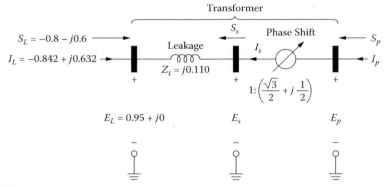

FIGURE E2.6

The input voltage to output voltage of the transformer has a phase shift due to the delta-to-wye construction. As specified in Table 2.6, the phase shift is standardized to be +30° to the high-voltage side of any wye-to-delta or delta-to-wye transformer, so that

$$E_p = \left(\frac{\sqrt{3}}{2} + j\frac{1}{2}\right)E_s = 1.024 \angle 35.21° \quad \text{p.u.}$$

and the primary current is

$$I_p = \left(\frac{\sqrt{3}}{2} + j\frac{1}{2}\right)I_s = \left(\frac{\sqrt{3}}{2} + j\frac{1}{2}\right)(0.842 - j0.632)^*$$

$$= 1.050 - j0.126 \quad \text{p.u.}$$

The input power to the transformer is

$$S_p = E_p I_p^* = (1.0195 + j0.093)(0.842 - j0.632)^*$$

$$= 0.80 + j0.723 \quad \text{p.u.}$$

Using the base power, $S_{BASE} = 50$ MVA, and the high-voltage base, $V_{BASE} = 80$ kV, the input to the transformer in engineering units is

$$S_p = 40 + j36.15(\text{MW, Mvar})$$

$$|E_p| = |E_p|(\text{p.u.}) | \times 80 = 81.92(\text{kV line-to-line})$$

TABLE 2.6

On-Nominal ($\alpha = 1$) Equivalent Symmetrical Component Per-Unit Circuits for Three-Phase Transformers (Excitation Currents Negligible)

Bus p	Bus p^a	Positive Sequence	Negative Sequence	Zero Sequence	Circuit Symbol
Wye, G	Wye, G	Z_{pq}	Z_{pq}	$Z_{pq} + 3Z_g + 3Z_g'$	Z_g $\quad$ Z_g'
Wye, G	Wye	Z_{pq}	Z_{pq}		Z_g
Wye, G	Delta	Z_{pq}, $1:(0.866 + j0.5)$	Z_{pq}, $1:(0.866 - j0.5)$	$Z_{pq} + 3Z_g$	Z_g
Wye	Wye	Z_{pq}	Z_{pq}		
Wye	Delta	Z_{pq}, $1:(0.866 + j0.5)$	Z_{pq}, $1:(0.866 - j0.5)$		
Delta	Delta	Z_{pq}	Z_{pq}		

Thus far only balanced operation of the transformer has been treated. If line currents are not balanced, the zero-sequence symmetrical component current,

$$I^o = \tfrac{1}{3}(I^a + I^b + I^c)$$

is nonzero. If the transformer has either a neutral conductor (three-phase, four-wire) or a grounded neutral, then a flow path exists for the zero-sequence current. A delta-connected set of terminals (primary, secondary, or tertiary) does not have a neutral; hence, an *open circuit* exists for a zero-sequence path through the delta. Similarly, an ungrounded wye (or star) transformer has an open circuit in the zero-sequence circuit if the neutral "floats." Table 2.6 summarizes the symmetrical component equivalent circuits [5, 6] for six common transformer connections. The following example demonstrates the method by which equivalent circuits are obtained.

Example 2.7

Derive the zero-sequence symmetrical component equivalent circuit for the impedance-grounded wye-to-impedance-grounded wye transformer shown in Table 2.6. A linearized magnetic core circuit is given in Figure E2.7.1, where Z_p and Z_s are primary and secondary leakage reactances plus winding resistances (in ohms). The excitation currents are negligible, so that n_1 and n_2 are primary and secondary turns, respectively, of ideal transformers. Express the equivalent circuit in per unit on the transformer power rating.

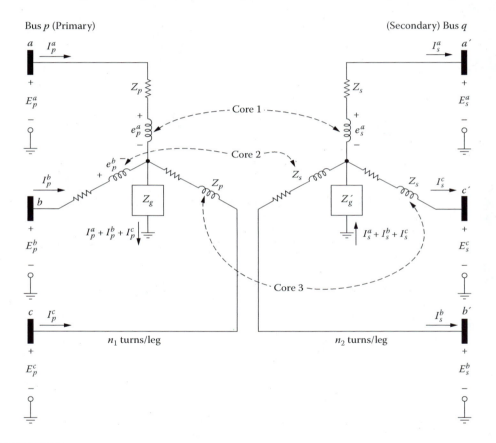

FIGURE E2.7.1

Solution

Short the secondary terminals a' b' c' to the ground and apply the same voltage $E°$ between primary terminals a, b, c and ground. The line currents are then $I^a = I^b = I^c = I^0$. For the phase a primary circuit, Kirchhoff's voltage law is

$$E^0 = Z_p I^0 + e_p^a + 3Z_g I^0 \tag{1}$$

where the current in the grounding impedance is $I_p^a + I_p^b + I_p^c = 3I^0$ and e_p^a is the voltage induced in the primary due to secondary current flow. Kirchhoff's voltage law for the phase a' (secondary) circuit is

$$e_s^a = Z_s I_s + 3Z_g' I_s = \frac{n_2}{n_1} e_p^a$$

$$= (Z_s + 3Z_g') \frac{n_1 I^0}{n_2} \tag{2}$$

where I_s is the secondary current and the turns ratio has been used. Substituting Equation 2 into 1 for e_p^a yields

$$E^0 = Z_p I^0 + 3Z_g I^0 + \left(\frac{n_1}{n_2}\right)^2 (Z_s + 3Z_g') I^0 \tag{3}$$

The zero-sequence impedance, with the secondary *shorted*, is the ratio

$$Z_0 = \frac{E^0}{I^0} = Z_p + \left(\frac{n_1}{n_2}\right)^2 Z_s + 3Z_g + 3\left(\frac{n_1}{n_2}\right)^2 Z_g' \quad \text{ohms} \tag{4}$$

If the impedances are expressed in per unit on the transformer MVA base (see Example 2.3), and the turns ratio is set for *rated* open-circuit voltage, then

$$Z_s(\text{p.u.}) = Z_s \left(\frac{E_{prated}}{E_{prated}}\right)^2 \left(\frac{S_{BASE}}{E_{srated}^2}\right) = \left(\frac{n_1}{n_2}\right)^2 Z_s \left(\frac{S_{BASE}}{E_{prated}^2}\right) \tag{5}$$

Therefore, Equation 4 expressed in per unit is

$$Z_0 = Z_p + Z_s + 3Z_g + 3Z_g'$$

$$= Z_{pq} + 3Z_g + 3Z_g' \quad (\text{p.u.}) \tag{6}$$

where primary and secondary leakage and resistance are combined in Z_{pq}.

The equivalent zero-sequence circuit is shown in Figure E2.7.2 for the case of open circuits on the abc and $a'\, b'\, c'$ terminals. Only if the p or q set of phase terminals are shorted to the reference (ground), or a load provides this connection, is there electrical continuity for unbalanced current. As a limiting case, if one neutral of the wye-to-wye transformer floats with respect to ground, $Z_g \to \infty$ or $Z_g' \to \infty$, the zero-sequence impedance is *open*, which is the wye, G-to-wye (open) connection of Table 2.6.

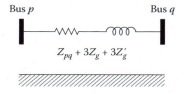

FIGURE E2.7.2

2.4.2 Multiple-Winding Transformers

Multiple-winding transformers are tested by pairs of windings, and the real or reactive losses are measured at *rated* current for one of the windings. Assuming that a phase-shift wye-to-delta exists on one winding, relative to the others, the equivalent circuit, referred to the primary, is as shown in Figure 2.28. Test data for the transformer are usually in the following form, which disregards phase shifting.

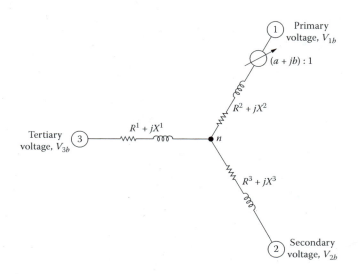

FIGURE 2.28
Equivalent for three-winding transformer employing one wye-to-delta coupling.

  Measurement 1: Primary to tertiary

  Rated current losses P_1 converted to resistance r_a

  Per-unit reactance x_a on primary power base S_{1b}, V_{1b}

  Measurement 2: Primary to secondary

  Rated current losses P_2 converted to resistance r_b

  Per-unit reactance x_b on primary power base S_{1b}, V_{1b}

  Measurement 3: Secondary to tertiary

  Rated current losses P_3' converted to resistance r_c'

  Per-unit reactance x_c' on secondary power base S_{2b}, V_{2b}

Each measurement involves a series combination of impedances from Figure 2.28. Measurement 3 has power base $S_{2b} \neq S_{1b}$, so it is converted to base S_{1b}:

$$r_c = R^1 + R^3 = \frac{P_3'}{I_{2b}^2}\left(\frac{S_{1b}}{V_{1b}^2}\right) \tag{2.87a}$$

$$x_c = X^1 + X^3 = x_c'\left(\frac{V_{2b}^2}{S_{2b}}\right)\frac{S_{1b}}{V_{1b}^2} \tag{2.87b}$$

Since measurements x_a and x_b are also series combinations of X^1, X^2, and X^3, the reactances are easily found by

$$X^1 = \frac{1}{2}(x_a - x_b + x_c) \tag{2.88a}$$

$$X^2 = \frac{1}{2}(x_a + x_b - x_c) \tag{2.88b}$$

$$X^3 = \frac{1}{2}(-x_a + x_b + x_c) \tag{2.88c}$$

and the resistances are found by a similar calculation.

The central node, n, of Figure 2.28 becomes a bus equivalent to bus t of Figure 2.27. If voltage V_{1b} is off-nominal because of a tap-setting or winding ratio, the π equivalent for this branch n to 1 is derived from Equation 2.85 and Equation 2.86. Similarly, if V_{2b} or V_{3b} are off-nominal, the branches n to 2 or n to 3 may also be converted to π-sections if necessary to obtain the per-unit equivalent.

Example 2.8

Factory tests on a 345/138/13.8 kV, 240 MVA wye–wye–wye autotransformer yield the following data for pairs of windings:

	Measurement 1	Measurement 2	Measurement 3
Winding pair	345 to 13.8	345 to 138	138 to 13.8
Load loss 75°C (total watts for three phases)	80,757	26,913	83,874
Percent impedance	6.86	7.51	4.46
At 75°C on 240 MVA base	(Referred to 345 kV)	(Referred to 345 kV)	(Referred to 138 kV)

The low-voltage winding has a rating of 58.7 MVA because of its construction. Find the positive-sequence equivalent circuit for the transformer as referred to a 345 kV line-to-line 240 MVA base. Express results in per unit on this base.

Solution

The transformer losses for measurement 1 are represented by resistors R^1 and R^2 in Figure 2.28. The losses expressed in terms of positive-sequence resistances referred to the HV base are

$$R^1 + R^2 = r_a = \left(\frac{0.080757}{I^2_{BASE}} \right) \frac{S_{BASE}}{V^2_{BASE}} = \frac{0.080757}{240} = 0.0003365 \quad \text{p.u.}$$

All impedance measurements are on a 240 MVA base, so measurement 1 is the sum of two impedances:

$$X^1 + X^2 = x_a = 0.0686 \text{ p.u.}$$

Define the impedance $z_a = r_a + jx_a$ for the measurement. For measurement 2, z_b is the sum of two elements:

$$z_b \begin{cases} R^2 + R^3 = \left(\dfrac{0.0269131}{I^2_{BASE}} \right) \dfrac{S_{BASE}}{V^2_{BASE}} = 0.0001121 \text{ p.u} \\[2ex] X^2 + X^3 = 0.0751 \text{ p.u.} \end{cases}$$

Measurement 3 is referred to the 138 kV winding, and on the 345 kV base becomes

$$z_c \begin{cases} R^1 + R^3 = 0.0003495 \text{ p.u.} \\[2ex] X^1 + X^3 = 0.0446 \left[\dfrac{Z_{BASE138}}{Z_{BASE345}} \right] \left(\dfrac{345}{138} \right)^2 = 0.0446 \text{ p.u.} \end{cases}$$

Solving the simultaneous equations for the resistances and reactances by means of Equation 2.88a to Equation 2.88c yields

$$R^1 = 0.000287 \quad R^2 = 0.000050 \quad R^3 = 0.000063$$

$$X^1 = 0.01905 \quad X^2 = 0.04955 \quad X^3 = 0.02555$$

The equivalent circuit on the 345 kV, 240 MVA base is shown in Figure E2.8. Often the resistances are neglected because they are small compared to the reactances. Depending on the magnitudes measured, it is possible to obtain *capacitive* reactances instead of inductive reactances in the three-winding transformer equivalent circuit.

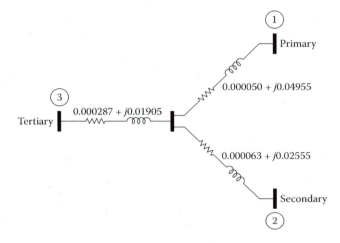

FIGURE E2.8

An extensive compilation of equivalent circuits for two- and three-winding transformers is given in reference [5]. This reference also includes excitation impedances and a large variety of grounding schemes. For the purposes of understanding the principles, the simple equivalents of Table 2.6 are sufficient to present here.

2.5   Synchronous Machine Representation

Synchronous generators, such as the one whose rotor is shown in Figure 2.29, are the only large-scale rotating machines used for power generation. There exist a large variety of analytical models for synchronous machines, both linear and nonlinear. Each model is derived to study particular aspects of a machine-network connection. A comprehensive model does not exist. Consequently, in this section a limited linear model, often called the *classical model*, is presented for the purpose of understanding steady-state, transient, and subtransient synchronous machine performance. Synchonous machine response is significant during a short circuit which may be several cycles of the line voltage, or even seconds before electromechanical protective devices "clear" the fault, or isolate the fault from the remaining part of the network. A synchronous machine has a typical response to an abrupt load change, as shown in Figure 2.30 for a generator.

Depending on the instant of time the load change is applied to the generator, the dc component in Figure 2.30 may or may not be present. In this figure, the subtransient region is

FIGURE 2.29
Salient pole rotor for a 167 MVA, 120 rpm (60-pole) umbrella-type hydrogenerator being lowered into the stator at the Robert Moses Generating Station, Niagara Falls, New York. (Photograph courtesy of Westinghouse Electric Corporation STGD, Orlando, FL)

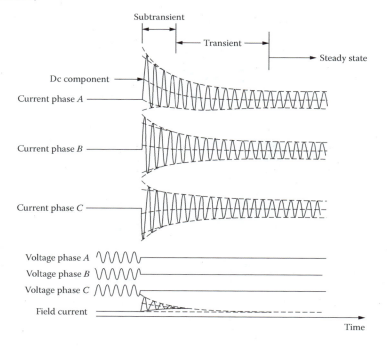

FIGURE 2.30
Typical oscillogram of a sudden three-phase short circuit for a synchronous machine.

approximated by an exponential rate of decay with time constant T_d''. The transient portion of the response is also approximated by exponential decay with a time constant T_d'. An experimental test procedure to measure these constants and other machine parameters is given in Section 7 of IEEE Standard 115 (Institute of Electrical and Electronic Engineers). The three regions of the machine response and their significance are:

1. *Steady state*: Approximately more than 6 sec after a disturbance to the machine's operating condition. As determined by the machine time constants, all transients have decayed to zero.

2. *Transient*: Approximately 0.1 to 6 sec after the disturbance. This is usually the operating time interval for electromechanical protective devices such as circuit breakers and interrupters, which sense abnormal voltages or currents.

3. *Subtransient*: Approximately 0 to 0.1 sec. The first peaks of the ac waveforms are in this time frame when passive circuit elements are affected but electromechanical devices are not capable of responding.

In order to understand the response of the synchronous machine in these three regions, an idealized model is examined. Consider the schematic of Figure 2.31, which represents a

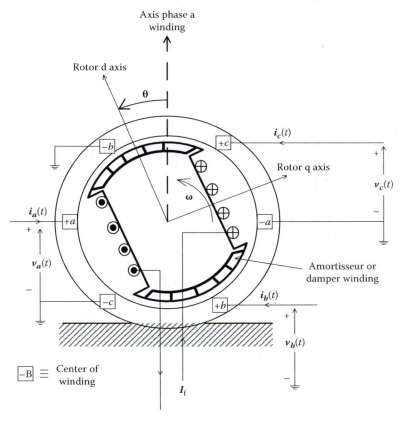

FIGURE 2.31
Three-phase, salient pole machine schematic for motor operation (Y-connected).

three-phase, two-pole synchronous machine with a salient rotor structure. There is no loss in generality in examining a two-pole as opposed to a multipole machine. The rotor has a field winding energized by dc current, I_f, and shorted copper bars called damper windings or amortisseur windings in the pole face. Further assume linear, lossless magnetic flux paths in both rotor and stator with an ideal sinusoidal air-gap flux density due to the field winding. The axis of the rotor magnetic flux density due to the field winding, called the *d* or direct axis, coincides with the principal axis of the rotor. The +*q*, or quadrature axis, is a half-pole pitch behind the +*d* axis for a counterclockwise rotation of the angle θ. θ is defined as the angle between the axis of the phase *a* winding and the +*d* axis. For short-circuit calculations, the rotor speed, ω, is assumed to be constant, such that

$$\frac{d\theta}{dt} = \omega$$

$$\theta = \omega t + \theta_0$$

where θ_0 is a phase angle of the rotor relative to a power system phase *a* reference at $t = 0$.

The three-phase windings are shown schematically in Figure 2.31 as concentrated coils *a*, *b*, *c*. Their self-inductances and mutual inductances depend on the rotor angle θ. For phase *a* and the field winding, the inductances are:

$L_{aa} = L_s + L_m \cos 2\theta \equiv$ phase *a* self-inductance

$L_{ab} = -M_s + L_m \cos \left(2\theta - \frac{2\pi}{3}\right) \equiv$ mutual inductance, phase *a* to *b*

$L_{ac} = -M_s + L_m \cos \left(2\theta + \frac{2\pi}{3}\right) \equiv$ mutual inductance, phase *a* to *c*

$L_{af} = M_f \cos \theta \equiv$ mutual inductance between phase *a* and the field

$L_{ff} = $ constant $\equiv$ self-inductance of rotor field winding

By means of shifting the angle $\theta \pm 2\pi/3$, the inductances of phases *b* and *c* are described by similar equations. Because linear magnetic material is assumed, the mutual inductances are reciprocal.

Let R_a be a phase winding resistance, R_f be the field winding resistance, sinusoidal voltages v_a, v_b, and v_c be applied to the phase windings, and sinusoidal currents i_a, i_b, and i_c flow into the windings for motor operation. Let V_f be a dc voltage applied to the rotor field winding and I_f be the dc current in this winding. Kirchhoff's voltage law for the field winding and phase *a*, respectively, employing motor notation (with current into the *a* winding), is

$$V_f = R_f I_f + \frac{d}{dt}\left[L_{ff} I_f + i_a M_f \cos\theta + i_b M_f \cos\left(\theta - \frac{2\pi}{3}\right) + i_c M_f \cos\left(\theta + \frac{2\pi}{3}\right) \right] \quad (2.89)$$

$$v_a = R_a i_a + \frac{d}{dt}\left[L_s i_a + i_a L_m \cos 2\theta - i_b M_s + i_b L_m \cos\left(2\theta - \frac{2\pi}{3}\right)\right.$$

$$\left. - i_c M_s + i_c L_m \cos\left(2\theta + \frac{2\pi}{3}\right) \right] - \frac{d}{dt}\left(I_f M_f \cos\theta\right) \quad (2.90)$$

By means of rotating subscripts, equations similar to Equation 2.90 may be written for phases *b* and *c*. Park's transformation [7, 8] (or Blondel's two-reaction theory), as modified by constant factors needed for field–*d*-*q* reciprocity, is used to convert the instantaneous phase currents into more convenient d, q, 0 currents. The transformation matrix and its inverse are

$$
\begin{bmatrix} i_0(t) \\ i_d(t) \\ i_q(t) \end{bmatrix} = \sqrt{\frac{2}{3}} \begin{bmatrix} \frac{1}{\sqrt{2}} & \frac{1}{\sqrt{2}} & \frac{1}{\sqrt{2}} \\ \cos\theta & \cos\left(\theta - \frac{2\pi}{3}\right) & \cos\left(\theta + \frac{2\pi}{3}\right) \\ \sin\theta & \sin\left(\theta - \frac{2\pi}{3}\right) & \sin\left(\theta + \frac{2\pi}{3}\right) \end{bmatrix} \begin{bmatrix} i_a(t) \\ i_b(t) \\ i_c(t) \end{bmatrix} = T_p \begin{bmatrix} i_a(t) \\ i_b(t) \\ i_c(t) \end{bmatrix}
\tag{2.91a}
$$

$$
\begin{bmatrix} i_a(t) \\ i_b(t) \\ i_c(t) \end{bmatrix} = \sqrt{\frac{2}{3}} \begin{bmatrix} \frac{1}{\sqrt{2}} & \cos\theta & \sin\theta \\ \frac{1}{\sqrt{2}} & \cos\left(\theta - \frac{2\pi}{3}\right) & \sin\left(\theta - \frac{2\pi}{3}\right) \\ \frac{1}{\sqrt{2}} & \cos\left(\theta + \frac{2\pi}{3}\right) & \sin\left(\theta + \frac{2\pi}{3}\right) \end{bmatrix} \begin{bmatrix} i_0(t) \\ i_d(t) \\ i_q(t) \end{bmatrix} = T_p^{-1} \begin{bmatrix} i_0(t) \\ i_d(t) \\ i_q(t) \end{bmatrix}
\tag{2.91b}
$$

This transformation is also applied to phase voltages v_a, v_b, and v_c. The result of transforming Kirchhoff's voltage law for the a, b, c windings is a two-axis system on the rotor turning at speed ω, equivalent to that in Equation 2.92:

$$
\begin{bmatrix} v_0(t) \\ v_d(t) \\ v_q(t) \end{bmatrix} = T_p \left\{ R_a + \frac{d}{dt} \begin{bmatrix} L_s + L_m\cos 2\theta & -M_s + L_m\cos\left(2\theta - \frac{2\pi}{3}\right) & -M_s + L_m\cos\left(2\theta + \frac{2\pi}{3}\right) \\ -M_s + L_m\cos\left(2\theta - \frac{2\pi}{3}\right) & L_s + L_m\cos\left(2\theta - \frac{2\pi}{3}\right) & -M_s + L_m\cos 2\theta \\ -M_s + L_m\cos\left(2\theta + \frac{2\pi}{3}\right) & -M_s + L_m\cos 2\theta & L_s + L_m\cos\left(2\theta + \frac{2\pi}{3}\right) \end{bmatrix} \right.
$$

$$
\left. \right\} T_p^{-1} \begin{bmatrix} i_0(t) \\ i_d(t) \\ i_q(t) \end{bmatrix} + T_p \frac{d}{dt} \begin{bmatrix} I_f M_f \cos\theta \\ I_f M_f \cos\left(\theta - \frac{2\pi}{3}\right) \\ I_f M_f \cos\left(\theta + \frac{2\pi}{3}\right) \end{bmatrix}
$$

$$
= \begin{bmatrix} R_a & 0 & 0 \\ 0 & R_a & -\omega L_q \\ 0 & \omega L_d & R_a \end{bmatrix} \begin{bmatrix} i_0(t) \\ i_d(t) \\ i_q(t) \end{bmatrix} + \begin{bmatrix} L_0 & 0 & 0 \\ 0 & L_d & 0 \\ 0 & 0 & L_q \end{bmatrix} \frac{d}{dt} \begin{bmatrix} i_0(t) \\ i_d(t) \\ i_q(t) \end{bmatrix} + \begin{bmatrix} 0 \\ \sqrt{\frac{3}{2}} M_f \frac{dI_f}{dt} \\ \sqrt{\frac{3}{2}}\omega M_f I_f \end{bmatrix}
\tag{2.92}
$$

On the right-hand side of Equation 2.92 several fundamental inductances have simplified the notation. These are defined to be the direct-axis, quadrature-axis, and zero-sequence inductances, respectively:

$$
L_d = L_s + M_s + \tfrac{3}{2} L_m
\tag{2.93a}
$$

$$
L_q = L_s + M_s - \tfrac{3}{2} L_m
\tag{2.93b}
$$

$$
L_0 = L_s - 2M_s
\tag{2.93c}
$$

Kirchhoff's voltage law for the field winding, Equation 2.89, is converted to the d, q, 0 axis system using the inverse of Park's transformation. The result is

$$
V_f = R_f I_f + L_{ff} \frac{dI_f}{dt} - \sqrt{\frac{3}{2}} M_f \frac{di_d}{dt}
\tag{2.94}
$$

Equation 2.92 and Equation 2.94 define the linear synchronous machine without damper windings in the d, q, 0 axis system. Observe that the zero-sequence component decouples from the other equations; hence, it can be solved separately with appropriate initial conditions. The θ angular dependence has been eliminated to yield a set of constant-coefficient linear equations when speed, ω, is constant. Even if ω does vary due to torques on the machine, its incremental change is small, so it may be approximated as a constant in Equation 2.92. Small ω perturbations are calculated using shaft torque dynamic equations.

Reciprocal coupling between the d-axis and the rotor field results from using the numerical coefficients of Equation 2.91. The mutual inductance between them is $\sqrt{3/2}\,M_f = L_{md}$ in Equation 2.92 and Equation 2.94. Since physical measurements on the nonideal machine are usually given in terms of armature reactance and leakage reactance, the following leakage reactances are defined for the d-axis, field, and q-axis, respectively:

$$\ell_d = L_d - \sqrt{\tfrac{3}{2}}\,M_f = L_d - L_{md} \tag{2.95a}$$

$$\ell_f = L_{ff} - \sqrt{\tfrac{3}{2}}\,M_f = L_{ff} - L_{md} \tag{2.95b}$$

$$\ell_q = L_q - L_{mq} \tag{2.95c}$$

The mutual reactance L_{mq} is the coupling of the damper winding in the quadrature axis.

These leakage reactances and the previous equations describe the circuit of the linear synchronous machine shown in Figure 2.32. The amortisseur or damper windings on the rotor, which are short-circuited windings, are shown dashed in Figure 2.32. Their mutual coupling and self-inductances are described later. The analytical model of the synchronous machine is valid for transient and steady-state analysis, and all variables are *dc quantities*.

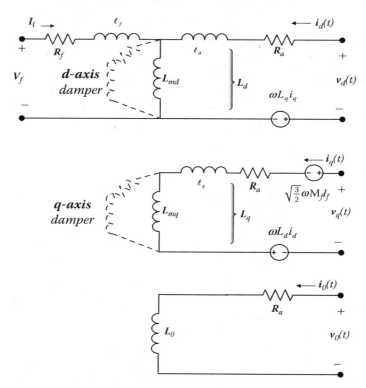

FIGURE 2.32
Equivalent circuit for synchronous machine in d, q, 0 rotating axes (motor operation).

FIGURE 2.33
Foreground: Assembly of a stator section for a 650 MVA, 72 rpm hydrogenerator. Background: Bore view of a 300 MVA, 1,800 rpm hydrogen-cooled generator. (Photograph courtesy of Westinghouse Electric Corporation STGD, Orlando, FL)

Figure 2.33 shows the distributed configuration of stator windings in synchronous generators, in contrast to the concentrated windings used for simple analysis.

2.5.1 Steady-State Synchronous Machine Equivalent

Several operating conditions of the machine are easily analyzed by applying terminal connections to the d, q, 0 circuits. For example, if the machine is operated with an open circuit on the output terminals and the field current I_f is constant, then $i_d = i_q = 0$. In this case, the networks of Figure 2.32 circuits show $v_d = 0$ and the v_q voltage is equal to

$$v_q(\text{open} - \text{circuit voltage}) = \omega \sqrt{\frac{3}{2}} \, M_f I_f \qquad (2.96)$$

The voltage v_q is a dc voltage on the rotating axis. When the salient pole (direct axis) points directly at the phase a winding in Figure 2.31, the ac voltage in phase a is at its peak value under no-load conditions.

When synchronous ω currents flow in the windings, the combination of field current and winding currents alters the air gap magnetic flux, and forces the internal generated voltage to an angle δ with respect to the phase a voltage. Therefore, the machine internal voltage with respect to the phase a reference is defined as the phasor:

$$\boxed{E_q = |E_q|\angle\delta = |\omega\sqrt{\tfrac{3}{2}}\,M_f I_f|\angle\delta = \text{internal voltage w/r to phase 'a'}}$$

The phase b and phase c voltages as computed from Park's Equation (Equation 2.91b) are at phase-shift angles $\delta - 120°$ and $\delta + 120°$, respectively.

Park's transformation of the a, b, c quantities yields two orthogonal voltages v_d and v_q. Currents i_d and i_q appear to be constant on the ω rotating reference and must be referred to the stator. Consider an equivalent y-x, two-phase circuit on the stator with y magnetic axis aligned with the $\theta = 0$ position (the same orientation as the stator phase a winding). The two-phase stator equivalent has orthogonal phase currents for balanced operation:

$$i_y = I\cos(\omega t) \tag{2.97a}$$

$$i_x = I\sin(\omega t) \tag{2.97b}$$

where I is the rms current in the winding. These y-x windings couple into the rotating d-q axis, which are at an angle $\theta = \omega t + \delta$, with respect to the stator:

$$i_d = i_y\cos(\theta) + i_x\sin(\theta) \tag{2.98a}$$

$$i_q = -i_y\sin(\theta) + i_x\cos(\theta) \tag{2.98b}$$

Stator currents from Equation 2.97a and b are substituted into these last equations to obtain

$$i_d = I\cos(\omega t)\cos(\theta) + I\sin(\omega t)\sin(\theta) = I\cos(\theta - \omega t) = I\cos(\delta) \tag{2.99a}$$

$$i_q = -I\cos(\omega t)\sin(\theta) + I\sin(\omega t)\cos(\theta) = I\sin(\theta - \omega t) = I\sin(\delta) \tag{2.99b}$$

Since the two-axis y-x equivalent current i_y on the stator is aligned with the a winding, the phase a current and the phase a voltage *on the stator* in phasor form are

$$I^1 = I_d + jI_q \quad \text{rms amperes} \tag{2.100a}$$

$$E^1 = V_d + jV_q \quad \text{rms volts} \tag{2.100b}$$

δ is the angle between the machine internal voltage, v_q, and the phase a terminal voltage as shown in Figure 2.34a for phasors. Observe that under no-load conditions, $v_d = 0$ on the rotor, and when reflected to the stator equivalent in phasor form, this is $V_d = 0$.

Direct, quadrature, and zero-axis inductances from the rotating frame are reflected to the stator fixed frame as reactances in the steady state. These reactances are

$$X_d = \omega L_d = \omega\left(L_s + M_s + \frac{3}{2}L_m\right)$$

$$X_q = \omega L_q = \omega\left(L_s + M_s - \frac{3}{2}L_m\right)$$

$$X_0 = \omega L_0 = \omega(L_s - 2M_s)$$

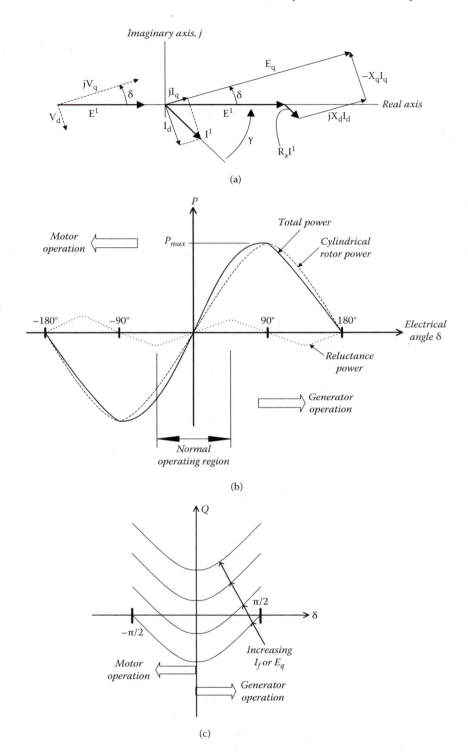

FIGURE 2.34
(a) Synchronous machine phasor diagram for generator operation. (b) Salient pole synchronous machine power characteristics. (c) Salient pole synchronous machine reactive power characteristics.

The two-phase equations referred to the stator (with currents out of the windings for generator operation) in steady-state phasor notation are

$$\begin{bmatrix} V_0 \\ V_d \\ V_q \end{bmatrix} = \begin{bmatrix} R_a & 0 & 0 \\ 0 & -R_a & X_q \\ 0 & -X_d & -R_a \end{bmatrix} \begin{bmatrix} I_0 \\ I_d \\ I_q \end{bmatrix} + \begin{bmatrix} 0 \\ 0 \\ \omega\sqrt{\frac{3}{2}}M_f I_f \end{bmatrix} \quad \text{(generator operation)} \tag{2.101}$$

Equations 2.100 to Equation 2.101 represent balanced steady-state operation of the synchronous machine derived from the general rotating frame. For motor operation, the algebraic signs of I_d and I_q are changed on the right side of Equation 2.101.

Let E_q be the internal voltage due to I_f, then Equation 2.101 is used to obtain

$$E^1 = V_d + jV_q = -R_a(I_d + jI_q) + X_q I_q - jX_d I_d + E_q = E^a \tag{2.102a}$$

Rearranging terms, the internal voltage E_q of the generator is calculated as

$$E_q = |E_q| \angle \delta = E^1 + R_a I^1 + jX_d I_d - X_q I_q \tag{2.102b}$$

This equation for generator operation is shown in the phasor diagram of Figure 2.34a for steady-state operation.

Current flow in the phase a winding is at an angle γ with respect to phase a voltage. The current induces magnetic fields in the salient pole direction (the easy d magnetic axis) and the q-axis (hard magnetic axis). The resultant air gap flux is due to three components of magnetizing currents. The easy axis magnetic flux changes very little, so the angle δ of the generated voltage changes mostly due to the component of current in the q-axis. The phase angle δ of E_q is the same as the angle of a fictitious voltage, V_q'', found from terminal conditions of the machine. The angle δ of the fictitious internal voltage is calculated from measured voltage E^1 and current I^1 terminal conditions [9]:

$$V_q'' = |V_q''| \angle \delta = E^1 + R_a I^1 + j\omega L_q I^1 = E^1 + (R_a + jX_q)I^1 \tag{2.103}$$

The reference is $E^1 = |E^1| \angle 0°$ for all phasors, and the angle γ is known from the operating condition (power factor).

After the angle δ is calculated from Equation 2.103, the d, q components of the currents are calculated from the expressions

$$|I_q| = |I^1| \cos(\gamma + \delta)$$

$$|I_d| = |I^1| \sin(\gamma + \delta)$$

As vector quantities, the d-q axis currents and voltages on the stator neglecting the small voltage drop due to R_a may be expressed as

$$I_d = -\left[\frac{|E_q| - E^1 \cos(\delta)}{X_d} \right] \tag{2.104a}$$

$$I_q = E^1 \sin(\delta)/X_q \tag{2.104b}$$

$$V_d = -E^1 \sin(\delta) \tag{2.104c}$$

$$V_q = E^1 \cos(\delta) \tag{2.104d}$$

The steady-state power calculated from the defined phasors, Equations 2.100a and b, is

$$\frac{S}{3} = E^1(I^1)^* = \frac{P+jQ}{3} = V_dI_d + V_qI_q + j(V_qI_d - V_dI_q) \tag{2.105a}$$

In the last equation, the small contribution of R_a has been neglected. Each term of Equation 2.104 may be substituted into the real and imaginary parts of Equation 2.105a to obtain the real power delivered by the salient pole generator:

$$\frac{P}{3} = V_dI_d + V_qI_q = E^1\sin(\delta)\left[\frac{|E_q|-E^1\cos(\delta)}{X_d}\right] + E^1\cos(\delta)\left[\frac{E^1\sin(\delta)}{X_q}\right]$$

$$= \frac{E^1|E_q|\sin(\delta)}{X_d} + (E^1)^2\sin(\delta)\cos(\delta)\left[\frac{1}{X_q} - \frac{1}{X_d}\right]$$

$$= \frac{E^1|E_q|\sin(\delta)}{X_d} + \frac{(E^1)^2\sin(2\delta)}{2}\left[\frac{1}{X_q} - \frac{1}{X_d}\right] \quad \text{salient pole} \tag{2.105b}$$

The first power term in Equation 105.b dependent upon $|E_q|$ and δ represents cylindrical rotor power. The second power term, $\sin(2\delta)$, which is independent of excitation $|E_q|$ and is dependent upon differences between the quadrature and direct-axis impedances, is called the magnetic reluctance power.

For a cylindrical rotor machine, $X_d = X_q = X_s$, where X_s is called the *synchronous reactance*. For the cylindrical type of rotor, Equation 2.105b simplifies to the power equation

$$\boxed{\frac{P}{3} = \frac{E^1|E_q|\sin(\delta)}{X_s} \quad \text{cylindrical rotor neglecting resistance}}$$

Figure 2.34b plots the power equation for a salient rotor machine as a function of the angle δ. Note for negative δ the power flow is into the synchronous machine operating as a motor and delivers power to a mechanical mover. If the peak power points are exceeded as either a generator or motor, the machine loses synchronization with the line frequency. For this maximum power consideration, synchronous machines are rated to operate at an angle $\delta \ll 90°$ at all times in order to absorb transient excursions.

For conventional synchronous machines the reluctance power or torque, $\tau = P/\omega_{mech}$, is always positive since $X_d \geq X_q$, but this reluctance power may be negative for permanent magnet machines because of the large ampere-turns force to magnetize the material of the permanent magnets. The reluctance torque represents the tendency of the d-axis of the salient pole rotor to align itself with the rotating magneto-motive force due to stator currents. The reluctance power is directly dependent upon the terminal voltage E^1 and not dependent on the current in the field winding that produces the internal voltage.

Equations 2.104a–d may also be substituted into the reactive part of Equation–2.105a to obtain the reactive power delivered by the machine:

$$\frac{Q}{3} = V_qI_d - V_dI_q = \frac{E^1|E_q|\cos(\delta)}{X_d} - \frac{(E^1\sin(\delta))^2}{X_q} - \frac{(E^1\cos(\delta))^2}{X_d} \tag{2.105c}$$

From Equation 2.96, it may be seen that the internal voltage E_q is dependent upon the field excitation current I_f. An increased excitation results in more reactive power delivered by the machine (Equation 2.105c). Figure 2.34c schematically plots the reactive power as a function of δ. The property of leading or lagging reactive power generation as dependent upon excitation, E_q or I_f, is also utilized for *synchronous condensors* that do not have real power capability. Synchronous condensors are employed as reactive compensators to supply or absorb reactive power. The magnetic material saturation effects due to the field current I_f are discussed in references [5, 9].

For a cylindrical rotor machine $X_d = X_q = X_s$, the reactive power delivered by the generator is

$$\boxed{\frac{Q}{3} = \frac{E^1 |E_q| \cos(\delta) - (E^1)^2}{X_s} \quad \text{cylindrical rotor neglecting resistance}}$$

Example 2.8.1

A 12-pole, 13.2 kV, Y-connected, three-phase, salient rotor generator is operated at rated voltage, and rated 100 MVA output at 0.8 lagging power factor. Neglect armature winding resistance R_a. The direct-axis reactance is 1.0 p.u., and the quadrature-axis reactance is 0.8 p.u.

(a) Find the internal voltage $|E_q|$ p.u. and the phase angle δ if the machine is assumed to be cylindrical, $X_q = X_d = X_s = 1.0$.
(b) Repeat part (a) if the machine has salient rotor characteristics.

Solution

(a) In per unit, numerical values can directly be substituted into Equation 1.102b with $R_a = 0$ to calculate the internal voltage:

$$|E_q| \angle \delta = E^1 + jX_d I_d - X_q I_q = E^1 + jX_d(I_d + jI_q) = E^1 + jX_s I^1$$

$$= 1.0 + j1.0(0.8 - j0.6) = 1.6 + j0.8 = 1.79 \angle 26.5°$$

(b) For a salient pole machine, the internal angle must be calculated from the terminal conditions according to Equation 2.103:

$$\left| V_q'' \right| \angle \delta = E^1 + (R_a + jX_q)I^1 \cong E^1 + jX_q I^1 = 1.0 + j0.8(0.8 - j0.6)$$

$$= 1.48 + j0.64 = 1.61 \angle 23.38°$$

With the angle $\delta = 23.38°$, the magnitude of the internal voltage E_q is calculated from the salient pole power Equation 2.105b.

$$|E_q| = \left[\frac{1}{\sin(\delta)} \right] \left[\frac{P}{3E^1} - \frac{E^1 \sin(2\delta)}{2} \left(\frac{1}{X_q} - \frac{1}{X_d} \right) \right] = 2.52 \left[0.8 - \frac{.728}{2}(1.25 - 1.0) \right] = 1.79$$

For this example, there is negligible difference in the magnitude of the internal voltage between the cylindrical and salient pole machine. However, saliency forces the salient pole machine to deliver the same power at 3.12° lesser angle than the cylindrical machine.

2.5.1.1 Short-Circuit Characteristics

The Thévenin equivalent for the machine is an open-circuit voltage, E_q, as derived in Equation 2.96 in series with an impedance. The Thévenin impedance is the open-circuit voltage divided by the short-circuit current. Equation 2.101 or the circuits of Figure 2.32, with currents out of the machine for generator operation, are used to obtain the shorted terminal currents:

$$V_d\Big|_{\text{circuit}}^{\text{short}} = 0 = -I_d R_a + X_q I_q \tag{2.106a}$$

$$V_q\Big|_{\text{circuit}}^{\text{short}} = 0 = -I_q R_a - X_d I_d + E_q \tag{2.106b}$$

Solving these simultaneous equations yields

$$I_{d(s.c)} = \frac{E_q}{X_d + R_a^2/X_q} \cong \frac{E_q}{X_d} = \frac{E_q}{\omega L_d} \tag{2.107a}$$

$$I_{q(s.c)} = \frac{R_a E_q}{X_d X_q + R_a^2} \cong 0 \tag{2.107b}$$

when the armature resistance $R_a << X_d$. Therefore, the Thévenin equivalent circuit for the synchronous machine is an induced voltage E_q (proportional to field current) behind a synchronous reactance $X_d = X_1$ for steady-state calculations. The negative-sequence imped-ance X_2 is given in the next section.

2.5.2 Transient Time-Frame Synchronous Machine Equivalent

If the generator experiences a balanced but abrupt change in line currents due to a transmis-sion line fault or load change, the rotor field flux momentarily remains constant. The phase current changes are balanced, so only a positive-sequence or direct-axis change occurs (see Example 2.1). Hence for a machine without damper windings, the rotor field current and direct-axis current changes (assuming negligible armature resistance) are related by incre-mental changes in the field winding loop of the d-axis (see Figure 2.32 or Equation 2.94):

$$\Delta \psi_f = 0 = -\sqrt{\tfrac{3}{2}}\, M_f\, \Delta i_d + L_{ff}\, \Delta I_f \tag{2.108}$$

where ψ_f is the field flux. In Equation 2.108, $\sqrt{3/2}\, M_f$ is the mutual coupling between the rotor circuit and the equivalent d-axis on the stator, and L_{ff} is the self-inductance of the rotor field winding. The change in field current may be substituted in the equation for direct-axis flux linkages (the stator loop of the d-axis in Figure 2.32) to obtain

$$\Delta \psi_d \simeq -L_d \Delta i_d + \sqrt{\frac{3}{2}}\, M_f\, \Delta I_f + = -\left(L_d - \frac{3M_f^2}{2L_{ff}} \right) \Delta i_d = -L_d' \Delta i_d \tag{2.109}$$

The quantity in parentheses is referred to as the direct-axis inductance in transient con-ditions. Multiplying by the frequency, the direct-axis transient reactance is given by

$$X_d' = X_1' = \omega L_d' = \omega \left(L_d - \frac{3M_f^2}{2L_{ff}} \right) \tag{2.110}$$

Since L_{ff} is a positive quantity, the direct-axis transient reactance is less than the syn-chronous value:

$$X_1' = X_d' < X_d = X_1 \tag{2.111}$$

In the quadrature axis, the change in flux linkages is also zero due to the abrupt phase current change, so that

$$\Delta\psi_f = 0 = L_q\,\Delta i_q \tag{2.112}$$

The inductance in the quadrature axis is constant, so the current change must be zero. Thus, the quadrature-axis transient reactance is identical to its steady-state value:

$$X_q' = \omega L_q' = \omega L_q = X_q \tag{2.113}$$

If reverse-phase, or negative-sequence, currents a, c, b,

$$i_a = I\cos\omega t \tag{2.114a}$$

$$i_b = I\cos\left(\omega t + \frac{2\pi}{3}\right) \tag{2.114b}$$

$$i_c = I\cos\left(\omega t - \frac{2\pi}{3}\right) \tag{2.114c}$$

are applied to the stator (armature) of the synchronous machine while rotation is in the positive sense, the ratio of phase a flux linkages to current is the negative-sequence inductance

$$\frac{\psi_a}{i_a} = L_2$$

In order to derive this reactance in terms of machine parameters, let the rotor be at position $\sigma = \omega t + \theta_0$ relative to the phase of the negative-sequence currents. Substituting Equation 2.114 into Park's transformation and by means of trigonometric identities, the resulting negative-sequence d, q, 0 currents are

$$\mathbf{I}_{odq}^{neg} = \begin{bmatrix} i_0 \\ i_d \\ i_q \end{bmatrix}^{neg} = \mathbf{T}_p \begin{bmatrix} i_a \\ i_b \\ i_c \end{bmatrix}^{neg} = \sqrt{\frac{3}{2}}\,I \begin{bmatrix} 0 \\ \cos(2\omega t + \theta_0) \\ \sin(2\omega t + \theta_0) \end{bmatrix} \tag{2.115}$$

In this operating condition, the magnetic flux wave due to the applied armature currents rotates at twice the synchronous speed relative to the field winding on the rotor (d-axis). Because the frequency is twice nominal, the field winding reactance dominates its circuit

$$2\omega\sqrt{\frac{3}{2}}\,M_f \gg R_f$$

such that Equations 2.108, 2.109, and 2.112 are again valid. The flux linkages in winding a are expressed using the inverse of Park's equation as

$$\begin{aligned}
\psi_a &= \sqrt{\frac{2}{3}}\,\cos\theta\,\psi_d + \sqrt{\frac{2}{3}}\,\sin\theta\,\psi_q \\
&= \sqrt{\frac{2}{3}}\,\cos\theta[-L_d'i_d] + \sqrt{\frac{2}{3}}\,\sin\theta\,[-L_q i_q] \\
&= -L_d'\,I\cos\theta\,\cos(2\omega t + \theta_0) - L_q I\,\sin\theta\,\sin(2\omega t + \theta_0) \\
&= -I\left(\frac{L_d' + L_q}{2}\right)\cos\omega t - I\left(\frac{L_d' - L_q}{2}\right)\cos(3\omega + 2\theta_0)
\end{aligned} \tag{2.116}$$

Fundamental plus third harmonic components of flux are present in the phase windings. The negative-sequence inductance is defined as the fundamental component of Equation 2.116:

$$L_2 = \frac{L_d' + L_q}{2} \tag{2.117}$$

such that the negative-sequence reactance is

$$X_2 = \frac{X_d' + X_q}{2} = \frac{X_1' + X_q}{2} \tag{2.118}$$

The transient reactance of both sides of this equation yields

$$X_2' = \frac{X_d'' + X_q'}{2} = \frac{X_1'' + X_q}{2} \tag{2.119}$$

where the subtransient reactance X_d'' is subsequently derived.

2.5.3 Subtransient Time-Frame Synchronous Machine Equivalent

The damper or amortisseur windings, which have thus far been neglected, contribute significantly to the synchronous machine behavior during the initial few cycles of the disturbance response. The d, q damper windings are assumed to be closely coupled to the other circuits in the axes. The flux linkage equations including damping circuits are

$$\psi_d = -L_d i_d + \sqrt{\frac{3}{2}} \, M_f I_f + \sqrt{\frac{3}{2}} \, M_{Dd} i_D \tag{2.120a}$$

$$\psi_q = -L_q i_q + \sqrt{\frac{3}{2}} \, M_{Qq} i_Q \tag{2.120b}$$

$$\psi_f = -\sqrt{\frac{3}{2}} \, M_f i_d + L_{ff} I_f + M_R i_D \tag{2.120c}$$

$$\psi_D = \sqrt{\frac{3}{2}} \, M_{Dd} i_d + M_R I_f + L_D i_D \tag{2.120d}$$

$$\psi_Q = \sqrt{\frac{3}{2}} \, M_{Qq} i_q + L_Q i_Q \tag{2.120e}$$

where the flux in the d-axis damper is ψ_D and in the q-axis damper the flux is ψ_Q. In these equations i_D and i_Q are the d- and q-axis damper winding currents, respectively. Mutual inductances M_{Dd} and M_{Qq} are defined between the damper circuits and the d, q, and field windings. L_D and L_Q are the self-inductances of the two damper windings.

Immediately following an abrupt change in the phase currents of the machine, the flux linkages in the field and direct-axis damper do not change instantly, so that $\Delta\psi_f = \Delta\psi_D = 0$.

Substituting incremental current changes ΔI_f and Δi_d from Equation 2.120c and Equation 2.120d, respectively, into Equation 2.120a, one obtains the subtransient direct-axis inductance

$$L_d'' = \frac{\Delta \psi_d}{\Delta i_d} = \left[L_d - \frac{3}{2}\left(\frac{M_f^2 + M_{Dd}^2 L_{ff}/L_D - 2M_f M_{Dd} M_R/L_D}{L_{ff} - M_R^2/L_D} \right) \right] \tag{2.121}$$

In a similar fashion, for constant q-axis damper flux linkage, $\Delta \psi_Q = 0$, the change in i_q leads to

$$L_q'' = \frac{\Delta \psi_q}{\Delta i_q} = L_q - \left(\frac{3}{2} \right) \frac{M_{Qq}^2}{L_Q} \tag{2.122}$$

Multiplying Equation 2.121 and Equation 2.122 by ω specifies the subtransient reactances X_d'' and X_q'', respectively. Observe that if $L_D \to \infty$ and $L_Q \to \infty$, or else the damper-to-axis mutual coupling inductance approaches zero, then Equation 2.121 and Equation 2.122 reduce to Equation 2.110 and Equation 2.113, respectively. Therefore, only if machine damper windings are present do the subtransient and transient reactances differ. Furthermore, only consideration of additional flux linkages or the inherent nonlinear magnetic material in the machine leads to deriving X_d''', the sub-subtransient reactance.

Figure 2.35 graphically shows the effect of the reactances and defines time constants for an abrupt short. The symmetrical component of the rms line current as a function of time for a terminal three-phase short circuit from no load [10] is approximated as

$$I_{ac} = E^1 \left[\frac{1}{X_d} + \left(\frac{1}{X_d'} - \frac{1}{X_d} \right) e^{-t/T_d'} + \left(\frac{1}{X_d''} - \frac{1}{X_d'} \right) e^{-t/T_d''} \right]$$

$$= i_d + (i_d' - i_d)\, e^{-t/T_d'} + (i_d'' - i_d')e^{-t/T_d''} \tag{2.123}$$

where E^1 is the *measured* (rated) open-circuit line-to-neutral voltage. Positive-sequence reactances may replace the direct-axis values in Equation 2.123. If the short circuit occurs

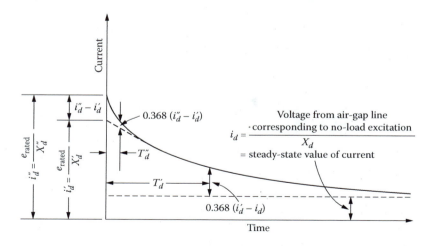

FIGURE 2.35
Symmetrical component of armature short-circuit current (three-phase short circuit from no-load rated voltage). Values are rms.

when the rate of change in voltage is nonzero, a dc component (see Figure 2.30)

$$I_{dc} = \frac{\sqrt{2} \, E^1 \cos \alpha}{X_d''} \, e^{-t/T_a}$$ (2.124)

is present in the measured current. T_a is the armature time constant (approximately 0.15 sec) and is indicative of negative-sequence currents [11] that force the terminal voltage to zero. Hence,

$$T_a = \frac{L_2}{R_a}$$ (2.125)

and the decay of the dc component is used to estimate $X_2 = \omega L_2$. The total current with a dc component present is

$$I_T = \sqrt{(I_{dc})^2 + (I_{ac})^2}$$ (2.126)

The dc component can be eliminated from a short-circuit record by using the mean value of successive ac peaks. The result is the symmetrical component shown in Figure 2.35.

To obtain sequentially $X_1 = X_d$, $X_1'' = X_d''$, T_d', $X_1' = X_d'$, then T_d'' from a test measurement as shown in Figure 2.35, the following steps [7] are used:

1. For long time durations, $t \rightarrow \infty$, Equation 2.123 yields

$$X_d = X_1 = \frac{E^1}{i_d}$$

2. At the onset of the transient, $t \sim 0$, then

$$X_d'' = X_1'' = \frac{E^1}{i_d''}$$

3. The time constants are such that $T_d' >> T_d''$, so for time $t' \sim T_d'$, the subtransient contribution is small:

$$e^{-t'/T_d''} \simeq 0$$

and the current envelope is determined only by the transient current t_d'. Subtract the steady-state current i_d from the envelope for several data points $t_1, t_2, \ldots, t_k$ near t', and then analytically fit a value of T_d' to these data points by means of taking *ratios* of successive points, $t_1, t_2, t_3, \ldots, t_k$:

$$e^{t_1/T_d'} = \{I_{ac}(t_1) - i_d\}/(i_d' - i_d)$$
$$e^{t_2/T_d'} = \{I_{ac}(t_2) - i_d\}/(i_d' - i_d)$$
$$\vdots \qquad \qquad \vdots$$
$$e^{t_k/T_d'} = \{I_{ac}(t_k) - i_d\}/(i_d' - i_d)$$

where the unknown value t_d' cancels in the ratios.

4. Let $t' = T_d'$, such that

$$e^{-t'/T_d'} = 0.368 \qquad e^{-t/T_d''} \simeq 0$$

and calculate $X_1' = X_d'$ and i_d' from Equation 2.123.

5. Subtract i_d' at $t = 0$ and estimate T_d'' from decay of the envelope for the difference $I_{ac} - i_d'$. The point $t'' = T_d''$ is indicated on Figure 2.35.

The following example shows how X_1 and X_1'' are measured.

Example 2.9

A wye-connected solidly grounded three-phase 60 Hz salient pole synchronous generator has a rating of 50 MVA, 13.2 kV rms line to line. The machine is tested by means of three types of *sustained short circuits* while operating at rated voltage, no load. The line rms currents measured for the sustained tests and conditions are as follows:

1. 2,000 A, three phases directly shorted
2. 2,900 A, line-to-line fault
3. 4,200 A, line to ground

Immediately after a sudden symmetrical three-phase short circuit, the current is found to be 20,000 A rms. The sustained and momentary measurements are used sequentially to determine the machine impedances X_d, X_2, X_0, and X_d''. Express the numerical value of these impedances in per unit using the machine ratings as base quantities. Neglect saturation effects and resistances.

Solution

The open-circuit schematic diagram for the generator is shown in Figure E2.9.1, where the internal reactances take on *different* values for the various tests. Currents I^a, I^b, I^c, and I^n are out of the terminals for generator convention.

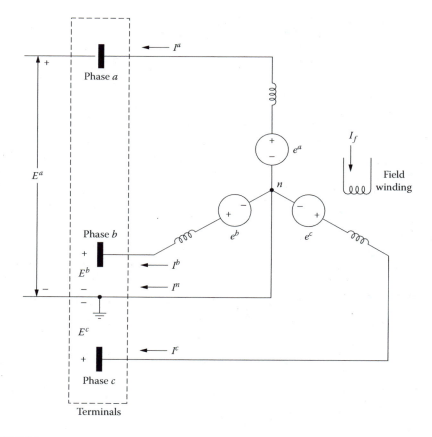

FIGURE E.2.9.1
Circuit diagram for wye-connected synchronous machine.

The internal line-to-neutral voltages e^a, e^b, and e^c are a balanced three-phase set that may be expressed as equivalent symmetrical component voltages:

$$\mathbf{E}^{0.12} = \mathbf{T}_s^{-1} \begin{bmatrix} e^a \\ e^b \\ e^c \end{bmatrix} = \mathbf{T}_s^{-1} \begin{bmatrix} 1 \\ a^2 \\ a \end{bmatrix} [e^a] = \begin{bmatrix} 0 \\ 1 \\ 0 \end{bmatrix} [e^a]$$

The measured line-to-line open-circuit voltage is 13.2 kV, so the numerical value of e^a is

$$|e^a| = \frac{13,200}{\sqrt{3}} = 7.621 \text{ kV rms}$$

This is also the equivalent voltage E_q of Equation 2.96.

Test 1: In this test terminals a, b, and c are directly connected together. This is also described by Equation 2.106 and Equation 2.107. The neutral connection is not needed because the resulting currents are balanced. Each of the currents is $I_{s.c.} = 2,000$ A, so the internal reactance per phase is

$$X_d = X_1 = \frac{|E_q|}{|I_{d(s.c.)}|} = \frac{|e^a|}{|I_{s.c.}|} = \frac{13,200/\sqrt{3}}{2000} = 3.810 \ \Omega$$

Therefore, for test 1, the generator positive-sequence circuit is as shown in Figure E2.9.2. Observe that the zero-sequence and negative-sequence circuits are as yet unknown because the machine response is unknown for unbalanced tests and opposite-phase rotations.

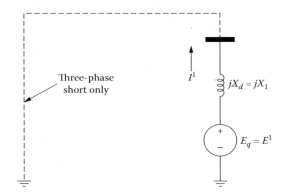

FIGURE E2.9.2
Balanced steady-state positive-sequence equivalent.

Test 2: In this test, let terminals b and c be connected together such that the phase currents are

$$I^c = -I^b = -I_{1.s.}$$

$$I^a = I^n = 0$$

When the symmetrical component transformation is applied to this set of currents,

$$\mathbf{I}^{012} \begin{bmatrix} I^0 \\ I^1 \\ I^2 \end{bmatrix} = \mathbf{T}_s^{-1} \mathbf{I}^{abc} = \mathbf{T}_s^{-1} \begin{bmatrix} 0 \\ I_{1.s.} \\ -I_{1.s.} \end{bmatrix} = \frac{j}{\sqrt{3}} \begin{bmatrix} 0 \\ 1 \\ -1 \end{bmatrix} [I_{1.s.}]$$

The resulting symmetrical component currents for this test are of opposite polarity, so the equivalent circuits are connected in series as shown in Figure E2.9.3. The measured current is 2,900 A rms, so Kirchhoff's voltage law around the loop in the circuit in Figure E2.9.3 is

$$E_q = |e^a| = 7621 = j(X_1 + X_2) I_{1.s} = j(X_1 + X_2) \frac{-j2900}{\sqrt{3}}$$

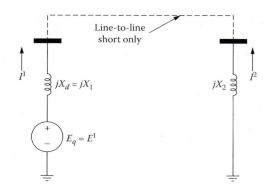

FIGURE E2.9.3

Since X_1 is known, this last equation may be solved for the unknown X_2:

$$X_2 = \frac{7621\sqrt{3}}{2900} - 3.810 = 0.742 \ \Omega$$

Test 3: In this test, let phase a be directly connected to neutral while phases b and c are open circuited:

$$I^a = -I_{1.n.} = -I^n$$
$$I^b = I^c = 0$$

This measurement yields the following set of symmetrical component currents, which have the same algebraic sign:

$$\mathbf{I}^{012} = \begin{bmatrix} I^0 \\ I^1 \\ I^2 \end{bmatrix} = \mathbf{T}_s^{-1} \mathbf{I}^{abc} = \mathbf{T}_s^{-1} \begin{bmatrix} I_{1.n.} \\ 0 \\ 0 \end{bmatrix} = \frac{1}{3} \begin{bmatrix} I_{1.n.} \\ I_{1.n.} \\ I_{1.n.} \end{bmatrix}$$

The symmetrical component equivalent circuits are in series for this test. The unknown impedance X_0, found using the circuit shown in Figure E2.9.4, is

$$X_0 = \frac{7261(3)}{4200} - X_1 - X_2 = 0.6344 \ \Omega$$

For the momentary-short-circuit test, the equivalent circuit is as shown in Figure E2.9.5 and the numerical value of the subtransient reactance is

$$X_d'' = \frac{13,200/\sqrt{3}}{20,000} = 0.381 \ \Omega$$

Using the machine ratings as the per-unit base, the base quantities are:

$$S_{\text{BASE}} = 50 \ \text{MVA}$$
$$V_{\text{BASE}} = 13.2 \ \text{MVA}$$
$$I_{\text{BASE}} = \frac{S_{\text{BASE}}}{V_{\text{BASE}}} = \frac{50 \ \text{MVA}}{13.2 \ \text{kV}} = 3.788 \ \text{kA}$$
$$Z_{\text{BASE}} = \frac{V_{\text{BASE}}}{I_{\text{BASE}}} = 3.4848 \ \Omega$$

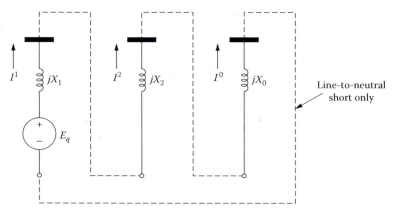

FIGURE E2.9.4

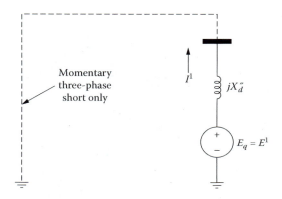

FIGURE E2.9.5

A summary of the machine reactances found using the four measurements is as follows:

Reactance	Value (p.u.)
Positive sequence	1.093
Negative sequence	0.213
Zero sequence	0.182
Subtransient	0.109

This generator's steady-state equivalent circuit is shown in Figure E2.9.6. With an *additional* measurement and δ specified for steady-state load conditions, it is possible to calculate X_q using the phasor diagram of Figure 2.34a. The calculated value of X_q, in turn, can be used in Equation 2.118 to obtain X_1' and subsequently specify all machine reactances.

Usually, the reactances X_0, X_1, X_2, X_1', X_2', and X_1'' are sufficient to calculate machine and network currents or voltages using phasor methods in the three time regions. These values can be obtained by fitting numerical values to experimental transient-time histories as outlined in IEEE Standard 115, "Test Procedures for Synchronous Machines." Another method

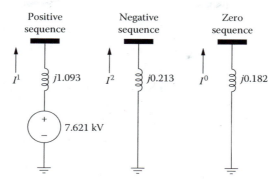

FIGURE E2.9.6
Generator equivalent circuit (p.u.)

is by means of examining machine experimental frequency response data as shown in Figure 2.36. Such frequency response data may be obtained with the machine stationary, off-line, or by perturbations introduced on the line waveform.

The classical synchronous machine model has been used to demonstrate the origin of transient and steady-state impedances required for power system network calculations. Because the model is "ideal," there is considerable error between theoretical values and experimental data. Consequently, manufacturer's data or typical numerical values are used for positive-, negative-, and zero-sequence parameters in subsequent sections rather than trying to derive these values from machine geometry, materials, and winding data. Table 2.7 gives typical values for the reactances as dependent on machine construction and rating. Included in Table 2.7 is a special case of a synchronous machine that does not have a prime-mover input-output drive shaft. This is a *synchronous condenser,* or simply *condenser*. It has three-phase winding and field excitation identical to a machine, but in the absence of a drive shaft, it essentially idles when on-line. By means of varying the excitation, the

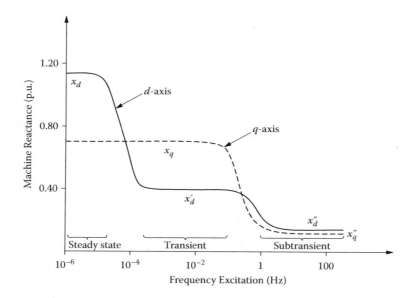

FIGURE 2.36
Frequency response of a synchronous machine stator.

TABLE 2.7

Typical Constants for Synchronous Machines[a]

	(1) X_d (Unsat.) X_d	(2) X_q Rated Current	(3) X'_d Rated Voltage	(4) X''_d Rated Voltage	(5) X_2 Rated Current	(6) X_0 Rated Current[b]	(7) R_a[c]	(8) T'_d	(9) T''_d
Two-pole turbine generators	1.20 / 0.95–1.45	1.16 / 0.92–1.42	0.15 / 0.12–0.21	0.09 / 0.07–0.14	$=x''_d$	0.03 / 0.01–0.08	0.001–0.007	0.6	0.035 / 0.02–0.05
Four-pole turbine generators	1.20 / 1.00–1.45	1.16 / 0.92–1.42	0.23 / 0.20–0.28	0.14 / 0.12–0.17	$=x''_d$	0.08 / 0.015–0.14	0.001–0.005	1.0	0.035 / 0.02–0.05
Salient pole generators and motors (with dampers)	1.25 / 1.60–1.50	0.70 / 0.40–1.80	0.30 / 0.20–0.50[d]	0.20 / 0.13–0.32[d]	0.20 / 0.13–0.32[d]	0.18 / 0.03–0.23	0.003–0.015	1.5 / 0.5–3.3	0.035 / 0.01–0.05
Salient pole generators (without dampers)	1.25 / 1.60–1.50	0.70 / 0.40–1.80	0.30 / 0.20–0.50[d]	0.30 / 0.20–0.50[d]	0.48 / 0.35–0.65	0.19 / 0.30–0.24	0.003–0.015	1.5 / 0.5–3.3	—
Condensers (air cooled)	1.85 / 1.25–2.20	1.15 / 0.95–1.30	0.40 / 0.30–0.50	0.27 / 0.19–0.30	0.26 / 0.18–0.40	0.12 / .025–0.15	0.0035 / 0.0025–0.008	2.0 / 1.2–2.8	0.035 / 0.02–0.04
Condensers (hydrogen cooled at ½ psi kVA rating)	2.20 / 1.50–2.65	1.35 / 1.10–1.55	0.48 / 0.36–0.60	0.32 / 0.23–0.36	0.31 / 0.22–0.48	0.14 / 0.030–0.18	0.0035 / 0.0025–0.005	2.0 / 1.2–2.8	0.035 / 0.02–0.04

Source: Reprinted with permission from reference [5].

[a] Values below the line give the normal range of values, while those above give an average value. Reactances are per unit; time constants are in seconds.

[b] X_0 varies so critically with armature winding pitch that an average value can hardly be given. Variations from 0.1 to 0.7 of X''_d. Low limit is for ⅔ pitch windings.

[c] R_a varies with machine rating; limiting values given are for about 50,000 and 500 kVA.

[d] High-speed units tend to have low resistance and low-speed units high resistance.

condenser operates at *leading* or *lagging* reactive power to provide rapid and continuously variable reactive compensation for power systems. Its reactances and reactive capacity are similar to those of a synchronous machine.

Problems

© Observe that many of the transmission line calculations can be performed by software programs *l.param.exe* and *c_param.exe*

2.1. Complete the following table by computing X_1 and X_1' for the blank spaces. Values for X_a and X_a' are available in Table 2.3 to Table 2.5.

Typical Single-Conductor ACSR Transmission Line Characteristics (without Ground and Shield Wire Effects)

Rated Voltage (kV line to line)	Conductor Code Name	Equivalent Spacing (ft)	Resistance at 50°C per Phase (Ω/mile)	X_1 Reactance per Phase (Ω/mile)	X_1' Shunt Capacitive Reactance per Phase (MΩ-mile)
69	Linnet	14	0.307		
115	Linnet	17	0.307		
138	Chickadee	18	0.259		
161	Chickadee	19	0.259		
230	Mallard	25	0.131		

2.2. A 345 kV line rated at 1,000 MVA is 14 miles long. The geometry of the line is shown in Figure P2.2. The conductor is aluminum conductor steel reinforced (ASCR) with the code name Bluebird. Only one conductor, 1.762 in. diameter, is used. The average height of the conductors above ground considering sag is 47 ft. Assume that the ground conductivity is 200 Ω-m, and there are no lighting shields on top of the line structure. Using the ACSR data in Table 2.3 to Table 2.5, calculate the impedances Z_1 and Z_0 and the shunt reactances X_1' and X_0'.

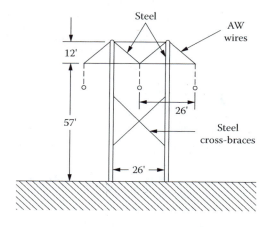

FIGURE P2.2

2.3. A 525 kV line rated at 2,000 MVA is 110 miles long and has steel construction with the geometry shown in Figure P2.3. Each phase uses two conductors with 18 in. spacing. The conductors are 1.88 in. in diameter, ACSR, with the code name Joree. Considering sag, the average height of the conductors above ground is 58 ft. The shield height is accordingly adjusted with respect to ground. Assume that the ground has 200 Ω-m resistivity. There are two lighting shield wires, 0.5 in. diameter (assume zero resistance and grounded at each tower). Using the Table 2.3 to Table 2.5, determine the impedances Z_1 and Z_0 and the shunt reactances X_1' and X_0'. Show the π-equivalent circuit for this line. Add to the equivalent circuit 100 MVA shunt reactors at both ends of the line that are used for voltage regulation (compensation).

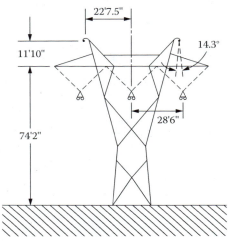

FIGURE P2.3

2.4. The ratings for a high-voltage circuit breaker are to be determined. The breaker is used to isolate a step-up transformer and two 50 MVA generators from the transmission network shown in Figure P2.4. The manufacturer's data on the generators and the 13.2:69 kV step-up transformer specified on a 13.2 kV, 50 MVA base are the following:

	Generator 1	Generator 2	Transformer
	$X_d = 1.20$	$X_d = 0.60$	$X_1 = X_2 = 0.06$
	$X_q = 0.70$	$X_q = 0.40$	
	$X_d' = 0.30$	$X_d' = 0.20$	
	$X_d'' = 0.30$	$X_d'' = 0.20$	
	$X_2 = 0.50$	$X_2 = 0.35$	

The resistances of the machines and transformer are negligible. Both generators are initially operating at full load, 50 MVA each, 0.8 power factor, with 13.2 kV output. Do the following:

(a) Calculate an equivalent voltage behind synchronous, transient, and subtransient reactances for both generators.

(b) Assume that both generators remain in synchronism, and the internal voltages from part (a) stay constant when a three-phase fault to ground occurs on

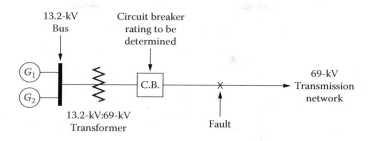

FIGURE P2.4

the 69 kV transmission line at the × indicated on the circuit diagram. Calculate the ac subtransient, transient, and steady-state short circuit through the 69 kV circuit breaker due to the generators.

2.5. Calculate the 60 Hz symmetrical component series impedances and shunt capacitor admittances for the 100-mile, 500 kV, 1,500 MVA horizontal transmission line, shown in Figure P2.5, with the following line geometry:

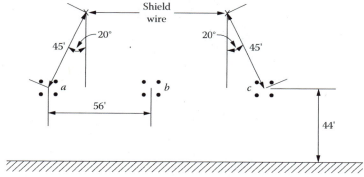

FIGURE P2.5

Phase spacing = 56 ft
Four conductors per phase, Pheasant ACSR, 1.382 in. diameter
Resistance per conductor = 0.0874 Ω/mile at 70°C
Bundle spacing = 18 in.
Two ground wires, $^{9}\!/_{16}$ in. diameter, at 20° shield angle and 45 ft from outside phases
Average conductor height above ground = 44 ft
The shield wires are Alumoweld with the following properties:

$$r_c = 1.1971 \,\Omega/\text{mile}$$

$$X_a = 0.707 \,\Omega/\text{mile}$$

$$\text{radius} = 0.023438\,\text{ft}$$

Assume that the ground has $\rho = 100 \,\Omega$-m resistivity. Perform the calculation for:

(a) An untransposed line
(b) A transposed line

Compare the zero-sequence impedance calculated in part (b) with the case of no shield wires (Equation 2.57).

2.6. Derive the symmetrical component equivalent circuit for the Z_g (p.u.) impedance-grounded wye-to-delta transformer shown in Table 2.6. A convenient starting point is a circuit for a three-core ideal transformer with leakage reactance shown in Figure P2.6. Let the leakage impedances shown in the primary and secondary be expressed in per unit on the transformer power base so that $Z_p + Z_s = Z_{pq}$ and the turns ratio cancels out. Excite the primary and short-circuit the secondary to obtain the positive- and negative-sequence equivalent circuit. For the zero-sequence current tests, let $I^a = I^b = I^c = I^0$ in the primary with the secondary shorted; then $Z_0 = E^0/I^0$.

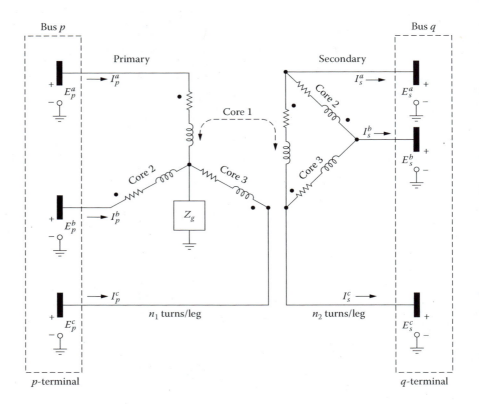

FIGURE P2.6

2.7. The construction of a 23 kV subtransmission line is shown in Figure P2.7. The unshielded phase conductors are 336 kcmil ACSR, code name Merlin, attached by insulators on the cross-arm of a wooden pole. Lightning arrestors (which are normally open circuits) are shown adjacent to the insulators. The average height above ground is 32 ft for the conductors, and they may be assumed to be *decoupled* from the 13.2/23 or 4.16 kV distribution circuit on the lower cross-arm. The subtransmission line is transposed. The conductor temperature is 50°C. Do the following:

(a) Find the subtransmission line positive- and zero-sequence series impedance (Ω/mile) in the presence of a 100 Ω-m ground.

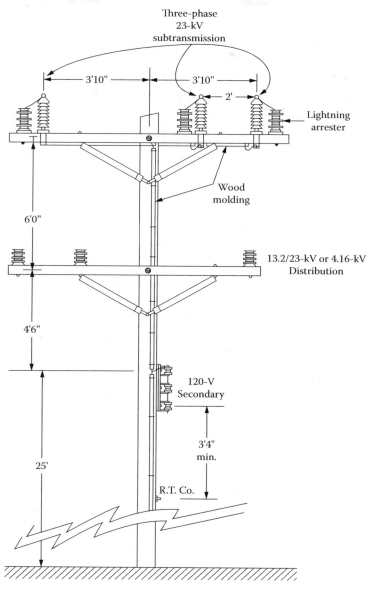

FIGURE P2.7

(b) Find the zero-sequence and positive-sequence shunt capacitive reactances considering ground.

(c) Draw the positive- and zero-sequence equivalent circuits.

2.8. The construction of a 115 to 230 kV compact transmission line is shown in Figure P2.8. The lightning shield wire is Alumoweld (7 no. 8) 0.385 in. diameter, GMR = 0.00209 ft, X_a = 0.749 Ω/mile, X'_a = 0.1224 MΩ-mile, and an ac resistance of 2.44 Ω/mile at 25°C. The average height above ground of the lowest phase conductor is 30 ft, and the ground resistivity is assumed to be 100 Ω-m. For ACSR phase

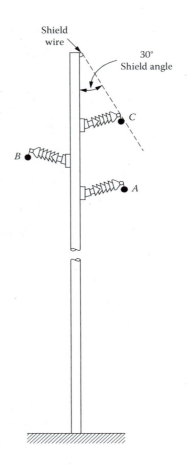

FIGURE P2.8

conductors at 25°C, complete the following table for any one conductor type: Dove, Drake, Cardinal, Grackle, or Falcon.

Conductor Spacing (in.)			$R_0 + jX_0$ (Ω/mile)	$R_1 + jX_1$ (Ω/mile)	X'_0 (MΩ – mile)	X'_1 (MΩ – mile)
$A - B$	$B - C$	$C - A$				
77	77	72				
95	95	84				
113	113	96				
130	130	108				
133	133	120				
139	139	144				

2.9. The graphical approach shown in Figure 2.35 may be used to find the machine reactances X''_d, X'_d, and X_d as well as the time constants T'_d and T''_d. The rms line current versus time was measured for a three-phase generator operating at nominal voltage, no load, when the machine was subjected to a sustained three-phase short circuit. Calculate these reactances and time constants using the following current envelop, which has the dc component removed.

Time (sec)	Line Current (p.u.)
0.00	4.48
0.05	3.41
0.10	2.96
0.15	2.72
0.20	2.56
0.25	2.47
0.40	2.19
0.50	2.04
0.70	1.80
1.00	1.55
1.50	1.27
2.00	1.13
2.50	1.05
3.00	1.01
$t = \infty$	0.96

2.10. Lightning strikes the phase *a* wire of a 60 Hz shielded transmission line as shown in the cross section in Figure P2.10. The *b* and *c* phases of the transmission line are essentially open circuit (zero current) to the lightning stroke because of their terminal loads. However, the shield wires are connected to earth at every tower, so they are at ground potential. The earth has a resistivity of 100 Ω – m. The phase conductors are code name Hawk ACSR single conductor, and the shield wires are .548 in. Alumoweld (GMR = .00296 ft, X_a = .707 Ω/mi, X'_a = .1121 Ω = mi). The resistance of the phase conductors, earth, and shield wires may be neglected, but earth corrections (see Equation 2.48 to Equation 2.51) are significant in answering the following questions.

(a) Considering the two shield wires, the phase *a* conductor and earth, what is the *series* impedance (inductive reactance) to the lightning stroke that must travel down the phase *a* conductor? (Ω/mile)

(b) What is the capacitive reactance (MΩ – mi) of the phase *a* conductor with respect to earth reference? Phases *b* and *c* may be neglected, but the shield wires must be considered.

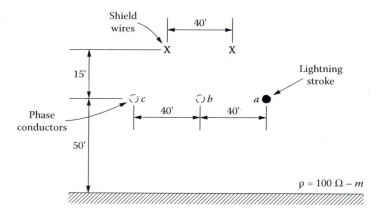

FIGURE P2.10

(c) What is the surge impedance presented to the lightning stroke

$$X_{surge} = \sqrt{X_{series}X_{Shunt}}$$

where X series and X shunt (both in ohms) are the result of parts (a) and (b)?

2.11. A three-phase transmission line runs parallel to a telephone line pair for 50 miles with the spacing shown in Figure P2.11. The transmission line currents are $I^a = 550 \angle 0°$, $I^b = 550 \angle -120°$, $I^c = 550 \angle +120°$ amperes. Assume zero current flows in the telephone wires. Observe the inductively coupled voltage line to ground in the telephone wires may be calculated using the method of Equation 2.54 to Equation 2.55. Because of the distances involved, the following calculation involves differences of large numbers, so numerical values should carry many decimal places.

(a) Calculate the voltage difference between the lines of the telephone pair, in other words, the voltage induced in the ungrounded telephone lines.
(b) How can the voltage induced in the telephone lines be eliminated?

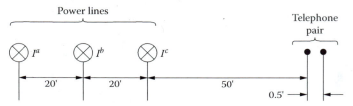

FIGURE P2.11

2.12. Two cylindrical rotor generators and a synchronous condenser feed a two-transmission-line network, shown in Figure P2.12. The loads are not shown because they are large impedances compared to the network parameters. Each component of the network is expressed in per unit on its own power and voltage rating.

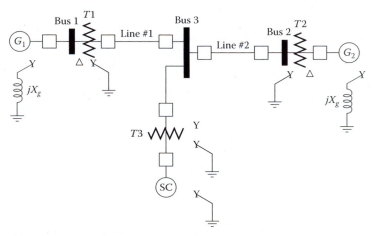

FIGURE P2.12

(a) Convert all components to a 100 MVA, 138 kV base by reflecting impedances through the transformers up to the transmission line voltage magnitude.

(b) Draw the *positive- and zero-sequence* diagrams for the network using π-equivalents for the transmission lines. Show numerical values of the components on this diagram.

Generator G1: 50 MVA, 13.8 KV, $X'_1 = X'_2 = .3$, $X_0 = .1$, $X_g = .1$

Generator G2: 100 MVA, 20 kV, $X'_1 = X'_2 = .8$, $X_0 = .2$, $X_g = .1$

Condenser SC: 20 MVA, 13.8 kV, $X'_1 = X'_2 = .2$, $X_0 = .1$

Primes in these elements represent transient reactances. Note that the effect of neutral grounding impedances is shown in Table 2.6.

Transformer T1: 50 MVA, 13.8:138 kV, $Z_{pq} = .033 + j.20$, Y – Δ

Transformer T2: 100 MVA, 20:138 kV, $Z_{pq} = j.10$, Y – Δ

Transformer T3: 20 MVA, 13.8:138 kV, $Z_{pq} = j.20$, Y – Y

Line 1: 100 MVA, 138 kV, $X_1 = X_2 = .03$, $X'_1 = X'_2 = -10$, $X_0 = .5$, $X'_0 = -20$

Line 2: 100 MVA, 138 kV, $X_1 = X_2 = .02$ $X'_1 = X'_2 = -8$, $X_0 = .4$, $X'_0 = -10$

2.13. A wye-connected, 480 V, 576 kvar, four-pole synchronous generator is delivering full load at 0.8 lagging power factor and rated voltage. The generator has a direct-axis reactance of $X_d = 0.25$ p.u. and a quadrature reactance of $X_q = 0.1875$ p.u.

(a) Find the internal voltage $|E_q|$ p.u. and the phase angle δ if the machine is assumed to be cylindrical, $X_q = X_d = X_s = 0.25$ p.u.

(b) Repeat part (a) if the machine has salient rotor characteristics $X_d = 0.25$, $X_q = 0.1875$.

2.14. A 345-kV line is 100 km long. Its series positive sequence reactance is $j0.15$ p.u. on a 100 MVA base. Its capacitive shunt susceptance is $j0.12$ p.u. Resistance of the line is neglected. If one end of the line is excited at rated voltage, and the other end is open-circuited, answer the following questions.

(a) What is the p.u. and kV line-to-line voltage measured at the open-circuit end of the line?

(b) What is the 3-phase reactive power input at the excited end?

(c) Why is it dangerous to equipment to operate the line open-circuited?

References

1. Magnusson, P. C. 1970. *Transmission lines and wave propagation*. Boston: Allyn and Bacon.
2. Kraus, J. D., and Carver, K. R. 1973. *Electromagnetics*. New York: McGraw-Hill Book Company.
3. Electric Power Research Institute. 1975. *Transmission line reference book—345 kV and above*. Palo Alto, CA: Author.
4. Hesse, M. H. 1963. Electromagnetic and electrostatic transmission-line parameters by digital computer. *IEEE Trans. Power Appar. Syst.*
5. Westinghouse Electric Corporation. 1964. *Electrical transmission and distribution reference book*. East Pittsburgh, PA: Author.
6. Chen, M.-S., and Dillon, W. E. 1974. Power system modeling. *Proc. IEEE* 62:901–15.

7. Kimbark, E. W. 1956. *Power system stability: Synchronous machines*. Vol. III. New York: John Wiley & Sons.

8. Bose, B. K. 2002. *Modern power electronics and AC drives*. Upper Saddle River, NJ: Prentice Hall PTR.

9. Chapman, S. J. 2005. *Electric machinery fundamentals*. 4th ed. New York: McGraw-Hill.

10. ANSI/IEEE. 1980. *Recommended practice for power systems analysis*. ANSI/IEEE Standard 399-1980.

11. Anderson, P. M. 1973. *Analysis of faulted power system*. Ames: Iowa State University Press.

3

Bus Reference Frame

Only linear networks are considered in this chapter. The balanced three-phase networks studied thus far are readily described by two types of formulation for Kirchhoff's voltage and current laws. The balanced aspect allows three-phase elements to be described by equivalent positive-sequence elements with respect to a potential neutral or ground point. Therefore, three-phase transmission lines between points may be drawn as single lines on a topological map of the terrain. An infinite, perfectly conducting plane below the lines serves as a ground potential surface. The electrical loads on the transmission line are connected between lines and the ground plane. Similarly, generation and shunt elements or line charging are connected between the transmission lines and ground plane. This single-line method will be continued.

There are few "islands" or isolated power systems that are not connected to neighbors by tie lines. For reliability—to aid each other in periods of power shortage, or economy—to buy or sell power to minimize operating costs, power systems have become interconnected. Through the interconnection, power may be sent from any member to another. Often small areas are coordinated in regulating frequency and interchanging power through "pool" control centers. The methods developed in this chapter are oriented toward digital computer calculations to describe steady-state conditions and study problems associated with power flow on the large number of buses involved in an area or a pool. At this time, large-memory-capacity computers can model power flow on the entire eastern U.S. interconnection, which is one of four in North America. Figure 3.1a shows the four synchronized, 60 Hz interconnections. Each interconnection maintains its own distinct 60 Hz frequency.

The separate interconnected areas developed historically, or are the result of very low MW power transmission capability between very large load-generation regions. Low MW interchange capability is why the Rocky Mountains in the United States break the eastern interconnection from the western. Between any interconnection there are only dc/dc links that only transmit or receive dc power converted from the ac system transmission lines. The dc/dc links between the interconnections, indicated by ►||◄, are shown in Figure 3.1b This figure also shows pool controllers (indicated by small circles in Figure 3.1b) in their approximate geographic location. The pool controllers often dispatch the generators and regulate ACE for areas within the interconnection.

Since the impact of an individual area diminishes with electrical distance into a large MW capacity interconnection, it is necessary to consider an area as an isolated power system with tie lines to neighbors replaced by equivalent power injections at the end of the tie lines. Selecting an appropriate equivalent for tie lines is a continuing research problem even though solutions are available for particular static or dynamic situations. The first step is to understand area power flow from a bus–line–load point of view within an area.

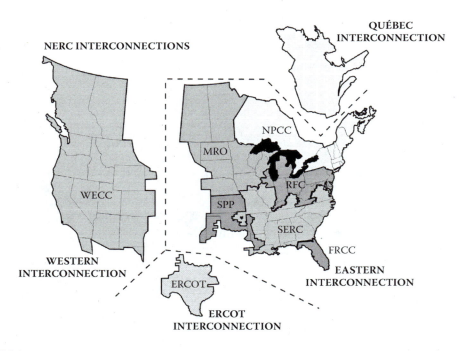

FIGURE 3.1
(a) North American separate 60 Hz interconnection regions. Each of the four regions regulates its own 60 Hz frequency. (Map courtesy of North American Electric Reliability Corporation (NERC)).

3.1 Linear Network Injections and Loads

Consider the positive sequence power system diagram with node-to-ground phasor voltages E_1, E_2, and E_3, shown in Figure 3.2. There are two generators supplying loads through simplified transmission lines Z_4 and Z_5. The network is linear and thus is described by Kirchhoff's voltage and current laws.

The connection point of generator 1, impedance Z_1, and impedance Z_4 is called a *bus* of the power system rather than a node as per Kirchhoff's laws. The current I_1 from generator 1, with positive sense into the network, is called an *injection current* and satisfies Kirchhoff's current law:

$$I_1 = I_{L1} + I_{13} \tag{3.1}$$

Real and reactive powers from generator 1 or generator 2 are called power injections into the network. The impedances Z_1, Z_2, and Z_3, which dissipate real and reactive power, are called loads, or negative power injections, on the system. Chapter 5 deals with the case where the load impedances vary with the applied voltage, a nonlinearity that is called the power flow problem.

In a linear network, such as the power system of Figure 3.2, the Thévenin equivalent at bus 1 has E_1 as the open-circuit voltage at the bus and a Thévenin impedance Z_{11} "looking into" the network. When all circuit elements are known, the Thévenin impedance at bus 1

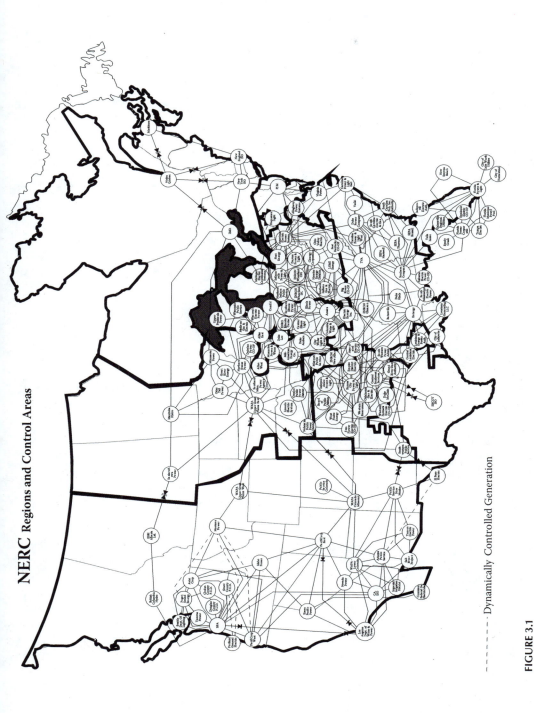

- - - - - - Dynamically Controlled Generation

FIGURE 3.1

(b) Major ACE regulation areas in North America. Small circles are the approximate location of the control centers. Regions are linked by ac–dc–ac (▶‖◀) conversion lines or devices. (Map courtesy of North American Electric Reliability Corporation (NERC)).

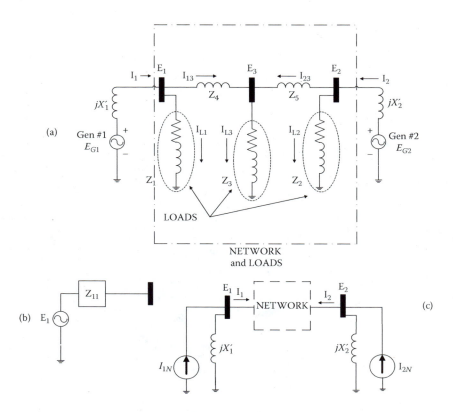

FIGURE 3.2
(a) A simple power system with two generators. (b) The Thevenin equivalent at bus 1. (c) Norton equivalents for generator injections.

into the network is found by shorting out all voltage sources and open-circuiting all current sources. For Figure 3.2, the driving-point impedance, or short-circuit impedance at bus 1, is a computed combination of $Z_1 \ldots Z_5$ and jX' of the generators.

$$Z_{11} = Z_{Thevenin} = \frac{Open-Circuit-Voltage}{Short-Circuit-Current} = \frac{E_1}{I_{SC}} \tag{3.2}$$

The resulting Thévenin equivalent at bus 1 is shown in Figure 3.2b. This equivalent holds for any impedance externally attached to bus 1 because Z_{11} is the result of all internal impedances, including the generators.

The dependent voltages E_1, E_2, and E_3, shown in Figure 3.2, are the result of the voltages E_{G1} and E_{G2} behind transient reactances. A better description for computations, of the generators at buses 1 and 2, is the Norton equivalent shown in Figure 3.2c. The Norton equivalent currents are I_{1N} and I_{2N} from the generators. From this consideration, a generalized Thévenin equivalent for the network of Figure 3.2, with two generators injecting currents, and no Norton injection current at bus 3, may be written in matrix form as

$$\begin{bmatrix} E_1 \\ E_2 \\ E_3 \end{bmatrix} = Z_{BUS} \begin{bmatrix} I_{1N} \\ I_{2N} \\ 0 \end{bmatrix} \tag{3.3}$$

The current into the network at bus 1 is

$$I_1 = I_{1N} - E_1/jX_1'$$ (3.4)

The Thévenin driving-point impedances Z_{11}, Z_{22}, and Z_{33} are on the diagonal of the 3×3 matrix Z_{BUS}. The off-diagonal terms of the Z_{BUS} matrix may be considered as mutual impedances found by superposition of the injection currents I_{1N} and I_{2N} taken one at a time.

Thus, the mutual impedance Z_{2N} may be considered the multiplier for the voltage induced at bus 1 caused by only the current I_{2N} injected at bus 2. Since the network is linear, the reciprocity property holds that $Z_{12} = Z_{21}$. In other words, the same mutual impedance is the multiplier for the voltage at bus 2 due to a current injected at bus 1.

For a general linear network, bus nodal voltages E_1, E_2, ..., E_M for an M bus network are calculated quantities due to injection currents I_{1N}, I_{2N}, ..., I_{MN} from generators or other sources. In matrix form this equation is called a generalized Thévenin equivalent, or a *bus impedance formulation*:

$$\begin{bmatrix} E_1 \\ E_2 \\ . \\ . \\ E_M \end{bmatrix} = \mathbf{E}_{BUS} = \mathbf{Z}_{BUS}\mathbf{I}_{BUS} = \mathbf{Z}_{BUS} \begin{bmatrix} I_{1N} \\ I_{2N} \\ . \\ . \\ I_{MN} \end{bmatrix} \quad \text{(generators included)}$$ (3.5)

The elements of $\mathbf{Z}_{BUS}$ are combinations of impedances from the network and the generators. The vector I_{BUS} is comprised of the Norton current injections when the generators are considered. When only the network terms are considered as in Figure 3.2a, $\mathbf{Z}_{BUS}$ does not contain generator impedances, and the form of the equations is

$$\begin{bmatrix} E_1 \\ E_2 \\ . \\ . \\ E_M \end{bmatrix} = \mathbf{I}_{BUS} = \mathbf{Z}'_{BUS}\mathbf{I}_{BUS} = \mathbf{Z}'_{BUS} \begin{bmatrix} I_1 \\ I_2 \\ . \\ . \\ I_M \end{bmatrix} \quad \text{(network only)}$$ (3.6)

Another form of the equations considering generator injections is found using the inverse of the bus impedance matrix:

$$\begin{bmatrix} I_{1N} \\ I_{2N} \\ . \\ . \\ I_{MN} \end{bmatrix} = \mathbf{I}_{BUS} = [\mathbf{Z}_{BUS}]^{-1}\mathbf{E}_{BUS} = \mathbf{Y}_{BUS}\mathbf{E}_{BUS} \quad \text{(generators included)}$$ (3.7)

$\mathbf{Y}_{BUS}$ is called the *bus admittance matrix*. Equation 3.7 is essentially a nodal formulation of Kirchhoff's current law [1]. With the increasing storage and computational speed of digital computers, except for very large networks, $\mathbf{Z}_{BUS}$ is almost always obtained by numerically inverting $\mathbf{Y}_{BUS}$ by direct methods or by the sparse matrix methods described later in this chapter.

For general power networks, the following definitions are used:

DEFINITION 3.1
The bus reference frame assumes complex voltages, $\mathbf{E}_{BUS}$, between every network bus and the reference, with bus injection currents, $\mathbf{I}_{BUS}$.

Even though this approach introduces extra bus voltages that must be eliminated in order to solve for network currents or voltages in terms of the independent variables (forcing functions), the bus reference frame subsequently is very suitable for large networks and power flow solutions. The fundamental matrices used in this formulation are defined as follows.

DEFINITION 3.2
$\mathbf{Z}_{BUS} \equiv$ bus impedance matrix. Its elements are the open-circuit driving-point and mutual (transfer) impedances between the bus and the reference.

DEFINITION 3.3
$\mathbf{Y}_{BUS} \equiv$ bus admittance matrix. Its elements are the short-circuit driving-point admittances and mutual (transfer) admittances between the bus and reference.

The number of equations in the bus reference frame is equal to the number of buses. The ground potential or neutral is often called an additional bus (or node) in circuit calculations, but here we refer only to points above ground potential as buses. It is necessary at times to designate an above-ground potential bus as a slack bus, swing bus, or reference bus, and use it as the common point for injections and voltages. These cases will be defined clearly later.

The driving-point and transfer impedances or admittances may be found using circuit methods as demonstrated in the following example.

Example 3.1

For the network in Figure E3.1.1, the generator internal voltages E_{G1} and E_{G2} establish potentials $E_1 = 1.05 + j0.0$ and $E_2 = 1.05 + j0.05$ at buses 1 and 2, respectively. For the numerical values of impedances indicated on Figure E3.1.1, do the following:

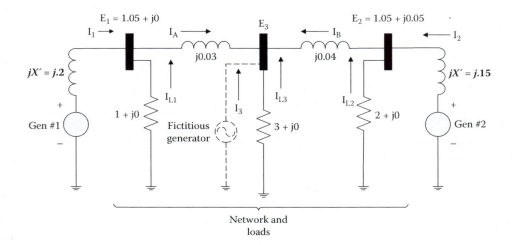

FIGURE E3.1.1

(a) Use Kirchhoff's voltage laws to calculate the potential E_3. Determine the Norton equivalent current injections from the generators.

(b) Use superposition of injection currents to determine the bus impedance matrix, $\mathbf{Z}_{BUS}$.

(c) Use superposition of voltage sources to find directly the bus admittance matrix, $\mathbf{Y}_{BUS}$.

(d) Use the bus admittance matrix to calculate E_3.

Solution

(a) Since the potentials E_1 and E_2 are fixed, $I_{L1} = -1.05$ and $I_{L2} = -(1.05 + j0.05)/2$. Two new branch current variables,

$$I_A = I_1 + I_{L1}$$

$$I_B = I_2 + I_{L2}$$

are used to express current I_{L3}:

$$I_{L3} = -I_A - I_B$$

Currents I_A and I_B are used to write Kirchhoff's voltage law for the two loops:

$$E_1 = j0.03I_A + 3(I_A + I_B)$$

$$E_2 = j0.04I_B + 3(I_A + I_B)$$

The solution for I_A and I_B is

$$\begin{bmatrix} I_A \\ I_B \end{bmatrix} = \begin{bmatrix} 3+j0.03 & 3 \\ 3 & 3+j0.04 \end{bmatrix}^{-1} \begin{bmatrix} E_1 \\ E_2 \end{bmatrix} = \frac{1}{j0.21 - 0.0012} \begin{bmatrix} 3+j0.04 & -3 \\ -3 & 3+j0.03 \end{bmatrix} \begin{bmatrix} 1.05 + j0 \\ 1.05 + j0.05 \end{bmatrix}$$

$$= \begin{bmatrix} -0.5142685 + j0.0029386 \\ 0.8642979 + j0.0022039 \end{bmatrix}$$

Hence, $I_{L3} = -I_A - I_B = -0.3500 - j0.0051425$ and the potential at bus 3 is

$$E_3 = -3I_{L3} = 1.05 + j0.0154$$

The currents into buses 1 and 2 of the network are

$$I_1 = I_A - I_{L1} = 0.5357315 + j0.0029386$$

and

$$I_2 = I_B - I_{L2} = 1.3892979 + j0.0272039$$

The Norton equivalent injection currents, as per Figure 3.2c, are

$$I_{1N} = E_1/j0.20 + I_1 = -j5.25 + I_1 = 0.5357315 - j5.2470614$$

$$I_{2N} = E_2/j0.15 + I_2 = 0.3333 - j7.0 + I_2 = 1.7226309 - j6.9727961$$

(b) The network bus impedance matrix may be found by injecting unit currents one at
a time. Thus, the potential at bus 1 is found by superposition of unit currents for I_1,
I_2, and a fictitious current I_3:

$$E_1 = E_1^1 + E_1^2 + E_1^3 = z_{11}I_1 + z_{12}I_2 + z_{13}I_3$$

where E_1^1 is the voltage measured at bus 1 due to a unit injection current with all
other bus injection currents open circuited. E_1^2 is due to a unit current injected at
bus 2 with the other bus injection currents open circuited, and so on. First, consider
bus 1 with an injection current $I_1^1 = 1.0$ as shown in Figure E3.1.2. This injection
yields the Thévenin impedance at bus 1, Z_{11}. Combining the resistors and inductive
reactances, the driving-point impedance is a combination of parallel impedances
indicated by the double lines:

$$\frac{E_1^1}{I_1^1} = z_{11} = j0.20 // 1.0 // [j0.03 + 3 //(j0.04 + 2 //j0.15)] = 0.01583 + j0.102181$$

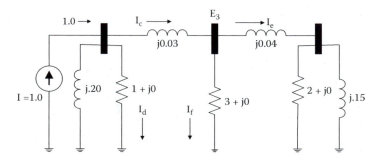

FIGURE E3.1.2

The injected current I_1^1 divides into I_c and I_d. Thus, the potential at bus 3 due to
injection current I_1^1 is used to calculate the transfer impedance.

$$\frac{E_3^1}{I_1^1} = z_{31} = z_{13} = I_c \left(\frac{3(j0.04 + 2//0.15)}{3 + j0.04 + 2//j0.15} \right) = 0.01514 + j0.087983$$

Using currents I_e and I_f defined in Figure E3.1.2 gives us

$$\frac{E_2^1}{I_1^1} = z_{21} = z_{12} = 0.01305 + j0.069254$$

$$I_c + I_d = I_1^1 = 1.0, \qquad I_e + I_f = I_c$$

The entire $\mathbf{Z}_{BUS}$ with generator currents as per Equation 3.5, calculated by injection
methods, is:

$$\mathbf{Z}_{BUS} = \begin{bmatrix} 0.01583 + j0.102181 & 0.01305 + j0.069254 & 0.01514 + j0.087983 \\ 0.01305 + j0.069254 & 0.01170 + j0.094579 & 0.01293 + j0.080034 \\ 0.01514 + j0.087983 & 0.01293 + j0.080034 & 0.01477 + j0.101635 \end{bmatrix}$$

The transient reactances, jX', for the generators are included in this computation. For each row, observe that the driving-point impedance on the diagonal is larger than any other mutual impedance on the row.

If the generators are not included in the Z_{BUS} formulation, the resulting equation for the network only is:

$$\begin{bmatrix} E_1 \\ E_2 \\ E_3 \end{bmatrix} = Z'_{BUS} \begin{bmatrix} I_1 \\ I_2 \\ 0 \end{bmatrix} = \begin{bmatrix} 0.5457 + j0.0092 & 0.5451 - j0.0154 & 0.5454 - j0.0045 \\ 0.5451 - j0.0154 & 0.5462 + j0.0301 & 0.5456 + j0.0010 \\ 0.5454 - j0.0045 & 0.5456 + j0.0001 & 0.5455 + j0.0119 \end{bmatrix} \begin{bmatrix} 0.5357 + j0.0029 \\ 1.3893 + j0.0272 \\ 0 \end{bmatrix}$$

(c) The bus admittance matrix may be found by applying unit voltages one at a time to each bus while short-circuiting the other bus voltage sources to ground. Thus, the injection current I_{1N} is found by superposition:

$$I_{1N} = I_1^1 + I_1^2 + I_1^3 = y_{11}E_1 + y_{12}E_2 + y_{13}E_3$$

where I_1^1 is the component of current resulting from a unit voltage applied at bus 1 with buses 2 and 3 shorted to ground. I_1^2 is the injection current (in the short circuit) resulting from a unit voltage applied to bus 2. Consider a unit voltage, $E = 1.0$, applied at bus 1, with the other potential sources shorted as shown in Figure E3.1.3.

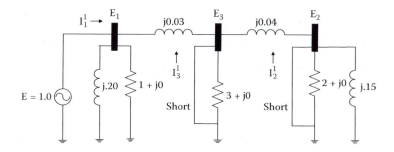

FIGURE E3.1.3

The current I_1^1 is readily calculated as

$$I_1^1 = \frac{E}{j0.20} + \frac{E}{1.0} + \frac{E}{j0.03} = 1.0 - j38.3 = y_{11}$$

and is numerically the same as the driving-point admittance. Since bus 3 is short-circuited to ground, the current I_3^1 is given by

$$I_3^1 = \frac{-E}{j0.03} = j33.3 = y_{13} = y_{31}$$

Notice that only buses directly connected by elements to bus 1 have current contributions due to the unit voltage applied at bus 1. Thus,

$$I_2^1 = 0 = y_{12} = y_{21}$$

When a unit voltage is applied to bus 2 with the other voltage sources shorted, the following driving-point and transfer admittances are easily calculated:

$$y_{22} = \frac{I_2^2}{E} = \frac{E/j0.15 + E/2 + E/j0.04}{E} = 0.5 - j31.666$$

$$y_{12} = y_{21} = 0$$

$$y_{23} = y_{32} = \frac{I_3^2}{E} = \frac{-E}{j0.04} = j25$$

The driving-point admittance at bus 3 is similarly calculated to be

$$y_{33} = 0.333 - j58.33$$

The complete bus admittance matrix for the power system of Figure E3.1.1 is

$$\mathbf{Y}_{BUS} = \begin{bmatrix} 1 - j38.33 & 0 & j33.33 \\ 0 & 0.5 - j31.666 & j25 \\ j33.33 & j0.25 & 0.333 - j58.33 \end{bmatrix}$$

where zeros appear at matrix locations (1, 2) and (2, 1) because there is no direct connection from bus 1 to bus 2. The generator transient reactances, jX', are included in this calculation.

The $\mathbf{Y}_{BUS}$ matrix is easily obtained for power systems of any size using a computer algorithm that searches an array of transmission lines to find those connected to each bus, and the mutual admittance from that bus to directly connected neighbors.

(d) The injection current I_3 is zero in the bus admittance formulation (see Figure E3.1.1), so $\mathbf{Y}_{BUS}$ may be partitioned as follows:

$$\begin{bmatrix} I_{1N} \\ I_{2N} \\ I_{3N} \end{bmatrix} = \begin{bmatrix} I_{1N} \\ I_{2N} \\ 0 \end{bmatrix} = \mathbf{Y}_{BUS} \ \mathbf{E}_{BUS} = \begin{bmatrix} \mathbf{Y}_{aa} & | & \mathbf{Y}_{ab} \\ \hline \mathbf{Y}_{ba} & | & \mathbf{Y}_{bb} \end{bmatrix} \begin{bmatrix} E_1 \\ E_2 \\ E_3 \end{bmatrix}$$

This equation may be verified by using the bus voltages to calculate I_{1N}, I_{2N}, and I_{3N}, which will agree with the results of part (a). The voltage E_3, or any bus voltage without Norton injection currents, may be found in terms of the other fixed bus voltages using the lower portion of the matrix:

$$E_3 = -[\mathbf{Y}_{bb}]^{-1} \mathbf{Y}_{ba} \begin{bmatrix} E_1 \\ E_2 \end{bmatrix} = \frac{-[j33.33 \quad j25]}{(0.333 - j58.33)} \begin{bmatrix} E_1 \\ E_2 \end{bmatrix}$$

$$= 1.05 + j0.0154$$

© Run the *zbus.exe* computer program to verify the $\mathbf{Z}_{BUS}$ matrix computed for the network of Figure E3.1.1 in Example 3.1. Note the impedances are specified as 100% p.u. The impedances are listed as either bus to bus or shunt elements to ground.

Next, edit the computed file *zbus* to set the bus-to-ground impedance at bus 1 to be 100% p.u. and the bus-to-ground impedance at bus 2 to be 200% p.u. With these changes, the result of another execution of **Zbus.exe** will match the $\mathbf{Z}_{BUS}$ matrix without generators.

Example 3.1 has demonstrated how elementary current injections or superposition of currents is a difficult approach to determine $\mathbf{Z}_{\text{BUS}}$. This method is unsuitable for large networks and does not easily generalize for computer applications.

Since $\mathbf{Z}_{\text{BUS}}$, or parts of it, is often required for network calculations, the next section introduces computer-oriented building algorithms to construct $\mathbf{Z}_{\text{BUS}}$ or modify an existing version. It is possible to formulate a version of $\mathbf{Z}_{\text{BUS}}$ that does, or does not, contain generator transient reactances. The following method can be used for the network-only version of $\mathbf{Z}_{\text{BUS}}$, or a version including the generator reactances.

3.2 Bus Impedance Matrix for Elements without Mutual Coupling

This section presents a systematic method to construct the bus impedance matrix one element at a time until all transmission lines, shunt elements, generators, and so on, that determine power flow have been included. Figure 3.3 shows reactors that can be added to

FIGURE 3.3
Air-cooled shunt reactors, each rated at 6.7 Mvar. Two in series are connected to the 13.8 kV winding of a transformer to shunt-compensate a high-voltage line. (Photograph courtesy of Westinghouse Electric Corporation, Pittsburgh, PA)

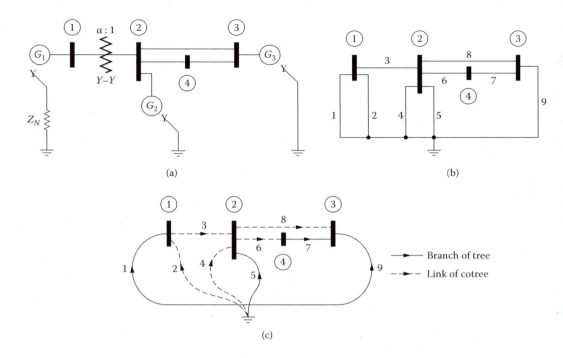

(a)   (b)

(c)

FIGURE 3.4
Isolated power system, impedance branches, and graph: (a) single-line diagram; (b) positive-sequence network for (a); (c) tree and cotree for the network. Part (b) replaces the transformer of (a) by a π-equivalent and shows the generator reactances as branches 1, 5, and 9 of a nine-branch network. Depending on the purpose of the network, either the generator synchronous or transient reactance values are used. The positive-sequence voltage of the generator is included in the element.

a network from bus to ground as separate elements. To begin the construction, it is necessary to borrow a few definitions from the formal discipline of graph theory for lumped networks [1] in order to establish a working vocabulary. Consider the four-bus, three-generator power system shown in the positive-sequence diagram in Figure 3.4a. Each lumped impedance between buses or bus to neutral is called a *branch* of the network if it contains a symmetrical component reactive element (capacitor or inductor) along with a dissipative element (resistor). Either the real or reactive element may be zero, but not both. Active sources (voltage or current sources) are considered to be a part of a branch. It is assumed that if mutual coupling exists between branches, a primitive impedance matrix is given to describe the coupling. Mutual coupling arises through common magnetic fluxes and common stored charge.

Every impedance shown in Figure 3.4b, unless of extreme magnitude, is significant for short-circuit studies. For power flow, the generator impedances are not used. The grounding impedance on the neutral of generator 1 does not appear in the positive-sequence diagram for balanced conditions. Generators 2 and 3 are directly grounded while the transformer neutrals float.

In Figure 3.4c a *tree* and its accompanying *cotree* have been defined for the network. A tree connects all the buses of the network without forming any closed paths. The buses of Figure 3.4b are connected by the tree using branches 1, 5, 7, and 9. There are many combinations of branches that could be used to connect all the buses. The remaining branches of the network are called *links* because each of them forms a loop when used in conjunction with tree branches. The selection of branches for the tree is not unique. The arrows on each branch of Figure 3.4c indicate the *branch current reference direction*. In the case of mutually coupled elements, the arrow defines positive or negative polarity of the coupling (discussed in Section 3.3) for the primitive elements. If no coupling exists, current direction is selected arbitrarily. The specified current direction arrows on the branches in Figure 3.4c are used to define a branch–bus reduced incidence matrix, **M**, which has the dimensions of *n* buses by *b* branches. The reference node is excluded, but at least one tree element should be connected to it. For Figure 3.4c this matrix has dimensions 4×9 and is

$$\mathbf{M} = \begin{matrix} & \text{elements} \rightarrow & \\ \begin{bmatrix} -1 & -1 & 1 & 0 & 0 & 0 & 0 & 0 & 0 \\ 0 & 0 & -1 & -1 & -1 & 1 & 0 & 1 & 0 \\ 0 & 0 & 0 & 0 & 0 & 0 & -1 & -1 & -1 \\ 0 & 0 & 0 & 0 & 0 & -1 & 1 & 0 & 0 \end{bmatrix} & \begin{matrix} \text{buses} \\ \downarrow \end{matrix} \end{matrix} \tag{3.8}$$

By convention, (i, j) is $+1$ if the branch j leaves bus i, and (k, j) is -1 if branch j terminates at bus k according to the reference directions assigned to the branch.

The tree branches and links defined for a network are next used to obtain $\mathbf{Z}_{BUS}$ for a network by progressively adding elements one at a time until all elements have been included. New matrices are formed as each element is added, with the order of the matrix increasing with each added tree branch. The four possible cases for adding an element are shown in Figure 3.5.

3.2.1 Adding a Tree Branch to Bus *p*

Assume that a partial network impedance $\mathbf{Z}^m_{BUS}$ is known for *m* buses and a branch, *without* mutual coupling $\mathbf{Z}_{new}$, is to be added at bus *p* as shown in Figure 3.5a. The new branch creates new bus $m + 1$. The partial network described by the equation

$$\mathbf{E}^m_{BUS} = \mathbf{Z}^m_{BUS} \mathbf{I}^m_{BUS} = \begin{bmatrix} z_{11} & z_{12} & \cdots & z_{1p} & \cdots & z_{1m} \\ \vdots & & & \vdots & & \vdots \\ z_{p1} & \cdots\cdots & & z_{pp} & \cdots & z_{pm} \\ \vdots & & & \vdots & & \vdots \\ z_{m1} & \cdots\cdots & & z_{mp} & \cdots & z_{mm} \end{bmatrix} \mathbf{I}^m_{BUS} \tag{3.9}$$

is known before the new branch is added. When the tree branch of impedance $\mathbf{Z}_{new}$ is added and a unit current is injected in bus $m + 1$ with respect to the reference, the voltage at this bus is

$$E_{m+1} = z_{m+1,m+1}I_{m+1} = (z_{pp} + Z_{new})I_{m+1} \tag{3.10}$$

In other words, the driving-point impedance at the new $m + 1$ bus is that of bus *p* plus the new element because they are in series. Because the unit current is injected into bus *p*, the voltages at all $k = 1, 2, 3, \ldots, p, \ldots, m \neq p$ buses, and hence the mutual impedances, are the same as those for bus *p*,

$$z_{k,m+1} = z_{m+1,k} = z_{k,p} = z_{p,k} \qquad k = 1, 2, 3, \ldots, p, \ldots, m \qquad k \neq m + 1$$

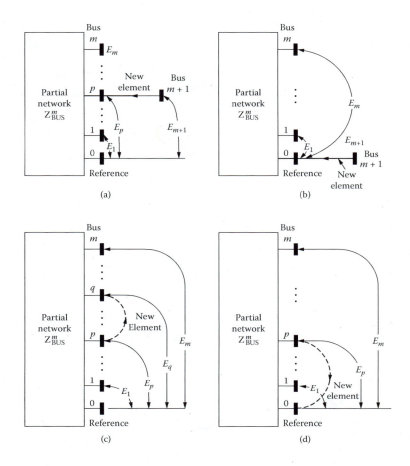

FIGURE 3.5
Four cases of adding a tree branch or cotree link: (a) tree branch added to bus p; (b) tree branch added to reference; (c) cotree link between p and q; (d) cotree link between p and reference.

because the same unit current injected at bus $m + 1$ also flows into bus p. Hence, the new bus impedance equations are

$$\mathbf{E}_{\text{BUS}}^{m+1} = \begin{bmatrix} E_1 \\ E_2 \\ \vdots \\ E_p \\ \vdots \\ E_m \\ \hline E_{m+1} \end{bmatrix} = \mathbf{Z}_{\text{BUS}}^{m+1}\mathbf{I}_{\text{BUS}}^{m+1} = \begin{array}{c} \textit{Add Tree Branch to Bus } p \\ \begin{bmatrix} z_{11} & \cdots & z_{1p} & \cdots & z_{1m} & \vdots & z_{p1} \\ & & & & & & \vdots \\ z_{p1} & \cdots & z_{pp} & \cdots & z_{pm} & \vdots & z_{pp} \\ \vdots & & & & & \vdots & \vdots \\ z_{m1} & \cdots & z_{mp} & \cdots & z_{mm} & \vdots & z_{pm} \\ \hline z_{p1} & \cdots & z_{pp} & \cdots & z_{pm} & \vdots & z_{pp} + Z_{\text{new}} \end{bmatrix} \end{array} \mathbf{I}_{\text{BUS}}^{m+1} \quad (3.11)$$

A column from the preceding matrix has been augmented with $z_{pp} + Z_{\text{new}}$ to generate the new bus impedance matrix.

3.2.2 Adding a Tree Branch to the Reference

When a new branch, $\mathbf{Z}_{new}$, is added to the network by connecting it to the reference as shown in Figure 3.5b, a new bus, $m + 1$, is created. The previous bus impedance is given by Equation 3.9. If it is desired to obtain $\mathbf{Z}_{BUS}^{m+1}$ by the superposition method of injecting currents into each bus, it is clear that there cannot be any voltages induced at $k = 1, 2, \ldots, m$ buses when current is injected at bus $m + 1$ because there is zero mutual impedance between the new bus and the previous partial network. The new bus impedance equation for a branch *without* mutual coupling is:

$$
\mathbf{E}_{BUS}^{m+1} = \begin{bmatrix} E_1 \\ \vdots \\ E_m \\ \hline E_{m+1} \end{bmatrix} = \begin{bmatrix} z_{11} & \cdots & z_{1m} & 0 \\ \vdots & & \vdots & 0 \\ & & & \vdots \\ z_{m1} & \cdots & z_{mm} & 0 \\ \hline 0 & \cdots & 0 & Z_{new} \end{bmatrix} \mathbf{I}_{BUS}^{m+1}
$$

Add Tree Branch to Reference

(3.12)

An example to demonstrate adding tree branches is:

Example 3.2

For the network shown in Figure E3.2, add the tree branches one at a time to obtain the network described by the tree elements. The element impedances are:

$$Z_1 = j0.01$$

$$Z_2 = j0.02$$

$$Z_3 = j0.03$$

$$Z_4 = j0.04$$

$$Z_5 = j0.05$$

$$Z_6 = j0.06$$

$$Z_7 = j0.07$$

$$Z_8 = j0.08$$

$$Z_9 = j0.09$$

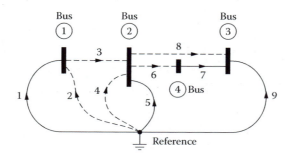

FIGURE E3.2

Solution

Let the tree branches be added in the sequence 1, 5, 9, 7 to avoid including bus 4 before bus 3 in the partial $\mathbf{Z}_{BUS}$ and allowing branch 7 to "float" with respect to the reference. The partial network using only branch 1 is

$$\mathbf{E}^1_{BUS} = j[0.01] \, \mathbf{I}^1_{BUS}$$

Branches 5 and 9 are connected to the reference, so the formulation using Equation 3.12 yields a diagonal partial bus impedance matrix

$$\mathbf{E}^3_{BUS} = \begin{bmatrix} E^3_1 \\ E^3_2 \\ E^3_3 \end{bmatrix} = j \begin{bmatrix} 0.01 & 0 & 0 \\ 0 & 0.05 & 0 \\ 0 & 0 & 0.09 \end{bmatrix} \mathbf{I}^3_{BUS}$$

Finally, when branch 7 is included, Equation 3.11 applies, so that

$$\mathbf{E}^4_{BUS} = \begin{bmatrix} E^4_1 \\ E^4_2 \\ E^4_3 \\ E^4_4 \end{bmatrix} = j \begin{bmatrix} 0.01 & 0 & 0 & 0 \\ 0 & 0.05 & 0 & 0 \\ 0 & 0 & 0.09 & 0.09 \\ 0 & 0 & 0.09 & 0.16 \end{bmatrix} \mathbf{I}^4_{BUS}$$

In Example 3.2 it was necessary to include tree branches directly connected to reference before the indirectly connected branches. This is a straightforward search process for a digital computer to select tree branches in this order to avoid an element floating with respect to ground.

3.2.3 Adding a Cotree Link between Buses *p* and *q*

In the case of a cotree link added between buses p and q as shown in Figure 3.5c, the driving-point impedances at buses p and q become parallel current paths due to the loop from the link. No new buses are introduced, but in the original impedance matrix every driving-point and mutual impedance may be modified when the link is added. Current flows due to the new element are shown in Figure 3.6. The current through the new element is

$$I_{new} = \frac{E_p - E_q}{Z_{new}} = Y_{new}(E_p - E_q) \tag{3.13}$$

which modifies the partial network injections. The m bus voltages of the original network due to the modified currents are

$$
\begin{aligned}
E_1 &= z_{11}I_1 & +++ & & z_{1p}(I_p - I_{new}) & +++ & & z_{1q}(I_q + I_{new}) & +++ \\
\vdots & & \vdots & & & \vdots & & & \vdots \\
E_p &= z_{p1}I_1 & +++ & & z_{pp}(I_p - I_{new}) & +++ & & z_{pq}(I_q + I_{new}) & +++ \\
\vdots & & \vdots & & & \vdots & & & \vdots \\
E_q &= z_{q1}I_1 & +++ & & z_{qp}(I_p - I_{new}) & +++ & & z_{qq}(I_q + I_{new}) & +++ \\
\vdots & & \vdots & & & \vdots & & & \vdots \\
E_m &= z_{m1}I_1 & +++ & & z_{mp}(I_p - I_{new}) & +++ & & z_{mq}(I_q + I_{new}) & +++
\end{aligned}
\tag{3.14}
$$

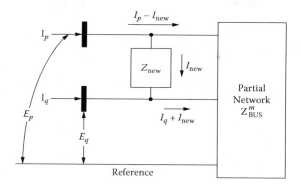

FIGURE 3.6
Current flow paths for adding a cotree link.

The new cotree link does not introduce another bus, so voltages E_p and E_q in Equation 3.13 may be expressed in terms of currents injected into the network

$$0 = E_p - E_q - Z_{new} I_{new}$$
$$= (z_{p1} - z_{q1})I_1 + (z_{p2} - z_{q2})I_2 + + + -(Z_{new} + z_{pp} - 2z_{pq} + z_{qq})I_{new} \tag{3.15}$$

The equivalent $m + 1$ simultaneous Equation 3.14 and Equation 3.15 may be written as

$$
\begin{bmatrix} E^m_{BUS} \\ \hline 0 \end{bmatrix} =
\begin{bmatrix}
 & Z^m_{BUS} & & \vdots & \begin{matrix} -(z_{1p} - z_{1q}) \\ \vdots \\ -(z_{mp} - z_{mq}) \end{matrix} \\
\hline
(z_{p1} - z_{q1}) & \cdots & (z_{pm} - z_{qm}) & \vdots & -(Z_{new} + z_{pp} - 2z_{pq} + z_{qq})
\end{bmatrix}
\begin{bmatrix} I^m_{BUS} \\ \hline I_{new} \end{bmatrix} \tag{3.16}
$$

The undesired current I_{new} can be eliminated from Equation 3.16,

$$I_{new} = [(z_{p1} - z_{q1})I_1 + (z_{p2} - z_{q2})I_2 + + + +]/(Z_{new} + z_{pp} - 2z_{pq} + z_{qq})$$

to yield entries of the modified bus impedance matrix:

$$z^{m+1}_{ij} = z^m_{ij} - \frac{\left(z^m_{ip} - z^m_{iq}\right)\left(z^m_{pj} - z^m_{qj}\right)}{Z_{new} + z^m_{pp} - 2z^m_{pq} + z^m_{qq}} \tag{3.17}$$

Since only one element at a time is to be added, matrix methods are applicable that obtain a modified inverse of a large matrix by means of inverting two small matrices. Equation 3.17 is a special case of one of these more general methods that is presented in Section 3.3.1. For uncoupled elements, the algorithm for adding a cotree link is summarized as

$$
\boxed{
\begin{array}{c}
\textit{Add Cotree Link Bus p to q} \\[4pt]
\mathbf{Z}^{m+1}_{BUS} = \mathbf{Z}^m_{BUS} - w\mathbf{X}\mathbf{X}^t \\[4pt]
\text{where } w = 1/(Z_{NEW} + z^m_{pp} - 2z^m_{pq} + z^m_{qq}) \\[4pt]
\mathbf{X} = \text{column vector which is the difference between} \\
\text{the } p\text{th and } q\text{th columns of } \mathbf{Z}^m_{BUS}
\end{array}
} \tag{3.18}
$$

In Equation 3.18, w is a scalar computed from the impedance of the cotree link and elements from the previous matrix $\mathbf{Z}_{\text{BUS}}^m$. The product of $\mathbf{X}$ and its transpose is an $m \times m$ matrix, but only the diagonal and upper-triangular portion need be computed or stored because the matrix is symmetric.

3.2.4 Adding a Cotree Link from Bus p to Reference

This is a special case of Equation 3.18 where bus q is the reference bus. Because bus q is the reference, $z_{qq}^m = 0$ and $z_{pq}^m = z_{qp}^m = 0$, Equation 3.15 becomes

$$I_{\text{new}} = \frac{\sum z_{pi}^m I_i}{\left(z_{pp}^m + Z_{\text{new}}\right)}$$

so the new bus impedance matrix is obtained from

$$\boxed{\begin{array}{c} \text{Add Cotree Link Bus } p \text{ to Reference} \\ \mathbf{Z}_{\text{BUS}}^{m+1} = \mathbf{Z}_{\text{BUS}}^m - \mu \mathbf{Y}\mathbf{Y}^t \\ \text{where } \mu = 1/(\mathbf{Z}_{\text{new}} + z_{pp}^m) \\ \mathbf{Y} = p\text{th column vector from } \mathbf{Z}_{\text{BUS}}^m \end{array}} \tag{3.19}$$

The product $\mathbf{YY}^t$ is symmetric, so only the diagonal and upper-triangular portion need be computed.

The following example is used to demonstrate how Equation 3.18 and Equation 3.19 add one link at a time to modify the bus impedance matrix.

Example 3.3

The partial $\mathbf{Z}_{\text{BUS}}^4$ matrix as derived in Example 3.2 has only tree branches in the network. Add the cotree link 6, which goes between two buses and cotree link 2 from bus 1 to the reference. The partial $\mathbf{Z}_{\text{BUS}}^4$ without link is

$$\mathbf{Z}_{\text{BUS}}^4 = j\begin{bmatrix} 0.01 & 0 & 0 & 0 \\ 0 & 0.05 & 0 & 0 \\ 0 & 0 & 0.09 & 0.09 \\ 0 & 0 & 0.09 & 0.16 \end{bmatrix}$$

Solution

The w and $\mathbf{X}$ vector of Equation 3.18 are computed for link 6 from $p = 2$ to $q = 4$ as follows:

$$w = \frac{1}{Z_{\text{new}} + z_{22}^4 - 2z_{24}^4 + z_{44}^4}$$

$$= \frac{-j}{0.06 + 0.05 - 2(0) + 0.16} = \frac{-j}{0.27}$$

$$\mathbf{X} = \begin{bmatrix} z_{12}^4 - z_{14}^4 \\ z_{22}^4 - z_{24}^4 \\ z_{32}^4 - z_{34}^4 \\ z_{42}^4 - z_{44}^4 \end{bmatrix} = j\begin{bmatrix} 0 \\ 0.05 \\ -0.09 \\ -0.16 \end{bmatrix}$$

Substituting values into Equation 3.18 gives

$$\mathbf{Z}^5_{BUS} = \mathbf{Z}^4_{BUS} - \frac{+j}{0.27}\begin{bmatrix} 0 \\ 0.05 \\ -0.09 \\ -0.16 \end{bmatrix}\begin{bmatrix} 0 & 0.05 & -0.09 & -0.16 \end{bmatrix}$$

$$= \mathbf{Z}^4_{BUS} - \frac{j}{2700}\begin{bmatrix} 0 & 0 & 0 & 0 \\ 0 & 25 & -45 & -80 \\ 0 & -45 & 81 & 144 \\ 0 & -80 & 144 & 256 \end{bmatrix}$$

$$= j\begin{bmatrix} 0.01 & 0 & 0 & 0 \\ 0 & 0.0407 & 0.0166 & 0.0296 \\ 0 & 0.0166 & 0.0600 & 0.0367 \\ 0 & 0.0296 & 0.0367 & 0.0652 \end{bmatrix}$$

For cotree link 2 from bus $p = 0$ to $q = 1$, the values of μ and the $\mathbf{Y}$ vector are computed as follows:

$$\mu = \frac{1}{Z_{new} + z^5_{00} - 2z^5_{10} + z^5_{11}}$$

$$= \frac{1}{j0.02 + j0.01} = \frac{-j}{0.03}$$

$$\mathbf{Y} = \begin{bmatrix} z^5_{11} \\ z^5_{21} \\ z^5_{31} \\ z^5_{41} \end{bmatrix} = j\begin{bmatrix} 0.01 \\ 0 \\ 0 \\ 0 \end{bmatrix}$$

Substituting these values into Equation 3.19, we have

$$\mathbf{Z}^6_{BUS} = \mathbf{Z}^5_{BUS} - \frac{j}{0.03}\mathbf{YY}^t = \mathbf{Z}^5_{BUS} - j\begin{bmatrix} 0.00333 & 0 & 0 & 0 \\ 0 & 0 & 0 & 0 \\ 0 & 0 & 0 & 0 \\ 0 & 0 & 0 & 0 \end{bmatrix}$$

$$= j\begin{bmatrix} 0.00666 & 0 & 0 & 0 \\ 0 & 0.0407 & 0.0166 & 0.0296 \\ 0 & 0.0166 & 0.0600 & 0.0367 \\ 0 & 0.0296 & 0.0367 & 0.0652 \end{bmatrix}$$

Observe that bus 1 is not yet connected to buses 2, 3, and 4, so that zeros appear in the mutual impedance locations. Cotree links 3, 4, and 8 must be added to complete the bus impedance matrix. When these final links are included, the network bus impedance matrix is

$$\mathbf{Z}_{BUS}^9 = j \begin{bmatrix} 0.005871 & 0.002288 & 0.001476 & 0.001913 \\ & 0.01259 & 0.008119 & 0.01052 \\ & \downarrow & 0.03719 & 0.02153 \\ & \text{Symmetry} & & 0.04793 \end{bmatrix}$$

© Copy the file example *33.sss* into **Zbus.sss**, then execute *zbus.exe* to duplicate the results of this example.

The algorithm to construct the $\mathbf{Z}_{BUS}$ matrix by adding one element at a time may be used to remove branches by using negative elements. The computation to add a tree branch is minimal, but for an n-bus system each cotree link added requires $\sim n \times n$ multiplications for all links $b > n$. Hence, on the order of n^3 multiplications are needed to obtain $\mathbf{Z}_{BUS}$ for an n-bus system. This total number of multiplications is the same as that required for a direct inverse of the $\mathbf{Y}_{BUS}$ matrix. Section 3.3 examines properties of the $\mathbf{Y}_{BUS}$ formulation, and Section 3.3.1 returns to constructing $\mathbf{Z}_{BUS}$ for the case of mutually coupled elements.

3.3   The Bus Admittance Matrix

Figure 3.7 shows a capacitor bank which would introduce an element ωC, bus-to-reference, in the bus admittance matrix.

The bus admittance matrix is comprised of the driving-point admittance for the buses and the transfer (or mutual) admittances between buses. If mutual coupling exists between parallel transmission lines or any two elements of the network, for a specified current sense, coupling is given by off-diagonal terms in either the primitive impedance matrix,

$$[\mathbf{Z}_{PRIM}] = \begin{bmatrix} Z_1 & 0 & 0 & 0 & 0\ldots \\ 0 & Z_{22} & Z_{23} & 0 & \\ 0 & Z_{32} & Z_{33} & 0 & \\ \vdots & 0 & 0 & Z_{4\cdots} & \end{bmatrix} = [\mathbf{Y}_{PRIM}]^{-1} \tag{3.20}$$

or its inverse, $\mathbf{Y}_{PRIM}$. The dimensions of these ordered matrices are $b \times b$, where b is the number of elements. If the concept of a directed graph of elements is again used, the elements of a power system (zero, positive, or negative sequence) may be separated into tree branches that connect each bus to the reference and cotree links that form closed loops, as shown in Figure 3.8.

Assigning an arbitrary direction to each uncoupled element of the network and maintaining mutual element current sign conventions, the reduced bus–branch matrix (without the reference) corresponding to Figure 3.8 is

$$\mathbf{M} = \begin{bmatrix} \overset{\text{elements}\rightarrow}{} & & & & & & & & & \overset{\text{buses}}{\downarrow} \\ -1 & 1 & 0 & 0 & 0 & 0 & 0 & 0 & \cdots\cdots \\ 0 & -1 & 1 & 1 & 0 & 0 & 0 & 0 & \cdots\cdots \\ 0 & 0 & 0 & -1 & -1 & 1 & 0 & 0 & \cdots \\ 0 & 0 & 0 & 0 & 0 & -1 & 1 & 1 & \cdots \end{bmatrix} \tag{3.21}$$

FIGURE 3.7
A 230 kV switched capacitor bank rated at 72 Mvar. Used as shunt compensation for voltage regulation and reactive support on 230 kV transmission system.

In the **M** matrix the *m* rows correspond to buses and the *b* columns correspond to the ordered *b* elements defined in Equation 3.20. The entries of the **M** matrix are +1 if an element leaves bus *i* and –1 if the element terminates on bus *j*, and are zero otherwise. The reference bus is not included in the **M** matrix, so that elements 1, 3, 5, 7, and so on, of Figure 3.8, which either emanate or terminate on the reference bus, have only one entry in their column.

Let **I** be a vector of complex current from the buses into the *b* elements of the network, **V** be a vector of voltage across each of the *b* elements, and **J** be the vector of source currents similar to Equation 2.21a. The voltage, current, and element direction of the *k*th element use the convention shown in Figure 3.9.

Both I_k and J_k flow in the Y_k element, so that Kirchhoff's current law for *b* elements is

$$\mathbf{I} + \mathbf{J} = \mathbf{Y}_{\text{PRIM}}\mathbf{V} \tag{3.22}$$

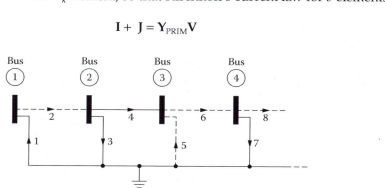

FIGURE 3.8
General network indicating tree branches and cotree links.

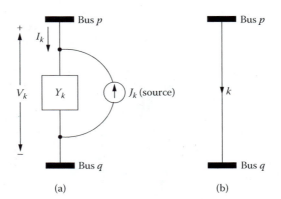

FIGURE 3.9
Element voltage, current, and direction: (a) detailed circuit; (b) directed graph symbol.

The reduced bus–branch incidence matrix indicates the branches terminating at each bus; thus,

$$\mathbf{MI} = 0 \qquad (3.23)$$

is a statement of Kirchhoff's current law at each bus. Multiplying Equation 3.22 by **M** yields

$$\mathbf{MJ} = \mathbf{I}_{BUS} = \mathbf{MY}_{PRIM}\mathbf{V} \qquad (3.24)$$

where $\mathbf{I}_{BUS}$ is defined to be the sum of the current sources at each bus. The element voltages may be expressed in terms of bus voltages as

$$\mathbf{V} = \mathbf{M}^{t}\mathbf{E}_{BUS} \qquad (3.25)$$

Therefore, the bus admittance formulation (Equation 3.24) may be written as

$$\mathbf{I}_{BUS} = \mathbf{MY}_{PRIM}\mathbf{M}^{t}\mathbf{E}_{BUS} = \mathbf{Y}_{BUS}\mathbf{E}_{BUS} \qquad (3.26)$$

The *b* elements include mutual coupling, and the *n* bus voltages are independent. An example is used next to demonstrate the application of Equation 3.20 to Equation 3.26.

Example 3.4

Derive the bus admittance matrix for the network described by the directed graph shown in Figure E3.4. Calculate the bus voltages in the network. Impedances and currents are expressed in per-unit values. Mutual coupling between elements 4 and 5 is given in terms of the directed graph. Elements 1 and 3 are admittance forms

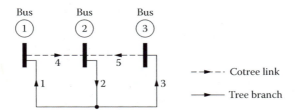

FIGURE E3.4

representing two generators. The primitive impedance matrix and source current vector are

$$
\mathbf{Z}_{\text{PRIM}} = \begin{bmatrix} j.3 & 0 & 0 & 0 & 0 \\ 0 & 1+j0 & 0 & 0 & 0 \\ 0 & 0 & j.2 & 0 & 0 \\ 0 & 0 & 0 & j.03 & -j0.01 \\ 0 & 0 & 0 & -j0.01 & j0.02 \end{bmatrix} \qquad J = \begin{bmatrix} -0.6+j3 \\ 0 \\ -0.4+j5 \\ 0 \\ 0 \end{bmatrix}
$$

Solution

The distinction of cotree links or tree branches is not significant for bus admittance equations. The branch–bus reduced incidence matrix for the defined element directions is

$$
\mathbf{M} = \begin{bmatrix} -1 & 0 & 0 & 1 & 0 \\ 0 & 1 & 0 & -1 & -1 \\ 0 & 0 & -1 & 0 & 1 \end{bmatrix}
$$

The mutual coupling between elements 4 and 5 is negative for their defined directions. The primitive admittance matrix is

$$
\mathbf{Y}_{\text{PRIM}} = [\mathbf{Z}_{\text{PRIM}}]^{-1} = \begin{bmatrix} -j3.33 & 0 & 0 & 0 & 0 \\ 0 & 1 & 0 & 0 & 0 \\ 0 & 0 & -j5 & 0 & 0 \\ 0 & 0 & 0 & -j40 & -j20 \\ 0 & 0 & 0 & -j20 & -j60 \end{bmatrix}
$$

The bus current injection vector is calculated as

$$
\mathbf{I}_{\text{BUS}} = \mathbf{MJ} = \begin{bmatrix} 0.6-j3 \\ 0 \\ 0.4-j5 \end{bmatrix}
$$

The bus admittance matrix is obtained using Equation 3.26:

$$
\mathbf{Y}_{\text{BUS}} = \mathbf{MY}_{\text{PRIM}}\mathbf{M}^t = \begin{bmatrix} -j43.33 & j60 & -j20 \\ j60 & 1-j140 & j80 \\ -j20 & j80 & -j65 \end{bmatrix}
$$

The inverse of the bus admittance matrix is calculated as

$$
\mathbf{Z}_{\text{BUS}} = [\mathbf{Y}_{\text{BUS}}]^{-1} = \begin{bmatrix} 0.01442+j0.1403 & 0.01515+j0.1191 & 0.01421+j0.1035 \\ 0.01515+j0.1191 & 0.01592+j0.1252 & 0.01493+j0.1174 \\ 0.01421+j0.1035 & 0.01493+j0.1174 & 0.01401+j0.1280 \end{bmatrix}
$$

© This result may be verified by copying the file example *34.sss* into *Zbus.sss*, executing *Zbus.exe*, and that branches are directed and mutual coupling is defined in terms of direction.

The resulting bus voltages as calculated by Equation 3.5 are slightly less than 1.0 p.u. in magnitude:

$$\mathbf{E}_{BUS} = \mathbf{Z}_{BUS}\mathbf{I}_{BUS} = \begin{bmatrix} 0.9527 + j0.01127 \\ 0.9593 - j0.00168 \\ 0.9646 + j0.00062 \end{bmatrix}$$

In Example 3.4, only the element from bus 2 to reference can dissipate power. Straight-forward calculations can show that it dissipates 0.925 p.u. power, which is the sum of the real power injected by the generators at buses 1 and 3.

If the elements connected to a bus *are not mutually coupled* to others, then:

1. The diagonal term, y_{ii}, of the bus admittance matrix is the sum of all admittances connected to bus i. This is the bus driving-point admittance.
2. The off-diagonal terms, y_{ij}, of the bus admittance matrix are the negative of the sum of admittances directly connected from bus i to bus j. These are called trans-fer admittances.

Therefore, in the case of no mutual coupling it is possible to write the $\mathbf{Y}_{BUS}$ matrix by inspection. In a computer algorithm, it is easily accomplished by a search of the endpoints of each line and transformer.

Another significant property of $\mathbf{Y}_{BUS}$ is that entries in the matrix are zero unless buses have a direct connection. For large power systems with many buses and few lines to each bus (few off-diagonal transfer admittances), there are far more zeros than entries in $\mathbf{Y}_{BUS}$, so it is called a *sparse matrix*. A matrix with less than 15% nonzero elements is considered to be sparse. Only the nonzero terms and their matrix location must be stored instead of $n(n+1)/2$ terms for a symmetric matrix. For example, consider the $\mathbf{Y}_{BUS}$ matrix for $n = 8$ buses that has the nonzero elements marked by ×'s in Figure 3.10. Since the matrix is symmetric, only the diagonal and upper-triangular portions of the matrix must be stored. This is $n(n+1)/2 = 36$ computer words for real-valued entries.

By means of integer pointer tables that are half-word length, it is possible to indicate where entries are located in each row, and in a parallel table to give the numerical value of each entry. Figure 3.11 shows the pointer arrays and storage for the matrix of Figure 3.10.

The diagonal elements require full-word storage. The row entry of the upper-triangular pointer table indicates the location. The entries for row 1 start at the 1 location in the

$\mathbf{Y}_{BUS} =$

	1	2	3	4	5	6	7	8
	×	×						×
	×	×					×	×
			×				×	×
				×		×		
					×	×		
				×	×	×	×	×
		×	×				×	×
	×	×	×				×		×

FIGURE 3.10
Entries for $\mathbf{Y}_{BUS}$ matrix.

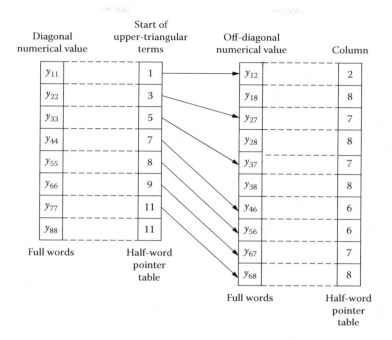

FIGURE 3.11
Pointers and storage for the matrix of Figure 3.10.

off-diagonal array. There are $3 - 1 = 2$ elements in the upper-triangular portion of the first row. Their values are y_{12} and y_{18} and are located in the second and eighth columns, respectively.

The entries for the second row start in the third storage location in the off-diagonal array. There are $5 - 3 = 2$ entries in the upper-triangular portion of the second row. The entries are in the seventh and eighth columns.

Counting the full words and half words in the Figure 3.11 arrays, there are 27 full words of equivalent storage when using the pointer method. This is a modest saving in storage 27/36, for the matrix of Figure 3.10, but the advantage increases for power systems studying a large number of buses.

Another advantage of using $\mathbf{Y}_{BUS}$ is that the sparsity makes it suitable to use efficient numerical techniques to compute the inverse. One of these methods has become a standard for the power industry. It is presented in Section 3.4.

3.3.1 Bus Impedance Matrix for Elements with Mutual Coupling

The $\mathbf{Y}_{BUS}$ matrix for a network is readily determined because network elements are modified by adding admittances to the diagonal of $\mathbf{Y}_{BUS}$ and subtracting from appropriate off-diagonal entries (unless the element is from bus to reference). Transmission network elements described by mutually coupled symmetrical component admittances are treated similarly. This additive property of admittances is now exploited to generalize the $\mathbf{Z}_{BUS}$ algorithm of Section 3.2 for the case of mutually coupled elements.

Let the $m \times m$ matrix $\mathbf{Z}_{BUS}^{m} = [\mathbf{Y}_{BUS}^{m}]^{-1}$ be calculated by progressively adding elements of the network (treated as tree branches or links) to partial networks. A mutually coupled link is to be added to calculate $\mathbf{Z}_{BUS}^{m+1}$. Only links are treated here because the $m \times m$ matrix $\mathbf{Z}_{BUS}^{m}$ can be constructed using uncoupled "dummy" branches until $\mathbf{Z}_{BUS}^{m}$ includes all desired buses (see Equation 3.11 and Equation 3.12). The new link to be added is assumed to have mutual coupling to one or more branches in the partial network. The method to add this link is based on the following identity from matrix theory [2]:

$$\mathbf{Z}_{BUS}^{m+1} = \left(\mathbf{Y}_{BUS}^m + \mathbf{C}_1^t \mathbf{Y}_c \mathbf{C}_1\right)^{-1}$$

$$= \mathbf{Z}_{BUS}^m - \mathbf{Z}_{BUS}^m \mathbf{C}_1^t \left(\mathbf{Y}_c^{-1} + \mathbf{C}_1 \mathbf{Z}_{BUS}^m \mathbf{C}_1^t\right)^{-1} \mathbf{C}_1 \mathbf{Z}_{BUS}^m \qquad (3.27)$$

Equation 3.27 can be considered as building the inverse of a large matrix $\mathbf{Z}_{BUS}^{m+1}$ by means of inverting two small matrices whose order equals the number of elements mutually coupled to the additional element. In this equation $\mathbf{Y}_c$ is a $c \times c$ matrix containing self-admittances (less dummy terms) and mutual admittances from the primitive admittance matrix, Equation 3.20. Matrix $\mathbf{C}_1$ is the bus-to-bus connection matrix of the added link and elements coupled to it. Matrix $\mathbf{C}_1$ has order $c \times m$ with entries $+1$, -1, or mostly zeros, similar to the incidence matrix, Equation 3.8. The product $\mathbf{C}_1^t \mathbf{Y}_c \mathbf{C}_1$ expresses modifications to $\mathbf{Y}_{BUS}^m$ due to the additional link. The cotree link formulas of Equation 3.18 and Equation 3.19 are special cases of Equation 3.27 when $\mathbf{C}_1^t$ is a column vector and $\mathbf{Y}_c$ is a scalar without mutual coupling. Example 3.5 demonstrates the application of the equation.

Example 3.5

Derive the bus impedance matrix for the three-bus network whose tree and cotree are shown in Figure E3.5. Numerical values for two coupled impedances and three uncoupled positive-sequence elements are given in the following table:

Element	Impedance	Mutual Coupling
1	$j0.01$	—
2	$j0.02$	—
3	$j0.04$	$(3-5)\,j0.02$
4	$j0.04$	—
5	$j0.04$	$(5-3)\,j0.02$

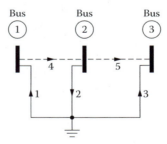

FIGURE E3.5

The mutual coupling between elements 3 and 5 is positive for the branch orientation specified.

Solution

The partial bus impedance matrix utilizing only elements 1 and 2 is

$$\mathbf{Z}_{BUS}^2 = j\begin{bmatrix} 0.01 & 0 \\ 0 & 0.02 \end{bmatrix}$$

A dummy uncoupled impedance $Z = 0 + j0.03$ from ground to bus 3 is employed to increase the order of the partial bus impedance matrix to 3×3:

$$\mathbf{Z}_{BUS}^3 = j \begin{bmatrix} 0.01 & 0 & 0 \\ 0 & 0.02 & 0 \\ 0 & 0 & 0.03 \end{bmatrix}$$

Next, element 4, with admittance $\mathbf{Y}_c = -j25$, is to be added using identity of Equation 3.27. The connection matrix $\mathbf{C}_1$ that adds admittance $-j25$ between buses 1 and 2 is

$$\mathbf{C}_1 = [1 \quad -1 \quad 0]$$

With these definitions of $\mathbf{C}_1$ and $\mathbf{Y}_{c'}$, Equation 3.27 to add element 4 is written as

$$\mathbf{Z}_{BUS}^4 = \mathbf{Z}_{BUS}^3 - \mathbf{Z}_{BUS}^3 \begin{bmatrix} 1 \\ -1 \\ 0 \end{bmatrix} \left\{ j.04 + [1 \quad -1 \quad 0]\mathbf{Z}_{BUS}^3 \begin{bmatrix} 1 \\ -1 \\ 0 \end{bmatrix} \right\}^{-1} [1 \quad -1 \quad 0]\mathbf{Z}_{BUS}^3$$

$$= \mathbf{Z}_{BUS}^3 - \mathbf{Z}_{BUS}^3 \begin{bmatrix} 1 \\ -1 \\ 0 \end{bmatrix} \left\{ j.04 + [1 \quad -1 \quad 0]\begin{bmatrix} j0.04 \\ -j0.02 \\ 0 \end{bmatrix} \right\}^{-1} [1 \quad -1 \quad 0]\mathbf{Z}_{BUS}^3$$

$$= \mathbf{Z}_{BUS}^3 - \mathbf{Z}_{BUS}^3 \begin{bmatrix} 1 \\ -1 \\ 0 \end{bmatrix} \{j.04 + j0.03\}^{-1} [1 \quad -1 \quad 0]\mathbf{Z}_{BUS}^3$$

$$= \mathbf{Z}_{BUS}^3 - \begin{bmatrix} j0.01 \\ -j0.02 \\ 0 \end{bmatrix} \frac{1}{j0.07} [j0.01 \quad -j0.02 \quad 0]$$

$$= \mathbf{Z}_{BUS}^3 - \frac{1}{j0.07} \begin{bmatrix} -10^{-4} & 2 \times 10^{-4} & 0 \\ 2 \times 10^{-4} & -4 \times 10^{-4} & 0 \\ 0 & 0 & 0 \end{bmatrix}$$

$$= j \begin{bmatrix} 0.008571 & 0.002857 & 0 \\ 0.002857 & 0.01429 & 0 \\ 0 & 0 & 0.03 \end{bmatrix}$$

It is next necessary to replace the dummy element by the coupled element. As a prior step, the primitive impedance matrix and its inverse are presented to show the mutual coupling:

$$[\mathbf{Z}_{PRIM}] = j \begin{bmatrix} 0.01 & 0 & 0 & 0 & 0 \\ 0 & 0.02 & 0 & 0 & 0 \\ 0 & 0 & 0.04 & 0 & 0.02 \\ 0 & 0 & 0 & 0.04 & 0 \\ 0 & 0 & 0.02 & 0 & 0.04 \end{bmatrix}$$

$$[\mathbf{Y}_{\text{PRIM}}] = -j\begin{bmatrix} 100 & 0 & 0 & 0 & 0 \\ 0 & 50 & 0 & 0 & 0 \\ 0 & 0 & \dfrac{100}{3} & 0 & \dfrac{-50}{3} \\ 0 & 0 & 0 & 25 & 0 \\ 0 & 0 & \dfrac{-50}{3} & 0 & \dfrac{100}{3} \end{bmatrix}$$

Consider elements 3 and 5 to be inserted simultaneously so that the new connection matrix $\mathbf{C}_1$ is of order 2×3. It connects element 5, directed from bus 2 to bus 3, and element 3, from bus 3 to reference.

$$\begin{array}{ccc} \text{Bus 1} & \text{Bus 2} & \text{Bus 3} \end{array}$$

$$\mathbf{C}_1 = \begin{bmatrix} 0 & 0 & -1 \\ 0 & 1 & -1 \end{bmatrix} \begin{array}{l} \text{element 3} \\ \text{element 5} \end{array}$$

The admittance matrix for the elements to be added, less the dummy element (from reference to bus 3), is:

$$\mathbf{Y}_c = -j\underbrace{\begin{bmatrix} \dfrac{100}{3} & \dfrac{-50}{3} \\ \dfrac{-50}{3} & \dfrac{100}{3} \end{bmatrix}}_{\text{From } \mathbf{Y}_{\text{PRIM}}} - \underbrace{\begin{bmatrix} \dfrac{-j100}{3} & 0 \\ 0 & 0 \end{bmatrix}}_{\substack{\text{Dummy} \\ \text{element}}} = j\begin{bmatrix} 0 & \dfrac{50}{3} \\ \dfrac{50}{3} & \dfrac{-100}{3} \end{bmatrix}$$

The final bus impedance matrix for the network is expressed in the form of Equation 3.27:

$$\mathbf{Z}_{\text{BUS}}^5 = \mathbf{Z}_{\text{BUS}}^4 - \mathbf{Z}_{\text{BUS}}^4 \begin{bmatrix} 0 & 0 \\ 0 & 1 \\ -1 & -1 \end{bmatrix} \left\{ \mathbf{Y}_c^{-1} + \begin{bmatrix} 0 & 0 & -1 \\ 0 & 1 & -1 \end{bmatrix} \mathbf{Z}_{\text{BUS}}^4 \begin{bmatrix} 0 & 0 \\ 0 & 1 \\ -1 & -1 \end{bmatrix} \right\}^{-1} \begin{bmatrix} 0 & 0 & -1 \\ 0 & 1 & -1 \end{bmatrix} \mathbf{Z}_{\text{BUS}}^4$$

$$= \mathbf{Z}_{\text{BUS}}^4 - \mathbf{Z}_{\text{BUS}}^4 \mathbf{C}_1^t \left\{ -j\begin{bmatrix} 0.12 & 0.06 \\ 0.06 & 0 \end{bmatrix} + j\begin{bmatrix} 0 & 0 & -1 \\ 0 & 1 & -1 \end{bmatrix}\begin{bmatrix} 0 & 0.002857 \\ 0 & 0.01429 \\ -0.03 & -0.03 \end{bmatrix} \right\}^{-1} \mathbf{C}_1 \mathbf{Z}_{\text{BUS}}^4$$

$$= \mathbf{Z}_{\text{BUS}}^4 - \mathbf{Z}_{\text{BUS}}^4 \mathbf{C}_1^t \begin{pmatrix} -j0.09 & -j0.0300 \\ -j0.0300 & j0.04429 \end{pmatrix}^{-1} \mathbf{C}_1 \mathbf{Z}_{\text{BUS}}^4$$

$$= \mathbf{Z}_{\text{BUS}}^4 - j\mathbf{Z}_{\text{BUS}}^4 \mathbf{C}_1^t \begin{bmatrix} 9.064 & 6.140 \\ 6.140 & -18.42 \end{bmatrix} \mathbf{C}_1 \mathbf{Z}_{\text{BUS}}^4$$

$$= \mathbf{Z}_{\text{BUS}}^4 - j \times 10^{-4} \begin{bmatrix} 1.504 & 7.520 & -10.53 \\ 7.520 & 37.61 & -52.64 \\ -10.53 & -52.64 & -26.32 \end{bmatrix}$$

$$= j\begin{bmatrix} 0.008421 & 0.002105 & 0.001053 \\ 0.002105 & 0.01053 & 0.005264 \\ 0.001053 & 0.005264 & 0.03263 \end{bmatrix}$$

Example 3.5 demonstrated that inversion of a matrix, whose order is equal to the mutually coupled elements, is necessary to construct $\mathbf{Z}_{\text{BUS}}$ for a general network. In this case, as in prior methods, the numerical calculations are more efficient if symmetry properties of the matrices are exploited, multiplications by unity and zero are avoided, and vectors are saved for repeated operations.

The *reference* for $\mathbf{Z}_{BUS}$ or $\mathbf{Y}_{BUS}$ may be ground, another bus, or change during the course of computations. If the reference changes, it is not necessary to construct a new $\mathbf{Z}_{BUS}$ with the building algorithm or invert a new $\mathbf{Y}_{BUS}$ matrix.

3.4 Inversion of the $\mathbf{Y}_{BUS}$ Matrix for Large Systems

To study power flow, short-circuit currents, or dynamic response for power systems with more than several buses, it is absolutely mandatory to use computers for assistance, as calculations are far beyond human capabilities in terms of length, accuracy, and number of operations. In recent years digital computers with constantly increasing capability and versatility have replaced almost all power system studies previously performed on analog computers, hybrid computers, or miniaturized models of the system. The methods introduced in this section are oriented toward digital computer implementation for large (in terms of number of buses and lines) power systems. They are for steady-state power system operation in the sense that phasor representation is used for voltages, currents, and power flow because the system has been operating for an extended time at this point and transients have decayed to zero.

An *n*-bus power system is generally described by linear representations for transmission line elements and transformers, but with loads that are nonlinear functions of applied voltage. The steady-state power flow is then a nonlinear set of *n* equations:

$$F(\mathbf{E}_{BUS}, \mathbf{I}_{BUS}) = P + jQ$$

where $P + jQ$ is the complex power vector, and $\mathbf{E}_{BUS}$ and $\mathbf{I}_{BUS}$ are also complex vectors. The solution of this equation for $\mathbf{E}_{BUS}$ or $\mathbf{I}_{BUS}$ requires an iterative method, and convergence of this method is based on linear equations and linearization about an operating point. When bus admittance (or nodal) equations are used the equations have properties related to $\mathbf{Y}_{BUS}$. Therefore, an examination of methods to invert $\mathbf{Y}_{BUS}$ will yield insight in solving the power flow problem.

The bus admittance matrix, $\mathbf{Y}_{BUS}$, is sparse for large networks if each bus of the system is connected to few others as is normal for power systems. A matrix is called sparse if it has at least 85% zeros, so that large-dimension (*n*) matrices must be encountered before benefits can be realized from special methods for sparse matrices. For small-matrix examples, Cramer's rule, Gauss-Jordan reduction, Crout's method, and so on, appear simpler and more direct, but they are inefficient on large, sparse matrices.

3.4.1 Tinney's Optimally Ordered Triangular Factorization [3]

Consider a linear set of *n* equations with coefficients a_{ij}:

$$\mathbf{AX} = \mathbf{b} = \begin{bmatrix} a_{11} & \cdots & a_{1n} \\ \vdots & & \\ a_{n1} & \cdots & a_{nn} \end{bmatrix} \begin{bmatrix} x_1 \\ x_2 \\ \vdots \\ x_n \end{bmatrix} = \begin{bmatrix} b_1 \\ \vdots \\ b_n \end{bmatrix} \tag{3.28}$$

where the rank of matrix $\mathbf{A}$ is n, and the right-hand-side vector, $\mathbf{b}$, is specified. Assuming that $\mathbf{A}$ is nonsingular, it is possible to solve for the dependent variable $\mathbf{X}$ vector by direct Gaussian elimination, or triangularization:

$$
\mathbf{GX} = \begin{bmatrix}
1 & g_{12} & g_{13} & \cdots & g_{1n} \\
0 & 1 & g_{23} & \cdots & \cdot \\
0 & 0 & 1 & \cdots & \cdot \\
& & & 1 & g_{n-1,n} \\
0 & \cdots\cdots\cdots & & 0 & 1
\end{bmatrix}
\begin{bmatrix} x_1 \\ \vdots \\ x_n \end{bmatrix} =
\begin{bmatrix} c_1^1 \\ \vdots \\ c_n^n \end{bmatrix} = \mathbf{c}
\tag{3.29}
$$

where x_1 has been eliminated from rows 2 to n, x_2 has been eliminated from rows 3 to n, and so on. The $\mathbf{c}$ vector is the result of combining terms from $\mathbf{b}$ and operations of the a_{ij} coefficients. The superscript on the $\mathbf{c}$ vector indicates the number of arithmetic operations (e.g., $c_1^1 = b_1/a_{11}$, etc.). The final variable $x_n = c_n^n$ is then back-substituted in the lower-order $n-1, n-2, \ldots$, equations to obtain values of x_{n-1}, then x_{n-2}, and so on, until all the dependent variables are calculated.

If a new right-hand-side vector, $\mathbf{d}$, is used and a new set of dependent variables are to be determined, the operations performed on the right-hand-side vector are saved. This is conveniently accomplished by means of storing the reciprocal

$$
\frac{1}{a_{jj}^{j-1}} \quad j = 1, 2, 3, \ldots, n
\tag{3.30}
$$

where the $j-1$ superscript refers to the number of numerical operations on the diagonal of the original matrix, and on factors in the lower-triangular portion of the matrix, L, which is used to obtain the $\mathbf{c}$ vector in Equation 3.29. The form of these records, called a *table of factors*, is shown in the equation

$$
[\mathbf{L\,D\,U}] = \begin{bmatrix}
\dfrac{1}{a_{11}^0} & g_{12} & g_{13} & \cdots\cdots & g_{1n} \\
a_{21}^0 & \dfrac{1}{a_{22}^1} & g_{23} & g_{24} \cdots & g_{2n} \\
a_{31}^0 & a_{32}^1 & \dfrac{1}{a_{33}^2} & g_{34} & \cdot \\
a_{41}^0 & a_{42}^1 & a_{43}^2 & & \cdot \\
a_{51}^0 & a_{52}^1 & a_{53}^2 & & \cdot \\
\vdots & & & & \\
a_{n1}^0 & a_{n2}^1 & a_{n3}^2 & \cdots & \dfrac{1}{a_{nn}^{n-1}}
\end{bmatrix}
\tag{3.31}
$$

The table of factors is a form of solution derived from a Crout reduction, LU decomposition, or Cholesky's method [3, 5].

The upper-triangular portion of the matrix, U, contains the g_{ij} Gaussian factors from Equation 3.29. In the lower triangle, L, the zero superscript refers to the original a_{ij} coefficients of Equation 3.28, and $j - 1$ operations of the form

$$a_{ij}^{j-1} = a_{ij}^{j-2} - a_{ij-1}^{j-2} a_{i-1,j}^{j-1} \quad i > 1$$

$$j = 1, 2, \ldots < i$$

(3.32)

are performed on each column in the L triangle.

To compute a new solution X for a vector, d, the d_1 element is multiplied by $1/a_{11}^0$ and substituted in subsequent rows according to

$$d_1^1 = \frac{d_1}{a_{11}}$$

$$d_2^1 = d_2 - a_{21}^0 d_1^1 = d_2 - \frac{a_{21} d_1}{a_{11}}$$

$$d_3^1 = d_3 - a_{31}^0 d_1^1 = d_3 - \frac{a_{31} d_1}{a_{11}}$$

$$\vdots$$

$$d_n^1 = d_n - a_{n1}^0 d_1^1 = d_n - \frac{a_{n1} d_1}{a_{11}}$$

(3.33)

The next step is to normalize, then substitute d_2^1 into equations 3, 4, …, n:

$$d_2^2 = \frac{d_2^1}{a_{22}^1}$$

$$d_3^2 = d_3^1 - a_{32}^1 d_2^2 = d_3^1 - \frac{a_{32}^1 d_2^1}{a_{22}^1}$$

$$d_4^2 = d_4^1 - a_{42}^1 d_2^2$$

$$\vdots$$

$$d_n^2 = d_n^1 - a_{n2}^1 d_2^2$$

(3.34)

This process is continued using each column of the lower-triangular matrix until d_n^{n-1} is calculated. It is called forward substitution. After x_n is calculated,

$$x_n = \frac{1}{a_{nn}^{n-1}} d_n^{n-1}$$

(3.35)

the back substitution is started using the upper-triangular portion of the matrix. Example 3.6 demonstrates deriving the table of factors for repeated solutions.

Example 3.6

Find the dependent variable **X** using Gaussian elimination and obtain the [L D U] table of factors for the linear equations.

$$\mathbf{AX} = \begin{bmatrix} a_{11} & a_{12} & a_{13} \\ a_{21} & a_{22} & a_{23} \\ a_{31} & a_{32} & a_{32} \end{bmatrix} \begin{bmatrix} x_1 \\ x_2 \\ x_3 \end{bmatrix} = \begin{bmatrix} 2 & 3 & 6 \\ 4 & 5 & 7 \\ 8 & 9 & 10 \end{bmatrix} \begin{bmatrix} x_1 \\ x_2 \\ x_3 \end{bmatrix} = \begin{bmatrix} 1 \\ 1 \\ 1 \end{bmatrix} = b$$

Solution

A Gaussian elimination starts with solving the first equation for x_1,

$$x_1 = \frac{1}{2}(-3x_2 - 6x_3 + b_1) = \frac{1}{a_{11}^0}(-a_{12}x_2 - a_{13}x_3 + b_1) \tag{1}$$

$$= -g_{12}x_2 - g_{13}x_3 + \frac{b_1}{a_{11}^0}$$

and substituting this relation into the second and third rows:

$$4\left(\frac{1}{2}\right)(-3x_2 - 6x_3 + b_1) + 5x_2 + 7x_3 = b_2 \tag{2}$$

$$8\left(\frac{1}{2}\right)(-3x_2 - 6x_3 + b_1) + 9x_2 + 10x_3 = b_3 \tag{3}$$

Observe that the original a_{21} and a_{31} are used as multipliers when x_1 is introduced. These original matrix variables are in the first column below the diagonal in Equation 3.31. Collecting terms in equations (2) and (3), we have

$$(5-6)x_2 + (7-12)x_3 = b_2 - 2b_1 = 1 - 2 = -1 \tag{4}$$

$$(9-12)x_2 + (10-24)x_3 = b_3 - 4b_1 = 1 - 4 = -3 \tag{5}$$

The next step is to normalize the second row, equation 4, and solve for x_2:

$$x_2 = \frac{1}{-1}(5x_3 + b_2 - 2b_1) = \frac{1}{a_{22}^1}(5x_3 + b_2 - 2b_1) \tag{6}$$

$$= -g_{23}x_3 + \frac{1}{a_{22}^1}(b_2 - 2b_1) = -g_{23}x_3 + c_2^2$$

The third row, equation 5, may be written as

$$a_{32}^1 x_2 + a_{33}^1 x_3 = b_3 - a_{31}^0 b_1 \tag{7}$$

To complete the triangularization the value of x_2 is substituted into equation 7:

$$-3(-5x_3 + 1) - 14x_3 = x_3 - 3 = a_{32}^1(-5x_3 + 1) - 14x_3 = -3 \tag{8}$$

which is solved for x_3:

$$x_3 = \frac{1}{a_{33}^2}(-3+3) = 0 = c_n^n \tag{9}$$

Since $x_3 = 0$ has been found, this is back-substituted into equation 6 to calculate

$$x_2 = -g_{23}x_3 + c_2^2 = -5x_3 + 1 = 1 \tag{10}$$

The values of x_3 and x_2 are then used in equation 1:

$$x_1 = -g_{12}x_2 - g_{13}x_3 + c_1^1 = \frac{-3x_2}{2} - 3x_3 + \tfrac{1}{2} = -1 \tag{11}$$

The table of factors is

$$[L\,D\,U] = \begin{bmatrix} \tfrac{1}{2} & \tfrac{3}{2} & 3 \\ 4 & -1 & 5 \\ 8 & -3 & 1 \end{bmatrix} \tag{12}$$

The coefficients in the upper-triangular part of the table of factors are from the Gaussian elimination. The diagonal, D, and lower-triangular form, L, were used in forward substitutions on the independent variables, b. Define the following matrices from the table of factors, where only nonzero terms are specified:

$$D_i = \begin{bmatrix} 1 & & & & & & & & \\ & 1 & & & & & & & \\ & & 1 & & & & & & \\ & & & \cdot & & & & & \\ & & & & \cdot & & & & \\ & & & & & d_i & & & \\ & & & & & & 1 & & \\ & & & & & & & \cdot & \\ & & & & & & & & \cdot \\ & & & & & & & & & 1 \end{bmatrix} \tag{3.36}$$

$$L_i = \begin{bmatrix} 1 & & & & & & & \\ & 1 & & & & & & \\ & & \cdot & & & & & \\ & & & \cdot & & & & \\ & & & & 1 & & & \\ & & & & -l_{i+1,i} & 1 & & \\ & & & & -l_{i+2,i} & & & \\ & & & & -l_{i+3,i} & & \cdot & \\ & & & & \cdot & & & \\ & & & & \cdot & & & \\ & & & & -l_{n,i} & \cdots\cdots & & 1 \end{bmatrix} \tag{3.37}$$

The matrix D_i is diagonal where only the ith entry from the table of factors is other than unity. The L_i matrix has negative values from L in the ith column below the diagonal. In a similar fashion, the upper-triangular variables of U are used in the definition

$$
U_i = \begin{bmatrix}
1 & & & & & & & & \\
 & 1 & & & & & & & \\
 & & \ddots & & & & & & \\
 & & & 1 & -g_{i,i+1} & -g_{i,i+2} & \cdots & & -g_{i,n} \\
 & & & & 1 & & & & \\
 & & & & & \ddots & & & \\
 & & & & & & & \ddots & \\
 & & & & & & & & 1
\end{bmatrix}
$$

The negative signs are necessary in L_i and U_i because quantities are subtracted from previously computed values or by transposing values to the other side of the equation, as expressed by Equation 3.29. D_i and L_i as defined are used to express operations on the independent vector—the forward substitution—as

$$
\begin{bmatrix} c_1^1 \\ \vdots \\ c_n^n \end{bmatrix} = D_n L_{n-1} D_{n-1} L_{n-2} \cdots L_1 D_1 \begin{bmatrix} b_1 \\ \vdots \\ b_n \end{bmatrix}
\tag{3.38}
$$

The back substitution of variables to obtain the solution is expressed as

$$
\begin{bmatrix} x_1 \\ \vdots \\ x_n \end{bmatrix} = U_1 U_2 \cdots U_{n-1} \begin{bmatrix} c_1^1 \\ \vdots \\ c_n^n \end{bmatrix}
\tag{3.39}
$$

It must be emphasized that for a large power system D_i, L_i, and U_i have the $n \times n$ dimensions of the number of buses, but only their nonunity, nonzero elements are stored or multiplied in Equation 3.38 and Equation 3.39. The table of factors is a convenient form for computer programming and used to express the inverse as a product of matrices:

$$
\mathbf{X} = \mathbf{A}^{-1} \mathbf{b} \rightarrow U_1 U_2 \cdots U_{n-1} D_n L_{n-1} D_{n-1} L_{n-2} \cdots L_1 D_1 \mathbf{b}
\tag{3.40}
$$

Alternative forms for the inverse using the table of factors with the transpose of U_i and L_i matrices or column-by-column triangularization may also be derived. The following example demonstrates use of the product form.

Example 3.7

Use the product form of the table of factors to calculate $\mathbf{X} = \mathbf{A}^{-1}\mathbf{b}$ for the $\mathbf{A}$ matrix of Example 3.6 and the independent vector

$$\mathbf{b} = \begin{bmatrix} 1 \\ 2 \\ 3 \end{bmatrix}$$

Solution

Introducing numerical values from the Example 3.6 table of factors into Equation 3.40, the solution is expressed as

$$\mathbf{X} = U_1 U_2 D_3 L_2 D_2 L_1 D_1 \mathbf{b}$$

$$= \underbrace{\begin{bmatrix} 1 & -\frac{3}{2} & -3 \\ 0 & 1 & 0 \\ 0 & 0 & 1 \end{bmatrix} \begin{bmatrix} 1 & 0 & 0 \\ 0 & 1 & -5 \\ 0 & 0 & 1 \end{bmatrix}}_{\text{Back Substitution}}$$

$$\times \underbrace{\begin{bmatrix} 1 & 0 & 0 \\ 0 & 1 & 0 \\ 0 & 0 & 1 \end{bmatrix} \begin{bmatrix} 1 & 0 & 0 \\ 0 & 1 & 0 \\ 0 & 3 & 1 \end{bmatrix} \begin{bmatrix} 1 & 0 & 0 \\ 0 & -1 & 0 \\ 0 & 0 & 1 \end{bmatrix} \begin{bmatrix} 1 & 0 & 0 \\ -4 & 1 & 0 \\ -8 & 0 & 1 \end{bmatrix} \begin{bmatrix} \frac{1}{2} & 0 & 0 \\ 0 & 1 & 0 \\ 0 & 0 & 1 \end{bmatrix} \begin{bmatrix} 1 \\ 2 \\ 3 \end{bmatrix}}_{\text{Forward Substitution}}$$

The sequence of vectors corresponding to the products above is

$$X = \underbrace{\begin{bmatrix} -4 \\ 5 \\ -1 \end{bmatrix}, \begin{bmatrix} \frac{1}{2} \\ 5 \\ -1 \end{bmatrix}}_{\text{Back Substitution}}, \underbrace{\begin{bmatrix} \frac{1}{2} \\ 0 \\ -1 \end{bmatrix}, \begin{bmatrix} \frac{1}{2} \\ 0 \\ -1 \end{bmatrix}, \begin{bmatrix} \frac{1}{2} \\ 0 \\ -1 \end{bmatrix}, \begin{bmatrix} \frac{1}{2} \\ 0 \\ -1 \end{bmatrix}, \begin{bmatrix} \frac{1}{2} \\ 2 \\ 3 \end{bmatrix}}_{\text{Forward Substitution}}$$

The solution is $x_1 = -4$, $x_2 = 5$, $x_3 = -1$. The evolution of the solution starts with multiplying the b_1 term by $1/a_{11}$ by means of the D_1 matrix, then using this value in modifying b_2 and b_3 by means of the L_1 matrix. After all the operations have been performed on the right-hand-side vector, the back substitution starts with obtaining $x_3 = -1$ by means of the D_3 matrix. Subsequent U_i multiplications each yield a value for x_i. It is apparent that as a x_k is determined it is unchanged during $k - 1, k - 2, \ldots, 1$ back substitutions.

For symmetric matrices, computer programming takes advantage of the property

$$a_{ij}^{j-1}\Big|_{j>i} = a_{ij}^{j-1}\Big|_{j<i} \tag{3.41}$$

so that a row of the U matrix is not normalized by the diagonal term until a_{ij}^{j-1} for $j < i$ is used in the next row. Also, because of the normalization and symmetry,

$$L_i = D_i U_i^t D_i^{-1} \tag{3.42}$$

This equation for L_i may be substituted into Equation 3.40 to obtain the simpler form for *symmetric* matrices:

$$\mathbf{X} = \mathbf{A}^{-1}\mathbf{b} \rightarrow U_1 U_2 \cdots U_{n-1} D U_{n-1}^t \ U_{n-2}^t \cdots U_1^t \mathbf{b} \tag{3.43}$$

where the product $U_i^t D_j$ for $i > j$ is commutative, so the D_i factors can be grouped into D, the diagonal of Equation 3.31.

The number of multiplications to compute a solution vector using Equation 3.43 is n^2 if the ith row of the U_i matrix has no zeros. This is the same number of multiplications as if the inverse were computed by some other method, such as Cramer's rule. One advantage is the table of factors is obtained in one-third the number of operations of the direct inverse.

For a sparse matrix, when it is triangularized, the order in which the dependent variables are processed affects the number of nonzero terms in the upper triangle. An optimal reordering of the matrix would result in the least number of terms in the U_i table of factors. There does not appear to be a true optimal reordering scheme, but Tinney [4] presented three near-optimal methods that have been implemented in many computer algorithms. The methods are applicable to sparse matrices with symmetric patterns of elements, which include Hermitian matrices, and are carried out by computer algorithms before triangualrization. In order of increasing complexity to implement and increasing optimality, the schemes are given below.

3.4.1.1   Tinney's Schemes for Near-Optimal Ordering

1. Number the rows of the matrix according to the number of off-diagonal terms before elimination. The row with the least number of terms is eliminated first.

2. Number the rows so that at each step of the process the next row to be operated on is the one with the fewest nonzero terms. If more than one row meets this criterion, select any one. This scheme requires a simulation of the effects on the accumulation of nonzero terms of the elimination process. Input information is a list by rows of the column numbers of the nonzero off-diagonal terms.

3. Number the rows so that at each step of the process the next row to be operated on is the one that will introduce the fewest new nonzero terms. If more than one row meets this criterion, select any one. This involves a trial simulation of every feasible alternative of the elimination process at each step. Input information is the same as for scheme 2.

In addition to considering the number of zeros and the "fills" introduced during the elimination, elements of the matrix that change may be ordered last so that only the last factors (e.g., U_{n-2}, D_{n-1}, U_{n-1}, D_n) must be modified. Nonsymmetric terms due to phase shifters can be taken last, and elements of the **b** vector that change can be used to make the corresponding row the last to be calculated. Example 3.8 is used to demonstrate reordering the matrix for near-optimal elimination.

Example 3.8

Using scheme 1, optimally order the buses of the five-bus power system shown in Figure E3.8, then calculate the table of factors using computationally efficient methods. Line admittances and the generator transient reactance are:

$$Y_{12} = -j1$$

$$Y_{15} = -j2$$

$$Y_{25} = -j3$$

$$Y_{24} = -j4$$

$$Y_{23} = -j5$$

$$Y_{45} = -j6$$

$$jX_5' = j0.5$$

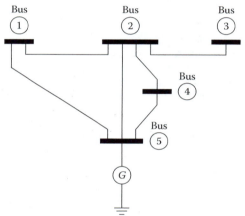

FIGURE E3.8

Solution

The bus admittance form of the injection equations has a symmetric $\mathbf{Y}_{BUS}$ matrix:

$$\mathbf{I}_{BUS} = \mathbf{Y}_{BUS}\mathbf{E}_{BUS} = -j\begin{bmatrix} 3 & -1 & 0 & 0 & -2 \\ -1 & 13 & -5 & -4 & -3 \\ 0 & -5 & 5 & 0 & 0 \\ 0 & -4 & 0 & 10 & -6 \\ -2 & -3 & 0 & -6 & 13 \end{bmatrix}\begin{bmatrix} E_1 \\ E_2 \\ E_3 \\ E_4 \\ E_5 \end{bmatrix} \qquad (1)$$

To apply scheme 1 for optimal ordering, inspection of the $\mathbf{Y}_{BUS}$ matrix yields:

One off-diagonal term, row 3
Two off-diagonal term, rows 1 and 4
Three off-diagonal terms, row 5
Four off-diagonal terms, row 2

Therefore, according to scheme 1, for a substitution to eliminate sequentially E_3, E_1, E_4, E_5, then E_2, the following change of variables is introduced:

$$x_1 = E_3$$
$$x_2 = E_1$$
$$x_3 = E_4 \tag{2}$$
$$x_4 = E_5$$
$$x_5 = E_2$$

The reordered system in matrix form is

$$\mathbf{b} = \mathbf{I}'_{BUS} = -j \begin{bmatrix} 5 & 0 & 0 & 0 & -5 \\ 0 & 3 & 0 & -2 & -1 \\ 0 & 0 & 10 & -6 & -4 \\ 0 & -2 & -6 & 13 & -3 \\ -5 & -1 & -4 & -3 & 13 \end{bmatrix} \begin{bmatrix} x_1 \\ x_2 \\ x_3 \\ x_4 \\ x_5 \end{bmatrix} = -j \begin{bmatrix} a_{11} & \cdots & a_{15} \\ \vdots & & \\ a_{51} & \cdots & a_{55} \end{bmatrix} \mathbf{X} \tag{3}$$

Observe that reordering locates the zeros so that they are the first terms involved in the elimination. If row 1 and row 4 of the original matrix, equation 1, are interchanged in the reordering scheme, it does not alter the nonzero element pattern in equation 3. As the elimination is developed, an asterisk will be used to indicate bypassed multiplications because zeros are present. Row-by-row elimination is used.

Observe that (3*) are avoided because of zeros in the first row. To eliminate x_1 from rows 2 to 4, the multiplication in the expression

12*
$$a^1_{ij} = a_{ij} - a_{i1}a^1_{1j} = a_{ij} \qquad i, j = 2, 3, 4 \tag{4}$$

is not necessary because of the zeros in the first column, $a_{21} = a_{31} = a_{41} = 0$. Also because of these zeros,

3*
$$a^1_{5j} = a_{5j} - a_{51}a^1_{1j} = a_{5j} \qquad j = 2, 3, 4 \tag{5}$$

Another multiplication (1*) is avoided in normalizing row 2. To eliminate x_2 from row 3, observe that a^1_{32} is zero, and hence

3*
$$a^2_{3j} = a^1_{3j} - a^1_{32}a^2_{2j} = a^1_{3j} = a_{3j} \qquad j = 3, 4, 5 \tag{6}$$

To eliminate x_2 from rows 4 and 5, one multiplication is avoided in each row because $a^2_{32} = 0$. A total of 25 multiplications are avoided by the reordering.

Successive eliminations do not have zeros, which can save multiplications. The resulting D and U terms in the table of factors are

$$[L\,D\,U] = j \begin{bmatrix} \frac{1}{5} & 0 & 0 & 0 & -1 \\ & \frac{1}{3} & 0 & -\frac{2}{3} & -\frac{1}{3} \\ & & \frac{1}{10} & -\frac{3}{5} & -\frac{2}{5} \\ & & & \frac{15}{121} & -\frac{91}{121} \\ & & & & \frac{121}{182} \end{bmatrix}$$

The **b** vector reorders a given bus injection vector $\mathbf{I}_{\text{BUS}}$ according to equation 2. The solution

$$\mathbf{X} = U_1 U_2 U_3 U_4 D U_4^t U_3^t U_2^t U_1^t \mathbf{b} \qquad (7)$$

also uses equation 2 to rearrange numerical values for $\mathbf{E}_{\text{BUS}}$.

Zero multiplications are easily determined during the elimination process by using pointer tables as shown in Figure 3.11. As fills, in other words, terms due to dependent variables from the previous rows, are introduced during the elimination, successive rows become progressively filled with nonzero terms. However, pointer tables are an efficient storage method for the U triangularization because the initial rows of U have the sparsity of $\mathbf{Y}_{\text{BUS}}$.

If the matrix $\mathbf{Z}_{\text{BUS}}$ is desired using the table of factors, it can be obtained column by column (bus by bus) by injection of unit currents. For example, if the column vector $\bar{Z}_{j2}$ is desired, a unit current injected into bus 2 calculates this column. Example 3.9 demonstrates this property.

Example 3.9

Use the network and table of factors of Example 3.8 to calculate the driving-point impedance and transfer impedances for bus 5 and bus 3.

Solution

A unit current injection at bus 5 is

$$\mathbf{I}_{\text{BUS}} = \begin{bmatrix} 0 \\ 0 \\ 0 \\ 0 \\ 1 \end{bmatrix}$$

The optimal ordering scheme renumbers the buses

$$b_2 = \mathbf{I}_1$$
$$b_5 = \mathbf{I}_2$$
$$b_1 = \mathbf{I}_3$$
$$b_3 = \mathbf{I}_4$$
$$b_4 = \mathbf{I}_5$$

such that $b = [0 \quad 0 \quad 0 \quad 1 \quad 0]^t$ is used as an injection in the renumbered system. The solution is

$$\mathbf{X} = \bar{Z}_{i5} = U_1 U_2 U_3 U_4 D U_4^t U_3^t U_2^t U_1^t \begin{bmatrix} 0 \\ 0 \\ 0 \\ 1 \\ 0 \end{bmatrix} = j \begin{bmatrix} 0.5 \\ 0.5 \\ 0.5 \\ 0.5 \\ 0.5 \end{bmatrix}$$

It is not surprising that the driving-point impedance and all transfer impedances are $j.5$ because the only connection from the transmission lines and buses to the reference is the transient reactance of the generator. In fact, $\mathbf{Y}_{BUS}$ is a singular matrix if there are no connections to the reference.

When a unit current is injected at bus 3, then $\mathbf{b} = [1 \quad 0 \quad 0 \quad 0 \quad 0]^t$ and the third column of $\mathbf{Z}_{BUS}$ is

$$\bar{Z}_{i3} = j \begin{bmatrix} 0.5549 \\ 0.6648 \\ 0.8648 \\ 0.5659 \\ 0.5000 \end{bmatrix}$$

If only a column of $\mathbf{Z}_{BUS}$ is needed, only this column has to be computed using the table of factors. It is unnecessary to store the entire $\mathbf{Z}_{BUS}$ matrix if it is not needed.

3.4.2   Several Iterative Methods for Linear Matrices

The electric power industry, service organizations to them, and power system researchers use many methods besides direct inverses or tables of factors to obtain an inverse for $\mathbf{Y}_{BUS}$. Several iterative methods for digital computers started with the early use of computers that had very limited active memories. At that time digital computers were the vacuum-tube type and were slow in comparison to analog computers. However, the coding versatility of the digital computers and the readout capabilities assured them of applications because they avoided the problems of hand-patched or hand-wired analog components.

The iterative methods used in early digital computer power systems programs continue to find applications. They are usually based on methods related to linear equations, and their proof of convergence is based on linear equations [5,6]. The principles of convergence for these methods are demonstrated by an elementary iterative scheme. Let $\mathbf{A}$ be an $n \times n$ matrix for which the dependent variable $\mathbf{X}$ is to be calculated as a function of the independent vector $\mathbf{b}$:

$$0 = \mathbf{A}\mathbf{X} - \mathbf{b} \tag{3.44}$$

If the unknown vector $\mathbf{X}$ is added to both sides of the equation, with $\mathbf{u}$ representing the identity matrix, then

$$\mathbf{X} = (\mathbf{A} + \mathbf{u})\mathbf{X} - \mathbf{b} \tag{3.45}$$

This equation is used for an elementary iterative scheme to solve for $\mathbf{X}$:

$$\mathbf{X}^{k+1} = (\mathbf{A} + \mathbf{u})\mathbf{X}^k - \mathbf{b} \tag{3.46}$$

where k is the iteration number. The sum of $\mathbf{A}$ and the identity matrix, $\mathbf{u}$, is called the iterative matrix. The numerical process is started with an arbitrary, random vector $\mathbf{X}^0$. As $k \to \infty$, it terminates with finite values

$$\mathbf{X}^{k+1}\Big|_{k \to \infty} = \mathbf{X}^k\Big|_{k \to \infty} \tag{3.47}$$

if the process is convergent, and only Equation 3.44 remains after cancellation of terms in Equation 3.46. Hence,

$$\mathbf{T} = \mathbf{A}^{-1}\mathbf{b} = \lim_{k \to \infty} \mathbf{X}^{k+1} \tag{3.48}$$

where **T** is called the true solution for the equations. To understand convergence to the solution, define the error vector ϵ Equation 3.89 after $k + 1$ iterative steps as

$$\epsilon^{k+1} = \mathbf{X}^{k+1} - \mathbf{T} = \mathbf{X}^{k+1} - \mathbf{A}^{-1}\mathbf{b} \tag{3.49}$$

Substituting Equation 3.46 into the error expression, a series of identities yields

$$
\begin{aligned}
\epsilon^{\kappa+1} &= (\mathbf{A} + \mathbf{u})\mathbf{X}^k - \mathbf{b} - \mathbf{A}^{-1}\mathbf{b} \\
&= (\mathbf{A} + \mathbf{u})\mathbf{X}^k - (\mathbf{A} + \mathbf{u})\mathbf{A}^{-1}\mathbf{b} \\
&= (\mathbf{A} + \mathbf{u})(\mathbf{X}^k - \mathbf{A}^{-1}\mathbf{b}) \\
&= (\mathbf{A} + \mathbf{u})\,\epsilon^k
\end{aligned}
\tag{3.50}
$$

But ϵ^κ can be expressed in terms of the error vector $\epsilon^{\kappa-1}$, so it follows inductively that

$$
\begin{aligned}
\epsilon^{k+1} &= (\mathbf{A} + \mathbf{u})(\mathbf{A} + \mathbf{u}) \cdots (\mathbf{A} + \mathbf{u})\,\epsilon^0 \\
&= (\mathbf{A} + \mathbf{u})^k\,\epsilon^0
\end{aligned}
\tag{3.51}
$$

Therefore, the error goes to zero as the number of iterations increases if the product

$$\lim_{k \to \infty} (\mathbf{A} + \mathbf{u})^k = \begin{bmatrix} 0 \\ 0 \\ \vdots \\ 0 \end{bmatrix} \tag{3.52}$$

It is a sufficient condition for convergence of the iterative scheme to the true solution (the error approaches zero) if the largest eigenvalue of the iterative matrix (in this case $\mathbf{A} + \mathbf{u}$) is less than 1.0 in absolute magnitude. In practice, the eigenvalues of the iterative method are rarely computed for large matrices or nonlinear iterative schemes, so other methods are used to ascertain convergence. Often, the true solution and inverse $\mathbf{A}^{-1}\mathbf{b}$ are unknown, so the iterative process is terminated at the kth iteration when

$$\| \mathbf{X}^{k+1} - \mathbf{X}^k \| < \bar{\epsilon} \tag{3.53}$$

where the double bars are absolute magnitude, and ε_i is some small-valued fraction of the largest $\|x_i\|$ expected. The accuracy of the method or rate of convergence can be determined by a check with a known solution calculated by other methods, or examination of residual values when the $\mathbf{X}^k$ vector is substituted into Equation 3.44.

3.4.2.1 Gaussian Iteration

The Gaussian iterative scheme also goes by several other names—the Jacobi, point iteration, or method of simultaneous displacements. It is essentially a line-by-line scheme to solve the simultaneous equations

$$\mathbf{AX} = \begin{bmatrix} a_{11} & \cdots & a_{1n} \\ \vdots & & \\ a_{n1} & \cdots & a_{nn} \end{bmatrix} \begin{bmatrix} x_1 \\ \vdots \\ x_n \end{bmatrix} = \begin{bmatrix} b_1 \\ \vdots \\ b_n \end{bmatrix} = \mathbf{b} \tag{3.54}$$

where the ith dependent variable is iteratively found at the $k + 1$ step in terms of the prior values of $\mathbf{X}$ and b_i:

$$x_i^{k+1} = \frac{1}{a_{ii}} \left\{ b_i - \sum_{\substack{j=1 \\ j \neq i}}^{n} a_{ij} x_j^k \right\} \quad i = 1, 2, 3, \ldots, n \tag{3.55}$$

The merit of this scheme is that only one line at a time of the matrix must be treated. If $\mathbf{A}$ is large and sparse, very few terms are used in the summation of Equation 3.55, and the inverse of $\mathbf{A}$ is never needed or computed. Pointer tables locate nonzero terms in each row.

To determine the iterative matrix for the scheme of Equation 3.55, let

$$\mathbf{A} = \mathbf{L} + \mathbf{D} + \mathbf{U}$$

where the matrix $\mathbf{A}$ is portioned into lower-triangular, diagonal, and upper-triangular portions. This is an equality, not a table of factors. Let $\mathbf{u}$ again be the identity matrix; then Equation 3.55 is written in matrix form as

$$\mathbf{X}^{k+1} = \{\mathbf{u} - \mathbf{D}^{-1}\mathbf{A}\}\mathbf{X}^k + \mathbf{D}^{-1}\mathbf{b} \tag{3.56}$$

where the iterative matrix is in braces.

The Gaussian iteration can be made to converge faster, using an acceleration factor, α, and old values from the solution vector. The accelerated scheme is

$$x_i^{k+1} = \frac{\alpha}{a_{ii}} \left\{ b_i - \sum_{j \neq i}^{n} a_{ij} x_j^k \right\} - (\alpha - 1) x_i^k \quad i = 1, 2, \ldots, n \tag{3.57}$$

A scalar acceleration factor, α, in the range $0 < \alpha \leq 2$ will improve the convergence but cannot change a divergent problem into one that converges. A vector formulation for Equation 3.57 is

$$\mathbf{X}^{k+1} = (\mathbf{u} - \alpha \mathbf{D}^{-1}\mathbf{A})\mathbf{X}^k + \alpha \mathbf{D}^{-1}\mathbf{b} \tag{3.58}$$

where $(\mathbf{u} - \alpha \mathbf{D}^{-1}\mathbf{A})$ is now the iterative matrix. When $\alpha = 1$, this becomes the basic Gaussian iteration; $0 < \alpha < 1$ is called underrelaxation and $1 < \alpha < 2$ is overrelaxation.

3.4.2.2 Gauss-Seidel Iteration

This is an extension of the preceding method, where the latest values of x_i^{k+1} are used in subsequent lines of iteration $m > i$. This is again a line-by-line iterative scheme to solve the equation $\mathbf{AX} = \mathbf{b}$ for the dependent variables x_i:

$$x_i^{k+1} = \frac{1}{a_{ii}} \left\{ b_i - \sum_{m<i} a_{im} x_m^{k+1} - \sum_{j>i}^{n} a_{ij} x_j^k \right\} \tag{3.59}$$

This is also a scheme where one line of the simultaneous equations is processed at a time and the inverse $\mathbf{A}^{-1}$ is not needed or computed. A scalar acceleration factor can be used to enhance convergence of this scheme as follows:

$$x_i^{k+1} = \frac{\alpha}{a_{ii}} \left\{ b_i - \sum_{m<i} a_{im} x_m^{k+1} - \sum_{j>i} a_{ij} x_j^k \right\} - (\alpha - 1) x_i^k \quad i = 1, 2, 3, \dots, m \quad (3.60)$$

When $0 < \alpha < 1$ this is underrelaxation, and $1 < \alpha < 2$ is called overrelaxation. Equation 3.60 expressed in matrix form is

$$\mathbf{X}^{k+1} = \left[\mathbf{u} - \left(\frac{\mathbf{D}}{\alpha} + \mathbf{L} \right)^{-1} \mathbf{A} \right] \mathbf{X}^k + \left(\frac{\mathbf{D}}{\alpha} + \mathbf{L} \right)^{-1} \mathbf{b} \quad (3.61)$$

where the iterative matrix is in brackets.

Example 3.10

Use the Gauss and Gauss-Seidel methods with and without an acceleration factor $\alpha = 0.5$ to find the driving-point and mutual impedances at bus 1 using $\mathbf{Y}_{BUS}$ for the three-bus system shown in Figure E3.10. Examine the convergence properties of all four

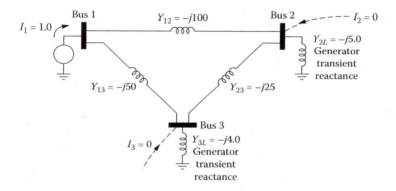

FIGURE E3.10

methods. Observe that the impedance vector is obtained by means of injecting a unit current in bus 1 to calculate the bus voltages (equivalent to a column 1 of $\mathbf{Z}_{BUS}$). Use a starting vector of

$$\mathbf{E}_{BUS}^t = [0 + j0.2 \quad 0 + j0.2 \quad 0 + j0.2]$$

Observe that only reactive elements are required for the circuit or impedance vector.

Solution

The $\mathbf{Y}_{BUS}$ formulation is

$$\mathbf{AX} = \mathbf{Y}_{BUS}\mathbf{E}_{BUS} = -j \begin{bmatrix} 150 & -100 & -50 \\ -100 & 130 & -25 \\ -50 & -25 & 79 \end{bmatrix} \begin{bmatrix} E_1 \\ E_2 \\ E_3 \end{bmatrix} = \mathbf{I}_{BUS} = \begin{bmatrix} 1 \\ 0 \\ 0 \end{bmatrix}$$

Let a_{ij} correspond to $\mathbf{Y}_{BUS}$ entries; then the Gauss-Seidel iterative scheme with an acceleration factor of $\alpha = 0.5$ is given by the equations

$$E_1^{k+1} = \frac{\alpha}{a_{11}}\left\{ I_1 - \sum_{j=2}^{3} a_{1j}E_j^k \right\} - (\alpha - 1)E_1^k$$

$$= \frac{0.5}{150}\left\{ j + 100E_2^k + 50E_3^k \right\} + 0.5E_1^k$$

$$E_2^{k+1} = \frac{\alpha}{a_{22}}\left\{ b_2 - a_{21}E_1^{k+1} - a_{23}E_3^k \right\} - (\alpha - 1)E_2^k$$

$$= \frac{50}{130}E_1^{k+1} + \frac{12.5}{130}E_3^k + 0.5E_2^k$$

$$E_3^{k+1} = \frac{\alpha}{a_{33}}\left\{ b_3 - a_{31}E_1^{k+1} - a_{32}E_2^{k+1} \right\} - (\alpha - 1)E_3^k$$

$$= \frac{25}{79}E_1^{k+1} + \frac{12.5}{79}E_2^{k+1} + 0.5E_3^k$$

To obtain the accelerated Gauss iterative scheme from these equations, replace $k + 1$ by k on the right-hand side of the equality. Let x_{11}, x_{12}, and x_{13} be the reactive components of the driving-point and mutual impedances at bus 1. The following table presents results of the four iterative methods to calculate these reactances. Observe as k iterations increase in the table, numerical values for the Gauss, accelerated Gauss, and Gauss-Seidel methods slowly converge toward similar final values. The accelerated Gauss-Seidel method shows *diverging* results—a consequence of an incorrect α.

Iterative Results for Example 3.10

	Gauss			Accelerated Gauss			Gauss-Seidel			Accelerated Gauss-Seidel		
Iteration	x_{11}	x_{12}	x_{13}	x_{11}	x_{12}	x_{13}	x_{11}	x_{12}	x_{13}	x_{11}	x_{12}	x_{13}
1	0.207	0.192	0.190	0.203	0.196	0.195	0.207	0.197	0.193	0.207	0.199	0.197
2	0.198	0.195	0.192	0.203	0.195	0.193	0.203	0.193	0.189	0.209	0.199	0.196
3	0.201	0.189	0.187	0.202	0.194	0.191	0.199	0.189	0.186	0.210	0.199	0.196
4	0.195	0.191	0.187	0.201	0.193	0.190	0.195	0.185	0.182	0.211	0.199	0.196
5	0.196	0.186	0.184	0.200	0.192	0.189	0.191	0.182	0.178	0.211	0.200	0.197
6	0.192	0.186	0.183	0.199	0.191	0.188	0.187	0.178	0.175	0.212	0.200	0.197
7	0.192	0.183	0.180	0.198	0.190	0.187	0.184	0.175	0.172	0.212	0.201	0.197
8	0.189	0.182	0.179	0.197	0.189	0.186	0.181	0.172	0.169	0.212	0.201	0.198
9	0.188	0.180	0.177	0.196	0.188	0.185	0.178	0.169	0.166	0.213	0.201	0.198
10	0.186	0.179	0.176	0.195	0.187	0.184	0.175	0.166	0.163	0.213	0.202	0.198
20	0.171	0.164	0.161	0.186	0.178	0.176	0.152	0.145	0.142	0.217	0.205	0.202
30	0.159	0.152	0.150	0.178	0.171	0.168	0.139	0.132	0.130	0.220	0.208	0.205
40	0.150	0.143	0.141	0.171	0.164	0.161	0.131	0.124	0.122	0.222	0.210	0.207
50	0.143	0.136	0.134	0.165	0.158	0.155	0.126	0.119	0.117	0.224	0.212	0.209
60	0.137	0.131	0.129	0.159	0.152	0.150	0.123	0.116	0.114	0.226	0.214	0.211
70	0.133	0.127	0.125	0.154	0.148	0.145	0.121	0.115	0.113	0.228	0.216	0.212
80	0.130	0.123	0.121	0.150	0.143	0.141	0.120	0.114	0.112	0.229	0.217	0.213
90	0.127	0.121	0.119	0.146	0.140	0.138	0.119	0.113	0.111	0.230	0.218	0.214
100	0.125	0.119	0.117	0.143	0.136	0.134	0.119	0.113	0.111	0.231	0.219	0.215

Three of the methods would converge to the true solution,

$$\bar{Z}_{il}^t = j[0.118 \quad 0.112 \quad 0.110] = j[X_{11} \quad X_{12} \quad X_{13}]$$

if enough iterations are performed. The Gauss-Seidel scheme without an accelerating factor exhibits the closest convergence to the true solution after 100 iterations.

Except for trial and error, there is no method to determine the best iterative method to use for a system of equations except for calculating the eigenvalues of the iterative matrix. The selection of an accelerating factor, α, is also based on trial and error. As Example 3.10 demonstrated, an accelerating factor may not enhance convergence, as the coefficients in the problem may be more suitable for one method than another. The starting vector is also important, and some methods may diverge for poor starting values, especially for nonlinear equations. If two iterative methods are to be compared for performance, convergence to within a tolerance or error band must be specified and the same starting vector must be used for both methods.

Problems

3.1. Designate bus 6 as the reference in the network shown in Figure P3.1. The line resistances and charging are neglected. The line data are as follows:

Line	X_1	X_0
1–2	0.04	0.12
2–5	0.05	Open
5–4	0.04	0.15
4–3	0.03	0.10
3–6	0.02	0.07
6–1	0.07	0.13
3–1	0.10	0.25

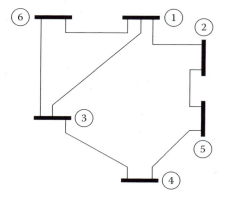

FIGURE P3.1

(a) Define a tree with transmission lines 1–3 and 2–5 as links. Use the $\mathbf{Z}_{BUS}$ building algorithm to obtain the positive- and zero-sequence impedance matrices.

(b) © Use the *zbus* computer program to verify the results of part (a) for both positive- and zero-sequence networks.

3.2.   For the network shown in Figure P3.2, use ground (neutral) as a reference. The transmission line positive-sequence impedances on a 100 MVA, 100 kV base are shown in Figure P3.2.

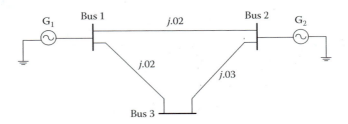

FIGURE P3.2

Transmission line resistance and shunt reactance are neglected. The generator reactances to ground on a 100 MVA, 100 kV base are

$$G_1, X_1' = X_2' = j0.1$$

$$G_2, X_1' = X_2' = j0.2$$

(a) Derive $\mathbf{Y}_{BUS}$ by inspection.

(b) Define a tree and cotree. Write the bus–branch incidence matrix and use it to obtain $\mathbf{Y}_{BUS}$.

(c) Derive $\mathbf{Z}_{BUS}$ by the building algorithm. Check the results by inverting the $\mathbf{Y}_{BUS}$ matrix.

3.3.   For the network shown in Figure P3.3, use the ground as the reference and determine the positive-sequence and zero-sequence bus impedance networks. Resistances are neglected. The transmission line parameters (p.u.) are:

	1–2	2–3	3–1
Positive- or negative-sequence reactance, X_1	0.02	0.03	0.02
Zero-sequence reactance, X_0	0.027	Open circuit	0.04
½ positive- or negative-sequence shunt susceptance, $\frac{\omega C_1}{2}$	0.5	0.25	0.333
Zero-sequence (capacitive) shunt susceptance, $\frac{\omega C_0}{2}$	0.5	Open circuit	1.0

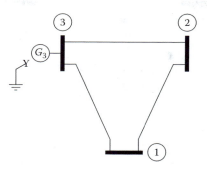

FIGURE P3.3

For the generator:

$$X_1' = X_2' = 0.50 \text{ p.u.}$$

$$X_0' = 0.15 \text{ p.u.}$$

3.4. Consider the network shown in Figure P3.4. The line positive-sequence imped-ances are:

Line	Impedance
1–2	$Z_{12} = j0.06$
2–3	$Z_{23} = j0.05$
3–1	$Z_{31} = j0.09$

The generator positive-sequence impedances are:

Generator 1, $Z_1' = j0.20$

Generator 2, $Z_1' = j0.10$

(a) Use ground (neutral) as a reference. Determine the positive-sequence bus admittance matrix, $\mathbf{Y}_{BUS}$, by means of inspection or by means of a reduced bus–branch incidence matrix for a tree.

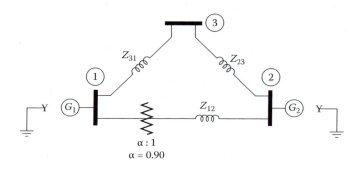

FIGURE P3.4

(b) Define a tree and cotree for the elements of the network. Apply the building algorithm to construct $\mathbf{Z}_{BUS}$ by successively adding branches or links, one element at a time.

3.5. For the network shown in Figure P3.5 the resistive parts of the elements are negligible. Positive mutual coupling between lines A and B is for the current flow directions shown on the network. The transmission line data (50 MVA base) are:

	Line C	Line A	Line B
Positive- or negative-sequence reactance	0.50	0.30	0.40
Zero-sequence reactance	1.50	1.0	1.20
½ shunt susceptance, $\frac{\omega C_1}{2}$	0.2	—	—
Positive- or negative-sequence mutual coupling	—	0.25 to B	0.25 to A

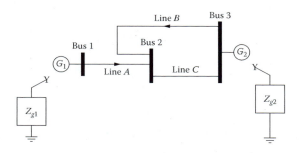

FIGURE P3.5

The generator data (50 MVA base) are:

Generator 1: $X'_1 = X'_2 = 0.30$ $\qquad X_0 = 0.25$ (includes $3Z_{g1}$)

Generator 2: $X'_1 = X'_2 = 0.50$ $\qquad X_0 = 0.70$ (includes $3Z_{g2}$)

Use *ground* as a reference and do the following:

(a) Draw the positive- and zero-sequence impedance circuit diagrams, labeling all reactances.

(b) Determine $\mathbf{Y}_{BUS}$ for the positive-sequence circuit.

(c) Use the $\mathbf{Z}_{BUS}$ building algorithm to calculate $\mathbf{Z}^1_{BUS}$, the positive-sequence bus impedance matrix.

(d) © Use the **zbus.exe** computer program to verify the results of part (c).

3.6. For the $\mathbf{Y}_{BUS}$ matrix given below, do the following:

(a) Reorder the rows to exploit the number of zeros in the admittance matrix.

$$\text{buses} \quad \rightarrow 1 \quad\quad 2 \quad\quad 3 \quad\quad 4 \quad\quad 5 \quad\quad 6$$

$$\mathbf{Y}_{BUS} = -j \begin{bmatrix} 35 & & & -15 & -20 & \\ & 20 & -20 & & & \\ & -20 & 60 & -15 & -10 & -10 \\ -15 & & -15 & 40 & & -10 \\ -20 & & -10 & & 35 & \\ & & -10 & -10 & & 20 \end{bmatrix}$$

(b) Use Tinney's matrix factorization to compute the inverse or [L D U] table of factors. Utilize the symmetry properties of the matrix.

(c) Compute the bus voltages $\mathbf{E}_{BUS} = \mathbf{Y}_{BUS}^{-1}\mathbf{I}_{BUS}$ when the current injection vector in the original numbering is

$$
\mathbf{I}_{BUS} = \begin{bmatrix} 0 \\ 0 \\ 1 \\ 0 \\ 0 \\ 0 \end{bmatrix} = \begin{bmatrix} I_1 \\ I_2 \\ I_3 \\ I_4 \\ I_5 \\ I_6 \end{bmatrix}
$$

(d) What is the driving-point impedance at bus 3?

(e) © Observe in $\mathbf{Y}_{BUS}$ that only buses 3 and 5 have admittances bus to ground. These shunt admittances are $y_{33} = y_{55} = -j5 = 1/z = \frac{1}{j0.2}$. For the network defined by $\mathbf{Y}_{BUS}$, convert all admittances to impedances and enter the impedances into the *zbus* computer program. The results of the *zbus* computer program can verify the results of part (d) or the complete $\mathbf{Z}_{BUS}$ matrix.

3.7. Shown in Figure P3.7 is the IEEE 14 bus network that is used as a standard test case for load-flow studies. Bus 1 is selected to be the reference bus, so it does not enter the $\mathbf{Y}_{BUS}$ matrix. Do not consider the generator reactances. To indicate only connections, every line is taken as 1 S in $\mathbf{Y}_{BUS}$ matrix, and transformers are on-nominal.

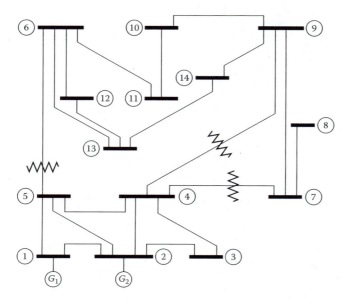

FIGURE P3.7

It is desired to perform an optimally ordered triangular factorization of the $\mathbf{Y}_{BUS}$ matrix for the network with respect to bus 1.

(a) Determine the order in which you would eliminate the rows according to Tinney's method 1.

(b) Of all buses with only two lines connected to them, which would you eliminate first, second, and so on? In other words, simulate the elimination process to select the next row with the minimum number of elements introduced by the substitution for the buses with two connecting lines.

3.8. It is intended to carry out a Gauss iterative scheme (not Gauss-Seidel) to solve the following linear, simultaneous equations:

$$\begin{bmatrix} 2 & 1 \\ 3 & 4 \end{bmatrix} \begin{bmatrix} x_1 \\ x_2 \end{bmatrix} = \begin{bmatrix} 30 \\ 70 \end{bmatrix}$$

Let the starting vector be $\begin{bmatrix} 0 \\ 0 \end{bmatrix}$ and do the following:

(a) Carry out five iterations.

(b) Observing your calculated values, will the iterations converge to a solution? If so, why?

(c) Prove whether or not the iterations will converge.

3.9. Solve the following simultaneous equations using the Gauss-Seidel iterative method with *all acceleration factors* equal to 1.5. Use the starting values $x_1^0 = 0, x_2^0 = 0$. Carry out iterations until both x_1 and x_2 change less than 0.0001 per iteration.

$$2x_1 + x_2 - 3 = 0$$

$$3x_1 + 4x_2 - 7 = 0$$

3.10. Use the Gauss-Seidel method with acceleration factors $\alpha = 0.5$ and $\alpha = 1.0$ to calculate the column vector $\overline{Z}_{i3}$ for Example 3.9. This is another method to calculate the driving-point and mutual impedances. Perform iterations until the largest impedance change for successive iterations is

$$\left| z_{i3}^{k+1} - z_{i3}^k \right| \le 0.0001 \quad i = 1, 2, 3$$

3.11 A 2-bus power system has a generator at bus 1 supplying a motor at bus 2 through 2 transmission lines as shown in Figure P3.11

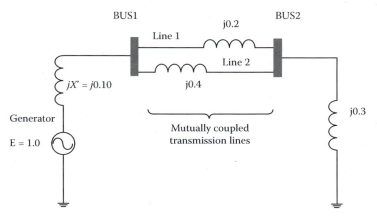

FIGURE P3.11

(a) Begin with the generator transient reactance and obtain the $\mathbf{Z}_{BUS}$ matrix with only line 1 included in the 2 bus system.

(b) Use the algorithm of Section 3.3.1 in order to add the mutually coupled transmission line 2 to the $\mathbf{Z}_{BUS}$ matrix of part (a). Line 2 has a mutual coupling of $j0.10$ with transmission line 1.

(c) Check the results of part (b) with the *zbus.exe* program.

3.12 The positive sequence reactance of a transmission line is $j0.05$ pu on a 100 MVA base. Another line was constructed on the same right-of-way with $j0.04$ pu positive sequence reactance. The pair of lines have $j0.03$ pu mutual reactance. The lines are operated in parallel, excited by the same sending generator as shown in Figure P3.12. What is the equivalent series positive sequence reactance when the lines are operated in parallel? **Hints:** (1) Analytically treat the parallel lines as coupled transformer windings with equal input voltage and shared current, I. (2) Use the *zbus.exe* program to connect both lines to a dummy bus with an impedance to ground, then introduce the mutual coupling.

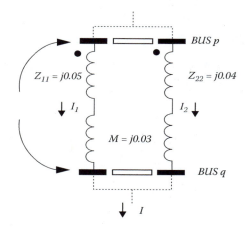

FIGURE P3.12

References

1. Chua, L. O., and Lin, P. 1975. *Computer-aided analysis of electronic circuits: Algorithms and computational techniques.* Englewood Cliffs, NJ: Prentice-Hall.
2. DeSoer, C. A., and Kuh, E. S. 1969. *Basic circuit theory.* New York: McGraw-Hill Book Company.
3. Khabaza, I. M. 1967. Introduction to matrix methods. In *The use of digital computers in electric power systems.* Stocksfield, Northumberland, England: Oriel Press Ltd.
4. Tinney, W. F., and Walker, J. W. 1967. Direct solutions of sparse network equations by optimally ordered triangular factorization. *Proc. IEEE* 55(11).
5. Ralston, A. 1965. *A first course in numerical analysis.* New York: McGraw-Hill Book Company.
6. Gerald, C. F. 1978. *Applied numerical analysis.* Reading, MA: Addison-Wesley Publishing Company.

4

Network Fault and Contingency Calculations

In designing protective circuits for generators, transformers, transmission lines, and other equipment, it is important to determine their terminal currents which flow due to local abnormal operating conditions such as line-to-line faults, or a fault from line to ground, and other device failures. If currents due to a fault are excessively high, accompanying electromagnetic forces could destroy equipment or fuse together the contacts of protective circuit breakers, thereby spreading the damage to many system components. Protective circuit breakers are selected according to voltage-withstanding capability, magnitude of fault currents to be interrupted, and response time from sensing the excessive current to the instant of opening the circuit. Other equipment, motors, for example, are voltage sensitive in the sense that if their supply voltage is too low due to an excessive current at another load point, the motor current increases to maintain the shaft power. This could cause motor overheating.

A general schematic for the three-phase network is shown in Figure 4.1. Two types of fault are shown in the figure, from lines to ground and a bus-to-bus impedance fault. The fault impedance between buses could represent a parallel line energized or an open circuit of the transmission line from bus p to bus q if negative values are used for $\mathbf{Z}_{pq}^{abc}$. Generators with grounding impedances in their neutral connections are shown in Figure 4.1, as the grounding impedance increases the zero-sequence impedance of the generator, which in turn limits unbalanced fault currents. Wye-connected transformers with impedance-grounded neutrals also aid in limiting unbalanced fault currents.

In general, all three of the symmetrical component impedances for network elements are required to determine fault currents and bus voltages during unbalanced fault conditions. The calculations are performed in the symmetrical component frame, then converted to *abc*-phase per-unit values, then ultimately to engineering units for physical voltages and currents, as shown in Figure 4.2.

4.1 Fault Calculations Using $\mathbf{Z}_{\mathrm{BUS}}$

In Chapter 2 the symmetrical component transformation was applied to three-phase *abc* elements. Whenever three-phase elements are balanced, the equivalent symmetrical component elements are decoupled, but unbalanced elements require all three components for an electrical computation. Therefore, in a general bus impedance formulation the linearized power network is described by the matrix equation

$$\mathbf{E}_{\mathrm{BUS}}^{012} = \mathbf{Z}_{\mathrm{BUS}}^{012}\mathbf{I}_{\mathrm{BUS}}^{012} \tag{4.1}$$

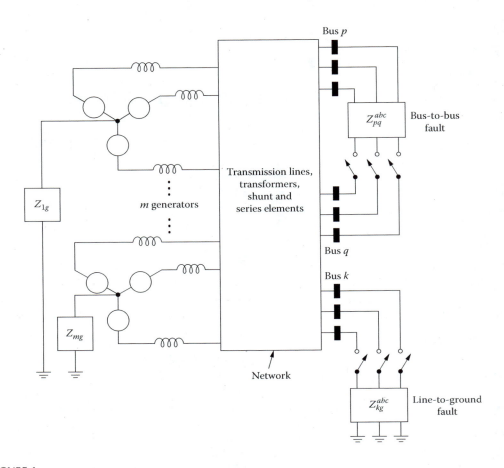

FIGURE 4
a,b,c circuit schematic of a power system with line-to-line and line-to-ground faults.

FIGURE 4.2
Faults and contingency calculations in the symmetric components frame.

In Equation 4.1 the following definitions are used for the ith voltage and current:

$$\mathbf{E}_i^{012} = \begin{bmatrix} E_i^0 \\ E_i^1 \\ E_i^2 \end{bmatrix} \qquad \mathbf{I}_i^{012} = \begin{bmatrix} I_i^0 \\ I_i^1 \\ I_i^2 \end{bmatrix} \qquad\qquad \text{(4.2a, b)}$$

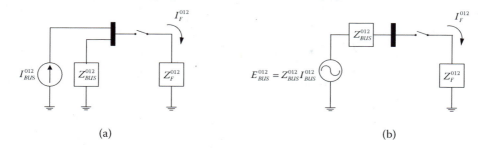

FIGURE 4.3
(a) Norton equivalent (b) Thévenin equivalent.

where 0, 1, and 2 are the zero-, positive-, and negative-sequence components, respectively. The (km)th impedance element is the 3×3 submatrix symmetrical component matrix

$$\mathbf{Z}_{km}^{012} = \begin{bmatrix} Z_{00} & Z_{01} & Z_{02} \\ Z_{10} & Z_{11} & Z_{12} \\ Z_{20} & Z_{21} & Z_{22} \end{bmatrix}_{k,m} \tag{4.3}$$

which reduces to diagonal elements for balanced networks. The elements of this submatrix are specified in Table 2.1 for several combinations of impedances. If there are n buses in the formulation, then $\mathbf{Z}_{BUS}^{012}$ is a $3n \times 3n$ matrix, and $\mathbf{I}_{BUS}^{012}$ and $\mathbf{E}_{BUS}^{012}$ are $3n$ vectors.

In the bus impedance formulation of Equation 4.1, the current vector is the forcing function or independent variable, such that if a fault impedance, $\mathbf{Z}_F^{012}$, were connected from a bus k to ground (reference), it alters $\mathbf{Z}_{BUS}^{012}$ as shown schematically in Figure 4.3a.

The fault impedance matrix, $\mathbf{Z}_F^{012}$, has only finite (or zero) entries for bus k, the faulted bus. Similarly, the fault current vector $\mathbf{I}_F^{012}$ is zero except for components corresponding to the faulted bus k. Because the bus impedance equations are linear, the bus voltages during the fault condition may be calculated by superposition from a Thévenin equivalent:

$$\mathbf{E}_{BUS}^{012}(F) = \mathbf{E}_{BUS}^{012}(0) - \mathbf{Z}_{BUS}^{012} \, \mathbf{I}_F^{012} \tag{4.4}$$

where $\mathbf{E}_{BUS}^{012}(0)$ is the specified or given initial voltage before the fault and $\mathbf{E}_{BUS}^{012}(F)$ is the voltage during the fault condition. The fault current vector is given by

$$\mathbf{I}_F^{012} = \begin{bmatrix} 0 \\ \vdots \\ \mathbf{I}_k^{012}(F) \\ \vdots \\ 0 \end{bmatrix} \tag{4.5}$$

such that the product $\mathbf{Z}_{BUS}^{012}\mathbf{I}_F^{012}$ utilizes only the kth submatrix of the impedance matrix. Also, the voltage at bus k is $\mathbf{Z}_F^{012}\mathbf{I}_k^{012}(F)$, so that Equation 4.4 is written as

$$
\mathbf{E}_{BUS}^{012}(F)=
\begin{bmatrix}
\mathbf{E}_1^{012}(F)\\
\vdots\\
\mathbf{E}_k^{012}(F)\\
\vdots\\
\mathbf{E}_n^{012}(F)
\end{bmatrix}
=
\begin{bmatrix}
\times\\
\times\\
\times\\
\mathbf{Z}_F^{012}\mathbf{I}_k^{012}(F)\\
\times\\
\times\\
\vdots
\end{bmatrix}
=
\begin{bmatrix}
\mathbf{E}_1^{012}(0)\\
\vdots\\
\mathbf{E}_k^{012}(0)\\
\vdots\\
\mathbf{E}_n^{012}(0)
\end{bmatrix}
-
\begin{bmatrix}
\mathbf{Z}_{1k}^{012}\mathbf{I}_k^{012}(F)\\
\vdots\\
\mathbf{Z}_{kk}^{012}\mathbf{I}_k^{012}(F)\\
\vdots\\
\mathbf{Z}_{kn}^{012}\mathbf{I}_k^{012}(F)
\end{bmatrix}
\tag{4.6}
$$

The $\times$'s indicate as-yet-unknown values, but the kth submatrix row is entirely known in terms of current $\mathbf{I}_k^{012}(F)$. Solving the kth submatrix for $\mathbf{I}_k^{012}(F)$, we have

$$
\mathbf{I}_k^{012}(F) = (\mathbf{Z}_F^{012} + \mathbf{Z}_{kk}^{012})^{-1}\mathbf{E}_k^{012}(0)
\tag{4.7}
$$

Observe that if the fault impedance $\mathbf{Z}_F^{012}$ is zero, the short-circuit current that flows is the open-circuit voltage divided by the driving-point (or input) impedance, as is well known from Thévenin's equivalent for linear circuits. After current $\mathbf{I}_k^{012}(F)$ is calculated using Equation 4.7, it is in turn substituted into submatrices 1, 2, ..., $k - 1$, $k + 1$, ... n of Equation 4.6 to find the voltages at these buses during the fault. Performing this substitution, the voltage at any bus i due to a fault $\mathbf{Z}_F^{012}$ from bus k to ground (reference) is given by

$$
\boxed{
\begin{array}{c}
\text{Voltage at Bus } i \text{ during Fault } \mathbf{Z}_F^{012} \text{ at Bus } k\\
\mathbf{E}_i^{012}(F) = \mathbf{E}_i^{012}(0) - \mathbf{Z}_{ik}^{012}(\mathbf{Z}_F^{012} + \mathbf{Z}_{kk}^{012})^{-1}\mathbf{E}_k^{012}(0)
\end{array}
}
\tag{4.8}
$$

When the network is balanced, the symmetrical component impedances are *diagonal*, so that it is possible to calculate $\mathbf{Z}_{BUS}$ separately for zero-, positive-, and negative-sequence impedances. There is no need to include the zero terms of a $3n \times 3n$ matrix, but only two or three matrices that are each $n \times n$. Each $\mathbf{Z}_{ik}^{012}$ element required for Equation 4.8 may be constructed from sequence components, and the respective form for the fault, either $\mathbf{Z}_F^{012}$ or $\mathbf{Y}_F^{012}$, may be used accordingly.

It is worth emphasizing that only the kth column of the bus impedance matrix is required in Equation 4.8 to determine the voltage at any bus during the fault. The entire $\mathbf{Z}_{BUS}^{012}$ matrix is not needed, so the methods of Chapter 3 that calculate the kth column $\overline{Z}_{ik}^{012}$ for $i = 1, 2, ...,$ n by means of inverting $\mathbf{Y}_{BUS}$ are applicable here, but all elements of $\overline{Z}_{ik}^{012}$ must be defined.

An example of an undefined element is a wye-connected transformer or generator with an ungrounded neutral that has an open circuit (infinite impedance) in the zero-sequence network, as shown in Table 2.6. If no connection exists through other parts of the network, these terminals "float" with respect to the ground reference, and the bus impedance matrix is undefined for zero-sequence elements. One method that ensures that the positive- and zero-sequence networks correspond to each other, element by element, is to create an artificial bus [1] at the center of the transformer as shown in Figure 4.4. H is a fictitious impedance of large numerical value.

The numerical value of H in Figure 4.4 is selected to be large enough so that it does not affect the engineering accuracy of the results more than, say, the fourth significant decimal place of the final $\mathbf{Z}_{BUS}^{012}$ matrix for the entire network, but small enough to permit normal

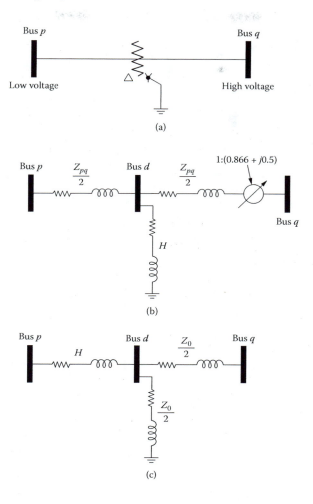

FIGURE 4.4
Artificial bus and elements to complete an open-circuit zero-sequence network: (a) circuit schematic delta-to-wye, G; (b) positive-sequence network, $H \to \infty$; (c) zero-sequence network, $H \to \infty$.

impedance computations. For example, a reasonable numerical value for H and its corresponding computer variable is

$$H = (1+j) \times 10^8 \to \text{COMPLX } (1.0E8, \ 1.0E8)$$

for a 32-bit computer whose largest floating-decimal-point number is approximately 10^{75}.

Example 4.1 is used to demonstrate the application of Z_{BUS}^{012} to short-circuit calculations and the use of the fictitious elements.

Example 4.1

A 100 MVA, 13.2 $\text{kv}_{\ell\text{-}\ell}$ cylindrical rotor generator as shown in Figure E4.1.1 is operated at rated terminal voltage. The generator is wye connected, impedance grounded, and driving a wye-to-wye-connected transformer exciting a 138 $\text{kv}_{\ell\text{-}\ell}$ transmission line. The sequence impedances for the elements are given in Figure E4.1.1. The transformer has 0.06 p.u. reactance and 0.1 p.u. grounding reactance on the generator side. Find the fault

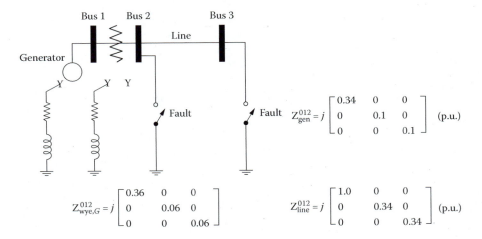

FIGURE E4.1.1

currents and bus voltages for a balanced three-phase fault to ground occurring at the transformer terminals and at the remote end of the transmission line.

Solution

The transformer has a floating neutral connection on the side connected to the transmission line, so the zero-sequence circuit is open from bus 1 to bus 2. A fictitious bus 4 is created at the center of the transformer to complete the zero-sequence diagram as shown in Figure E4.1.2.

The positive-sequence diagram shown in Figure E4.1.3 is complete up to bus 3, and bus 4 is at the center of the transformer. The corresponding bus impedance matrices are

$$
\begin{array}{cccc}
\phantom{Z^0_{BUS}=j}\;\;1 & \quad 4 & \quad 2 & \quad 3
\end{array}
$$

$$
Z^0_{BUS} = j
\begin{bmatrix}
0.1749 & 0.0874 & 0.0874 & 0.0874 \\
0.0874 & 0.1337 & 0.1337 & 0.1337 \\
0.0874 & 0.1337 & H & H \\
0.0874 & 0.1337 & H & H
\end{bmatrix}
\qquad
Z^{1,2}_{BUS} = j
\begin{bmatrix}
0.10 & 0.10 & 0.10 & 0.10 \\
0.10 & 0.13 & 0.13 & 0.13 \\
0.10 & 0.13 & 0.16 & 0.16 \\
0.10 & 0.13 & 0.16 & 0.50
\end{bmatrix}
$$

$$
\begin{array}{cccc}
\phantom{Z^{1,2}_{BUS}=j}\;\;1 & \quad 4 & \quad 2 & \quad 3
\end{array}
$$

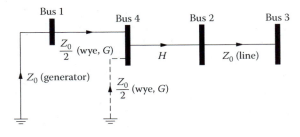

FIGURE E4.1.2
Zero sequence network.

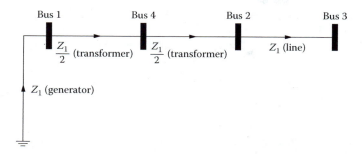

FIGURE E4.1.3
Positive and negative sequences.

The initial voltage at all points of the network is 1.0 p.u. For a balanced, three-phase fault to ground at bus 2, the fault current as per Equation 4.7 is

$$\mathbf{I}_2^{012}(F) = (\mathbf{Z}_F^{012} + \mathbf{Z}_{22}^{012})^{-1}\mathbf{E}_2^{012}(0)$$

$$= \left[\begin{bmatrix} 0 & 0 & 0 \\ 0 & 0 & 0 \\ 0 & 0 & 0 \end{bmatrix} + j \begin{bmatrix} H & 0 & 0 \\ 0 & 0.16 & 0 \\ 0 & 0 & 0.16 \end{bmatrix} \right]^{-1} \begin{bmatrix} 0 \\ E_2^1 \\ 0 \end{bmatrix} \quad \text{p.u.}$$

$$= -j \begin{bmatrix} 0 & 0 & 0 \\ 0 & 6.25 & 0 \\ 0 & 0 & 6.25 \end{bmatrix} \begin{bmatrix} 0 \\ E_2^1 \\ 0 \end{bmatrix} = \begin{bmatrix} 0 \\ -j6.25 \\ 0 \end{bmatrix} \quad \text{p.u.}$$

where $1/H$ is approximated as zero compared to nonzero matrix entries. At the 138 kV transformer terminals the base current is

$$I_{\text{BASE}} = \frac{S_{\text{BASE}}}{V_{\text{BASE}}} = \frac{100 \text{ MVA}}{138 \text{ kV}} = 0.7246 \text{ kA} = \sqrt{3} \; I_{\text{rated}}^a$$

which is used to convert to engineering units.
   The phase currents at the short circuit are found as

$$\mathbf{I}_2^{abc}(F) = \mathbf{T}_s \mathbf{I}_2^{012}(F) = \begin{bmatrix} 1 & 1 & 1 \\ 1 & a^2 & a \\ 1 & a & a^2 \end{bmatrix} \begin{bmatrix} 0 \\ -j6.25 \\ 0 \end{bmatrix} = -j \begin{bmatrix} 6.25 & \angle 0° \\ 6.25 & \angle -120° \\ 6.25 & \angle +120° \end{bmatrix} \quad \text{p.u.}$$

$$\rightarrow -j \begin{bmatrix} 2.615 & \angle 0° \\ 2.615 & \angle -120° \\ 2.615 & \angle +120° \end{bmatrix} \quad \text{kA}$$

The corresponding generator line currents are 27.34 kA. The voltage at bus i for a zero impedance fault at bus k (Equation 4.8) is given by

$$\mathbf{E}_i^{012}(F) = \mathbf{E}_i^{012}(0) - \mathbf{Z}_{ik}^{012}\mathbf{I}_k^{012}(F)$$

For bus 1, the voltage, using subscripts $i = 1$ and $k = 2$, is

$$\mathbf{E}_1^{012}(F) = \begin{bmatrix} 0 \\ E_1 \\ 0 \end{bmatrix} - j \begin{bmatrix} 0.0874 & 0 & 0 \\ 0 & 0.10 & 0 \\ 0 & 0 & 0.10 \end{bmatrix} \begin{bmatrix} 0 \\ -j6.25 \\ 0 \end{bmatrix} = \begin{bmatrix} 0 \\ 1 \\ 0 \end{bmatrix} - \begin{bmatrix} 0 \\ 0.625 \\ 0 \end{bmatrix} = \begin{bmatrix} 0 \\ 0.375 \\ 0 \end{bmatrix}$$

It is clear that only the positive-sequence network elements are necessary to calculate the voltages due to a balanced fault. The voltages at buses 2 and 3 are easily found to be

$$E_2^1 = E_2^1(0) - Z_{22}^1 I_2^1(F) = 0.0$$

$$E_3^1 = E_3^1(0) - Z_{23}^1 I_2^1(F) = 1.0 - (j0.16)(-j6.25) = 0.0$$

The zero-sequence and negative-sequence voltages at all buses are zero.

For a balanced three-phase fault to ground at bus 3, the fault current is calculated using positive-sequence values:

$$I_3^1(F) = \left(Z_F^1 + Z_{33}^1 \right)^{-1} E_3^1(0)$$

$$= (0 + j0.50)^{-1}(1.0) = -j2.0 \text{ p.u.} \rightarrow -j836.7 \text{ A line current}$$

The voltage at bus 3 is zero, of course, during the fault condition. At the other buses the voltages are

$$E_1^1(F) = E_1^1(0) - Z_{13}^1 \left(\frac{1}{Z_{33}^1} \right) E_3^1(0) = 1.0 - \frac{0.1}{0.5} = 0.800 \text{ p.u.}$$

$$E_2^1(F) = E_2^1(0) - Z_{23}^1 \left(\frac{1}{Z_{33}^1} \right) E_3^1(0) = 1.0 - \frac{0.16}{0.50} = 0.680 \text{ p.u.}$$

$$E_3^1(F) = 0.0 \text{ p.u.}$$

These per-unit values are easily converted into volts.

© The results calculated in this example may be verified using the software. Copy the file *example41.sss* and paste it into *zbus.sss* to verify the positive-sequence bus impedance matrix. Similarly, the file *example41Z.sss* may be used to check the zero-sequence results.

Often the 30° phase shift across a wye-to-delta connection is neglected for fault calculations using the positive- and negative-sequence symmetrical components. The transformer is treated as an ungrounded wye to wye, and the line-to-line voltage is used to obtain per-unit values of impedance. However, it must be recognized that the delta connection permits zero-sequence current to flow in the grounded wye, so the elements of the zero-sequence circuit must be retained.

In cases of a single-phase fault to ground, the fault matrix $\mathbf{Z}_F^{012}$ is not defined (e.g., open circuits are infinite impedances), but using the fault admittance form, the current may be expressed as

$$I_k^{012}(F) = Y_F^{012} E_k^{012}(F) \qquad (4.9)$$

and zeros replace the open-circuit conditions. The substitution of Equation 4.9 into the right side of Equation 4.6 yields the following result at the faulted bus:

$$E_k^{012}(F) = (u + Z_{kk}^{012} Y_F^{012})^{-1} E_k^{012}(0) \qquad (4.10)$$

Substituting Equation 4.9 into the i remaining rows of Equation 4.6 yields

$$\boxed{\begin{array}{c} \text{Voltage at Bus } i \text{ During Fault } Y_F^{012} \text{ at Bus } k \\ E_i^{012}(F) = E_i^{012}(0) - Z_{ik}^{012} Y_F^{012} (u + Z_{kk}^{012} Y_F^{012})^{-1} E_k^{012}(0) \end{array}} \qquad (4.11)$$

where u is the identity matrix.

Thus, with the Y_F^{012} form of the fault admittance, the voltage at any bus may be found. Table 4.1 summarizes single- and three-phase faults to ground.

The following example shows the derivation and application of the single-phase fault-to-ground results in Table 4.1.

Example 4.2

Consider Figure E4.2.1, where a 100 MVA 13.2 kv$_{\ell\text{-}\ell}$ cylindrical rotor wye-connected generator with an impedance-grounded neutral is operated at rated voltage. The generator characteristics are:

$$Z_{\text{gen}}^{012} = j \begin{bmatrix} 0.34 & 0 & 0 \\ 0 & 0.1 & 0 \\ 0 & 0 & 0.1 \end{bmatrix} = \begin{bmatrix} Z_0 & 0 & 0 \\ 0 & Z_1 & 0 \\ 0 & 0 & Z_2 \end{bmatrix}$$

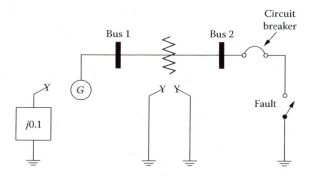

FIGURE E4.2.1

It is exciting an unloaded wye, G-to-wye, G 13.2:138 kV transformer. The transformer leakage reactance is 0.06 p.u. on a 100 MVA base. A single line-to-ground fault occurs on the high-voltage side of the transformer. Calculate the symmetrical component fault currents to be interrupted by a circuit breaker between the fault and the transformer.

TABLE 4.1

Single- and Three-Phase Faults to Ground

Fault Schematic	Fault Impedance Symmetrical Components	Fault Current (Fault at Bus k)	Bus Voltage at Bus i Due to Fault
Three-phase	$$Z_F^{012} = \begin{bmatrix} Z+3Z_g & 0 & 0 \\ 0 & Z & 0 \\ 0 & 0 & Z \end{bmatrix}$$	$$I_k^{012} = \begin{bmatrix} \dfrac{1}{Z_{kk}^0 + Z_F^0} & 0 & 0 \\[2mm] 0 & \dfrac{1}{Z_{kk}^1 + Z_F^1} & 0 \\[2mm] 0 & 0 & \dfrac{1}{Z_{kk}^2 + Z_F^2} \end{bmatrix}\begin{bmatrix} E_k^0(0) \\ E_k^1(0) \\ E_k^2(0) \end{bmatrix}$$	$$E_i^{012}(F) = E_i^{012}(0) - \begin{bmatrix} \dfrac{Z_{ik}^0}{Z_{kk}^0 + Z_F^0} & 0 & 0 \\[2mm] 0 & \dfrac{Z_{ik}^1}{Z_{kk}^1 + Z_F^1} & 0 \\[2mm] 0 & 0 & \dfrac{Z_{ik}^2}{Z_{kk}^2 + Z_F^2} \end{bmatrix} E_k^{012}(0)$$
Single-phase	$$Y_F^{012} = \frac{Y}{3}\begin{bmatrix} 1 & 1 & 1 \\ 1 & 1 & 1 \\ 1 & 1 & 1 \end{bmatrix}$$	$$I_k^{012} = \frac{E_k^0(0) + E_k^1(0) + E_k^2(0)}{Z_{kk}^0 + 2Z_{kk}^1 + 3/Y}\begin{bmatrix} 1 \\ 1 \\ 1 \end{bmatrix}$$ $$Z_{kk}^1 = Z_{kk}^2$$	$$E_i^{012}(F) = E_i^{012}(0) - \frac{(E_k^0(0) + E_k^1(0) + E_k^2(0))}{Z_{kk}^0 + 2Z_{kk}^1 + 3/Y}\begin{bmatrix} Z_{ik}^0 \\ Z_{ik}^1 \\ Z_{ik}^2 \end{bmatrix}$$

Solution

For a single line-to-ground fault at the output of the transformer, the impedance schematic is shown in Figure E4.2.2. Here $Z = j0.06$ is the single phase to ground impedance and Table 2.2 gives the 3-phase admittance as:

$$\mathbf{Y}_{1\phi}^{012} = \frac{1}{j0.18}\begin{bmatrix} 1 & 1 & 1 \\ 1 & 1 & 1 \\ 1 & 1 & 1 \end{bmatrix} = \frac{1}{3Z}\begin{bmatrix} 1 & 1 & 1 \\ 1 & 1 & 1 \\ 1 & 1 & 1 \end{bmatrix}$$

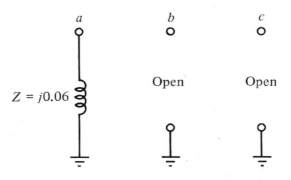

FIGURE E4.2.2

The inverse of the fault admittance matrix does not exist—a problem that may be circumvented by substituting $\mathbf{I}_k^{012}(F) = \mathbf{Y}_{1\phi}^{012}\mathbf{E}_k^{012}(F)$. The current vector $\mathbf{I}_F^{012}$ is zero everywhere except for the faulted bus, and submatrix k of Equation 4.6 is

$$\mathbf{E}_k^{012}(F) = \mathbf{E}_k^{012}(0) - \mathbf{Z}_{kk}^{012}\mathbf{Y}_{1\phi}^{012}\mathbf{E}_k^{012}(F)$$

which is solved for the fault voltage

$$\mathbf{E}_k^{012}(F) = (\mathbf{u} + \mathbf{Z}_{kk}^{012}\mathbf{Y}_{1\phi}^{012})^{-1}\mathbf{E}_k^{012}(0)$$

where equation $\mathbf{u}$ is the identity matrix. Premultiplying by $\mathbf{Y}_{1\phi}^{012}$ as per Equation 4.9 and substituting numerical values into this expression for the single-phase faulted transformer, we have

$$\mathbf{I}_{1\phi}^{012} = \frac{1}{j0.18}\begin{bmatrix} 1 & 1 & 1 \\ 1 & 1 & 1 \\ 1 & 1 & 1 \end{bmatrix}\left\{\begin{bmatrix} 1 & 0 & 0 \\ 0 & 1 & 0 \\ 0 & 0 & 1 \end{bmatrix} + \frac{1}{0.18}\begin{bmatrix} 0.34 & 0 & 0 \\ 0 & 0.1 & 0 \\ 0 & 0 & 0.1 \end{bmatrix}\begin{bmatrix} 1 & 1 & 1 \\ 1 & 1 & 1 \\ 1 & 1 & 1 \end{bmatrix}\right\}^{-1}\mathbf{E}_{gen}^{012}(0)$$

The inverse has a simple algebraic form because $Z_1 = Z_2$ for the generator. Using Z_0 and Z_1 symbols for the generator and Z for the fault impedance, the preceding equation can be algebraically manipulated to obtain

$$
\mathbf{I}_{1\phi}^{012} = \frac{1}{j0.18}
\begin{bmatrix} 1 & 1 & 1 \\ 1 & 1 & 1 \\ 1 & 1 & 1 \end{bmatrix}
\begin{bmatrix}
1+\dfrac{Z_0}{3Z} & \dfrac{Z_0}{3Z} & \dfrac{Z_0}{3Z} \\[2mm]
\dfrac{Z_1}{3Z} & 1+\dfrac{Z_1}{3Z} & \dfrac{Z_1}{3Z} \\[2mm]
\dfrac{Z_1}{3Z} & \dfrac{Z_1}{3Z} & 1+\dfrac{Z_1}{3Z}
\end{bmatrix}^{-1}
\mathbf{E}_{\text{gen}}^{012}(0)
$$

$$
= \frac{1}{3Z}
\begin{bmatrix} 1 & 1 & 1 \\ 1 & 1 & 1 \\ 1 & 1 & 1 \end{bmatrix}
\frac{1}{3Z+Z_0+2Z_1}
\begin{bmatrix}
3Z+2Z_1 & -Z_0 & -Z_0 \\
-Z_1 & 3Z+Z_0+Z_1 & -Z_1 \\
-Z_1 & -Z_1 & 3Z+Z_0+Z_1
\end{bmatrix}
\mathbf{E}_{\text{gen}}^{012}(0)
$$

$$
= \frac{1}{3Z+Z_0+2Z_1}
\begin{bmatrix} 1 & 1 & 1 \\ 1 & 1 & 1 \\ 1 & 1 & 1 \end{bmatrix}
\mathbf{E}_{\text{gen}}^{012} = \frac{E^0+E^1+E^2}{3Z+Z_0+2Z_1}
\begin{bmatrix} 1 \\ 1 \\ 1 \end{bmatrix}
= -jE^1 \begin{bmatrix} 1.39 \\ 1.39 \\ 1.39 \end{bmatrix} \quad \text{p.u.}
$$

The system was initially in balanced operation, so only $E^1(0)$ is nonzero. The line currents are $I^a = -j\,E^1\,(1.39)\,I_{\text{BASE}}/\sqrt{3} = -j\,4{,}373$, $I^b = 0$, $I^c = 0$.

4.1.1 Approximations Common to Short-Circuit Studies

There are several approximations that may be made to the power system to simplify the fault calculations. These are as follows:

1. Equivalent impedances of loads on the power system are much greater than the transmission line impedances. Therefore, bus-to-ground connections are open circuited at the load buses.

2. All transformers are at nominal voltage so that $\alpha = 1.0$ (off-nominal tap ratio) and all shunt terms, bus to ground, are open circuited in the π-equivalent circuit for the transformers. Tap ratios are usually near unity, so this approximation influences short-circuit currents to a very small extent.

3. The voltage at all buses before the fault is 1.0 p.u., *without any phase-angle difference* between buses. Usually, the bus voltages are 0.95 to 1.05 with less than 15° phase difference, so this is a good approximation.

4. The generators are represented by constant voltages behind transient reactances. Most circuit breakers on the power system such as shown in Figure 4.5 operate after a few line frequency cycles, when the subtransient response of the synchronous generators has decayed and the transient reactance is limiting the current. Magnetic flux linkages in the machine change slowly, thus maintaining the internal voltage behind the reactance.

Usually, power flow injections into a network are calculated using the output of generators as in Figure 4.6a, which must be converted to an equivalent current injection as shown in Figure 4.6b.

For fault calculations, all generators are often treated as cylindrical rotor machines in order to use the same circuit for positive- and negative-sequence calculations. The transient voltage behind the reactance, $X'_d = X'_1$, is

$$
E'_q = E_t + jX'_d I_t \qquad \text{p.u.} \tag{4.12}
$$

(a)

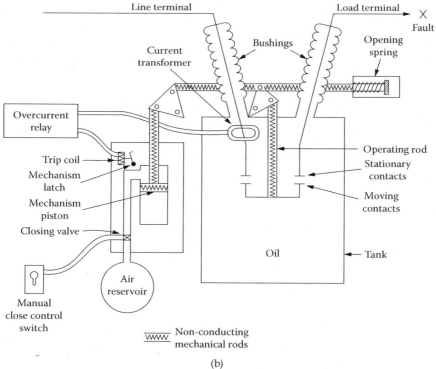

(b)

FIGURE 4.5
(a) Separate phase circuit breakers with SF_6 gas insulation, rated for 345 kV, 4 kA continuous duty, 50 kA interruption capability. (Photograph used with permission of, and copyrighted by, Mitsubishi Electric and its affiliates) (b) Schematic for one pole of an oil circuit breaker employing stored spring energy to interrupt fault currents. (Spring forces operating rod downward to open the circuit.)

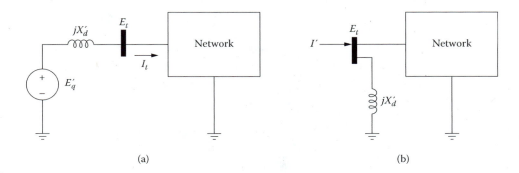

jX_d'
E_t
Network
I_t
E_q'
(a)
E_t
I'
Network
jX_d'
(b)

FIGURE 4.6
Including generator reactance: (a) generator injection; (b) equivalent.

where I_t and E_t are known from power flow calculations. The equivalent current injection is then

$$I' = I_t + \frac{E_t}{jX_d'} \qquad \text{p.u.} \tag{4.13}$$

This equivalence of generator injections plus approximations 1 through 3 stated above imply that only generators contribute to fault currents. Generator transient reactances plus transmission line impedances limit the current in a fault. Synchronous condensors are treated similarly to generators. Large motor loads, which are on the same order of magnitude as the generators and operate at a low power factor, contribute to fault currents because of their inductance. Induction motors have magnetic flux changes that last for more than a few cycles, so in the transient time frame they should be replaced by an equivalent impedance

$$Z_{eq} = \frac{|V_k|^2}{(P_k + jQ_k)^*} \tag{4.14}$$

where V_k, P_k, and Q_k are, respectively, the steady-state terminal voltage, and real and reactive power into the motor at bus k.

A suitable model for induction motors during the subtransient time frame—the first few cycles—is the locked rotor impedance in series with an internal source, E_i, as shown in Figure 4.7. The following example demonstrates approximations 1 through 4.

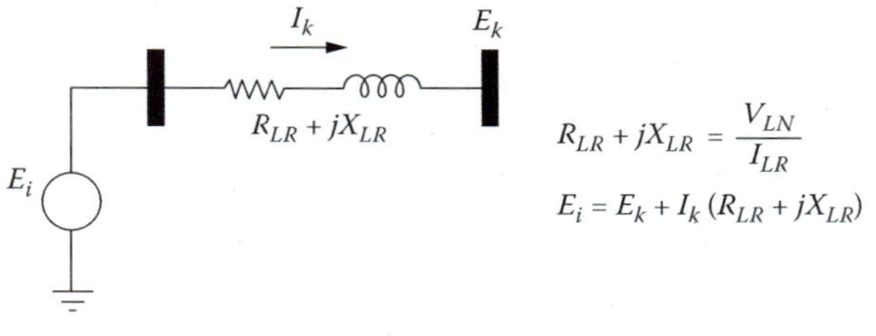

I_k
E_k
$R_{LR} + jX_{LR}$
E_i
$R_{LR} + jX_{LR} = \frac{V_{LN}}{I_{LR}}$
$E_i = E_k + I_k (R_{LR} + jX_{LR})$

FIGURE 4.7
Subtransient time frame equivalent for an induction motor. V_{LN} is the rated line-to-neutral voltage of the motor and I_{LR} is the locked rotor line current. The motor steady-state current I_k is determined from the steady-state operating conditions, $P_k + jQ_k$ and E_k. Often the resistance R_{LR} is neglected.

Example 4.3

For the sample power system shown in Figure E4.3.1, the loads at buses 2 and 4 are removed and the transformer tap ratios are $\alpha = 1.0$. All voltages are 1.0 p.u. before the fault occurs. All impedances and voltages are in per unit on a 100 MVA, 100 kV base. The data are:

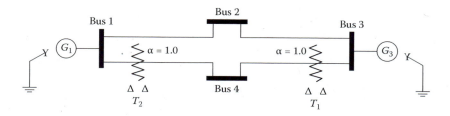

FIGURE E4.3.1

Lines:

$$1\text{--}2: X_1 = X_2 = 0.1 \qquad X_0 = 0.3$$

$$2\text{--}3: X_1 = X_2 = 0.2 \qquad X_0 = 0.4$$

Transformer T1:

$$X_1 = X_2 = 0.3$$

Transformer T2:

$$X_1 = X_2 = 0.4$$

Generators:

$$1: X_1 = X_2 = 0.1 \qquad X_0 = 0.04$$

$$3: X_1 = X_2 = 0.19 \qquad X_0 = 0.06$$

(a) Derive the zero-sequence and positive-sequence bus impedance matrices.
(b) Find the fault current for a three-phase-to-ground fault at bus 2. Also find the other bus voltages during the fault.

Solution

(a) A directed graph for the positive- and negative-sequence network, indicating tree branches and cotree links, is shown in Figure E4.3.2. It is easily seen that at bus 1 current injection should be $I_1' = 1.0/j0.1 = -j10$ and $I_3' = 1.0/j0.19 = -j5.26$ in order to establish the 1.0 p.u. voltage in the network. Rated voltage is $V^a = 100 \text{ kV}/\sqrt{3}$ line to neutral, and the base impedance is $X_{BASE} = 100 \ \Omega$.

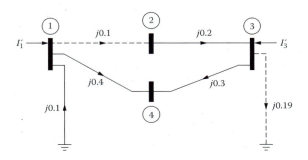

FIGURE E4.3.2

Using the Chapter 3 algorithm for construction of the bus impedance matrix, we obtain

$$\mathbf{Z}_{BUS}^{12} = j \begin{bmatrix} 0.080 & 0.066 & 0.038 & 0.056 \\ & 0.132 & 0.054 & 0.065 \\ \text{symmetry} & \swarrow & 0.118 & 0.084 \\ & & & 0.243 \end{bmatrix}$$

A directed graph for the zero-sequence network is shown in Figure E4.3.3. The Δ connected transformers are open for unbalanced currents so for buses 1,2,3 the corresponding zero-sequence bus impedance matrix is

$$\mathbf{Z}_{BUS}^{0} = j \begin{bmatrix} 0.038 & 0.023 & 0.0030 \\ \swarrow & 0.1955 & 0.0255 \\ \text{symmetry} & & 0.0555 \end{bmatrix}$$

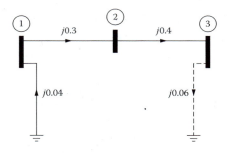

FIGURE E4.3.3

(b) The fault current at bus 2 is given by

$$\mathbf{I}_0^{012}(F) = (\mathbf{Z}_F^{012} + \mathbf{Z}_{22}^{012})^{-1} \mathbf{E}_2^{012}(0)$$

$$= \left\{ \begin{bmatrix} 0 & 0 & 0 \\ 0 & 0 & 0 \\ 0 & 0 & 0 \end{bmatrix} + j \begin{bmatrix} 0.1955 & 0 & 0 \\ 0 & 0.132 & 0 \\ 0 & 0 & 0.132 \end{bmatrix} \right\}^{-1} \begin{bmatrix} 0 \\ E^1 \\ 0 \end{bmatrix}$$

$$= -j \begin{bmatrix} 5.12 & 0 & 0 \\ 0 & 7.58 & 0 \\ 0 & 0 & 7.58 \end{bmatrix} \begin{bmatrix} 0 \\ E^1 \\ 0 \end{bmatrix}$$

Voltage before the fault is $\mathbf{E}_2 = 1.0$ p.u. so the fault current is 7.58 p.u. or 7.58 kA/$\sqrt{3}$ in terms of positive-sequence current. The voltages at bus $i = 1, 3, 4$ are given by

$$\mathbf{E}_i^{012}(F) = \mathbf{E}_i^{012}(0) - \mathbf{Z}_{i2}^{012}\left(\mathbf{Z}_F^{012} + \mathbf{Z}_{22}^{012}\right)^{-1}\mathbf{E}_2^{012}(0) = \mathbf{E}_i^{012} - \mathbf{Z}_{i2}^{012}\mathbf{I}_2^{012}(F)$$

At bus 1, the voltage using entries from $\mathbf{Z}_{\mathrm{BUS}}^0$ and $\mathbf{Z}_{\mathrm{BUS}}^1$ to express $\mathbf{Z}_{i2}^{012}$ is as follows:

$$\mathbf{E}_1^{012}(F) = \begin{bmatrix} 0 \\ E_1 \\ 0 \end{bmatrix} - j\begin{bmatrix} 0.023 & 0 & 0 \\ 0 & 0.066 & 0 \\ 0 & 0 & 0.066 \end{bmatrix}\mathbf{I}_2^{012}(F)$$

$$= \begin{bmatrix} 0 \\ E^1 \\ 0 \end{bmatrix}\left(1.0 - \frac{0.066}{0.132}\right) = 0.5\begin{bmatrix} 0 \\ E^1 \\ 0 \end{bmatrix} \quad \text{p.u.}$$

At bus 3, the voltage is

$$\mathbf{E}_3^{012}(F) = \begin{bmatrix} 0 \\ E_1 \\ 0 \end{bmatrix} - j\begin{bmatrix} 0.0255 & 0 & 0 \\ 0 & 0.054 & 0 \\ 0 & 0 & 0.054 \end{bmatrix}\mathbf{I}_2^{012}(F)$$

$$= \begin{bmatrix} 0 \\ E^1 \\ 0 \end{bmatrix}\left(1 - \frac{0.054}{0.132}\right) = 0.591\begin{bmatrix} 0 \\ E^1 \\ 0 \end{bmatrix} \quad \text{p.u.}$$

At bus 4, the zero sequence does not exist and the positive sequence voltage is

$$\mathbf{E}_4^{012}(F) = \begin{bmatrix} 0 \\ E^1 \\ 0 \end{bmatrix}\left(1 - \frac{0.065}{0.132}\right) = 0.508\begin{bmatrix} 0 \\ E^1 \\ 0 \end{bmatrix} \quad \text{p.u.}$$

It is seen that only the positive-sequence impedances between all buses and the faulted bus are needed to calculate balanced fault effects.

4.2 Fault Calculations Using the $\mathbf{Y}_{\mathrm{BUS}}$ Table of Factors

In Section 4.1 we used the property that in a balanced network the symmetrical component impedances decouple into the positive-, negative-, and zero-sequence values without mutual impedances. As a result, the bus impedance matrix $\mathbf{Z}_{\mathrm{BUS}}$, as well as the LDU table of factors for $\mathbf{Y}_{\mathrm{BUS}}$, can be obtained for a network considering sequence elements separately. From the $\mathbf{Y}_{\mathrm{BUS}}$ table of factors and the fault impedance characteristics, the voltages and currents are determined for the fault condition. Example 4.4 demonstrates this method.

Example 4.4

Calculate the three-phase short-circuit-to-ground fault current at bus 3 in the network shown in Figure E4.4, using the LDU table of factors. The network is initially at 1 p.u. voltage. The positive-sequence impedances are specified as follows:

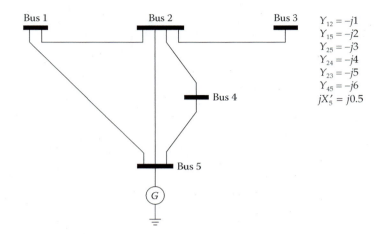

FIGURE E4.4

Solution

The reordered bus numbering to take advantage of the zeros in the $\mathbf{Y}_{\text{BUS}}$ matrix is 3, 1, 4, 5, 2. The corresponding table of factors calculated in Example 3.8 is

$$[\text{L D U}] = j \begin{bmatrix} \frac{1}{5} & 0 & 0 & 0 & -1 \\ & \frac{1}{3} & 0 & -\frac{2}{3} & -\frac{1}{3} \\ & & \frac{1}{10} & -\frac{3}{5} & -\frac{2}{5} \\ & \mathbf{L} & & \frac{15}{121} & -\frac{91}{121} \\ & & & & \frac{121}{182} \end{bmatrix}$$

Since bus 3 corresponds to the first row of the reordered system, the driving-point and mutual impedances $\overline{Z}_{i3}$ may be calculated using the unit current injection vector

$$\mathbf{b} = [1 \quad 0 \quad 0 \quad 0 \quad 0]^t$$

The impedances are calculated as

$$\begin{bmatrix} z_{33}^1 \\ z_{31}^1 \\ z_{34}^1 \\ z_{35}^1 \\ z_{32}^1 \end{bmatrix} = U_1 U_2 U_3 U_4 D U_4^t U_3^t U_2^t U_1^t \begin{bmatrix} 1 \\ 0 \\ 0 \\ 0 \\ 0 \end{bmatrix} = j \begin{bmatrix} 0.8648 \\ 0.5549 \\ 0.5659 \\ 0.5000 \\ 0.6648 \end{bmatrix}$$

Using Table 4.1, the short-circuit current at bus 3 is

$$I_3^1 = \frac{E_3^1(0)}{z_{33}^1} = \frac{E_3^1(0)}{j0.8648} = -j1.156E_3^1(0) \qquad \text{p.u.}$$

and the phase currents are

$$\left|I_3^a\right| = \left|I_3^b\right| = \left|I_3^c\right| = 1.156 I_{\text{BASE}}/\sqrt{3} \qquad \text{A}$$

Bus voltages are also found using Table 4.1.

$$E_i^1 = E_i^1(0) - \frac{z_{i3}E_3^1(0)}{z_{33}^1 + z_F^1} = E_i^1(0) - \frac{z_{i3}E_3^1}{z_{33}^1}$$

If all voltages are 1 p.u. before the fault, then bus 2, for example, is

$$E_2(F) = 1.0\left(1 - \frac{0.6648}{0.8648}\right) = 0.231 \qquad \text{p.u.}$$

If a fault is unbalanced, the zero-sequence bus impedance matrix must also be calculated. In general, the topology of the zero-sequence network is different from the positive-sequence because of the delta-connected transformers or floating wye-connected elements. Hence, a different optimal bus-ordering scheme is advantageous for zero-sequence elements. For unbalanced faults, the admittance form may be used if the impedance matrix does not exist.

A convenient way to summarize three-phase results of a short-circuit study is in terms of a *short-circuit level*. The short-circuit level at bus k is defined as:

$$\boxed{\begin{array}{c} \text{Three-Phase Short-Circuit Level} \\[6pt] (\text{MVA})_{\text{s.c.}} = \dfrac{V_{\text{line-to-line}}^2}{x_{kk}} = \dfrac{V_{\text{BASE}}^2}{x_{kk}} \end{array}} \qquad (4.15)$$

where the line-to-line voltage is expressed in megavolts and x_{kk} is the positive-sequence driving-point reactance at bus k expressed in ohms. Bus k is assumed to be electrically remote from generators, so the positive- and negative-sequence reactances are equal. The resistive component of z_{kk} is considered negligible (often $\mathbf{Z}_{\text{BUS}}^{012}$ disregards resistances). The three-phase short-circuit at bus k is given by

$$I_{\text{s.c.}} = \frac{V_{\text{BASE}}}{\sqrt{3}\ x_{kk}} = \frac{(\text{MVA})_{\text{s.c.}}}{\sqrt{3}\ V_{\text{BASE}}} \qquad (4.16)$$

where an X_F fault reactance may be added to x_{kk}.

The short-circuit level often specifies the range of values expected for different network operating conditions—lines or generators in and out of operation—and anticipated expansion of the power system in future years. As new equipment is added to a bus of the network, the current interrupting capacity of circuit breakers or switchgear must surpass the short-circuit current level in order to clear a fault occurring in the equipment. A typical summary is shown in Table 4.2 for connection points to a 132 kV transmission network.

The voltages at other buses on the system during faults or unbalanced fault currents cannot be obtained using short-circuit-level data. Therefore, the application of this type of data is limited to determining circuit breaker ratings, network voltage regulation characteristics, and simple source equivalents for harmonic current injections.

TABLE 4.2

Typical Short-Circuit Levels for Several Buses on a Power System

132 kV Busbar Name	Short-Circuit Level (MVA)			
	Minimum	Maximum	Normal	Future
Grantleigh	430	550	490	1,700
Dingo	290	480	430	1,700
Blackwater/Rangal	280	730	600	1,700
Gregory	480	1,440	580	1,700
Yukan	280	540	420	1,700
Cappabella	420	590	450	1,700
Moranbah	620	1,060	750	1,700
Dysart	400	870	650	1,700
Mt. McLaren	310	390	340	1,700

4.3   Contingency Analysis for Power Systems

The *state* of a power system is defined to be the voltage and its phase angle at every bus in the power system. Voltages are usually measured by step-down potential transformers on the a, b, and c lines and the average of the three is used as the magnitude of the positive sequence voltage. Until recently, phase angles of voltages are always calculated quantities from solving the power flow problem (Chapter 5) or a state estimator (Chapter 7). The advent of a synchronized, precise time reference (within 100 Nanoseconds at points separated by hundreds of miles) from earth-orbiting satellites in the Global Positioning System (GPS) has made it possible to directly measure voltage phase angles. GPS timing is restricted to a very few installations, such that power flow and state estimation remain the only viable method to determine voltage phase angles.

The voltage phase difference which determines power flow between adjacent buses, is very small, and an angle difference on the order of 0.1 degree can force large power flows on the transmission line. Communication links between the central computers of the power system and the Remote Terminal Units (RTUs) in the field can synchronize a timing reference pulse within 16 milliseconds. This is not adequate because in order to obtain a resolution of 1 electrical degree, a 60-Hz period of $\frac{1}{60}$ sec must be subdivided by

$$\frac{1}{60} \times \frac{1}{360} = 46.2 \quad \mu\text{sec}$$

Therefore the state of the power system must be a power flow computation, a state estimator, or a linearized network solved by applying Kirchhoff's Laws. After the application of one of these three methods, the state and the network parameters are known, so the linearized form of the network bus equations (positive sequence only) is

$$\mathbf{E}^1_{\text{BUS}} = \mathbf{Z}^1_{\text{BUS}} \, \mathbf{I}^1_{\text{BUS}} \tag{4.17}$$

$$\mathbf{I}^1_{\text{BUS}} = \mathbf{Y}^1_{\text{BUS}} \, \mathbf{E}^1_{\text{BUS}} \tag{4.18}$$

There is little experimental verification of calculated phase angles calculations with data from the physical network, although voltage magnitudes are usually in close agreement. With linear equivalents to replace the loads on the power system, Equation 4.17 and

TABLE 4.3

Periodically Executed Programs for an On-Line Digital Control Computer

2 Sec	4 Sec	30 Sec	3 Min	5 Min
System scan, topology check, data collection	Automatic generation control	State estimation, energy logging	Load flow, contingency calculation	Economic dispatch

Equation 4.18 are used for *contingency analysis*, a periodically executed calculation in the central digital computers that monitor and control power systems. The central computers are typically connected in the redundant configuration of Figure 1.5, called a SCADA (supervisory control and data acquisition) system to process network information. The software programs within the computers are usually executed periodically, as outlined in Table 4.3.

The purpose of both the state estimator and power flow programs is to accept periodic real-time data, essentially tracking the operating point of the system, and provide a calculated state for contingency cases. Basic definitions for contingencies are given below.

4.3.1 Contingency Analysis for Power Systems

Purpose: To alert the operator or dispatcher of the power system as to each vulnerable condition of the system on a real-time basis as load and generation change.

Method: Calculate the loss of critical transmission line, generator, or load increases taken one at a time (or possibly in combinations) in order to determine the loading on other system elements. Notice that there may be tens of generators or hundreds of transmission and tie lines to be considered. Linear Equation 4.17 and Equation 4.18 are employed.

Alarms: Determine bus voltages that are excessively high (above 1.05) or low (below 0.95). Find power flows that are above short-term and long-term overload limits for generators and transmission lines.

Remedial action: Construct tables or sets of instructions for the dispatcher to take corrective action *if* the contingency case should actually occur. Automatic corrective actions such as load-shedding tables can be constructed by the program and executed on command by the dispatcher.

A contingency case represents a hypothetical situation. The situation may describe the loss of a large generator or a heavily loaded tie line to neighbors, and is based on a dispatcher or system engineer's knowledge of the critical elements in the system. Extremely accurate results are not necessary since it is only a possible situation, so linear approximations are adequate. The system must be adequately described, and an equivalent should be used for neighbors connected through tie lines. Only balanced or positive-sequence changes are considered.

4.3.2 Contingencies Using $\mathbf{Z}_{BUS}$ in a Superposition Method

The bus impedance matrix $\mathbf{Z}_{BUS}$ is used with linearized loads replaced by current injections. If a generator or load changes at bus k by an amount ΔI_k, voltages after the change are

$$\mathbf{E}_{BUS}(F) = \mathbf{E}_{BUS}(0) + \mathbf{Z}_{BUS}\Delta\mathbf{I} = \mathbf{E}_{BUS}(0) + \begin{bmatrix} z_{1k} \\ z_{2k} \\ \vdots \\ z_{mk} \end{bmatrix}\Delta I_k \tag{4.19}$$

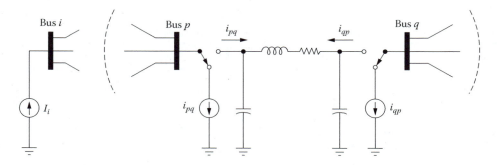

FIGURE 4.8
Loss of a transmission line as replaced by current source injections at buses.

If a $\Delta \mathbf{I}_k$ change is determined by an external impedance, this method is very similar to fault calculations at a bus. If a line from bus p to bus q was to open-circuit, a method [2] to estimate the system voltages is as follows.

4.3.3 $\mathbf{Z}_{BUS}$ Line Contingency Method

1. Initial voltage magnitudes and phase angles are specified at all buses of the system.
2. Replace all loads and generators by constant-current sources. The line from bus p to bus q is replaced by current sources i_{pq} and i_{qp}, which are inputs to the line π-equivalent (Figure 4.8). Thus, all elements to ground are removed locally.
3. Calculate $\mathbf{Z}'_{BUS}$ with the line removed by adding $-Z_{pq}$ as a link by means of the algorithms of Chapter 3.
4. Calculate the effect of reversing the current flow in line pq by injections at buses p and q.

$$\Delta \mathbf{E}_{BUS} = \begin{bmatrix} \Delta E_1 \\ \Delta E_2 \\ \vdots \\ \cdot \\ \cdot \\ \cdot \\ \cdot \\ \Delta E_m \end{bmatrix} = \mathbf{Z}'_{BUS}\, \Delta \mathbf{I} = \mathbf{Z}'_{BUS} \begin{bmatrix} 0 \\ 0 \\ \vdots \\ -i_{pq} \\ 0 \\ -i_{qp} \\ 0 \\ 0 \end{bmatrix} \tag{4.20}$$

Notice that only two columns of $\mathbf{Z}'_{BUS}$ are necessary for this step. All bus voltages are affected by the contingency.

5. Determine the new bus voltages and element currents after the fault.

$$\mathbf{E}_{BUS}(F) = \mathbf{E}_{BUS}(0) - \Delta \mathbf{E}_{BUS} \tag{4.21}$$

$$\mathbf{I} = \mathbf{Y}_{PRIM}\mathbf{V} = \mathbf{Y}_{PRIM}\mathbf{M}^t\mathbf{E}_{BUS}(F) \tag{4.22}$$

In Equations 4.20 to Equation 4.22 **I** is the vector of b element currents, $\mathbf{Y}_{PRIM}$ a $b \times b$ admittance matrix for the elements, and $\mathbf{M}^t$ the transpose of the reduced bus-branch incidence matrix (see Section 3.3). Transmission line currents are combinations of branch currents. Individual bus voltages in Equation 4.21 are

$$E_i(F) = E_i(0) + \Delta E_i = E_i(0) - z'_{ip} i_{pq} - z'_{iq} i_{qp} \tag{4.23}$$

where the impedances are from columns of $\mathbf{Z}'_{BUS}$ (positive sequence). By means of this method, the changes in voltages at all buses and resulting line flows are calculated. Comparing the flows or voltages with high or low limits determines the alarms that must be corrected by operator action. Often, the remedial action is complicated, such as shifting real and reactive power out of all generators, or switching transmission line compensation, so another program must be called to determine the remedial action.

4.4 Using the $\mathbf{Y}_{BUS}$ Table of Factors for Contingencies

The similarity between fault calculations and line contingency studies has been shown. For $\mathbf{Y}_{BUS}$ and the accompanying table of factors there is no need to retriangularize due to a line modification [3] because the network has been linearized and superposition of solutions holds. The following simple example demonstrates the principles.

Example 4.5

In the network shown in Figure E4.5.1 a unit current is injected into bus 2 to establish the initial voltage distribution. Calculate the effect of an open circuit of the line from bus 1 to bus 2.

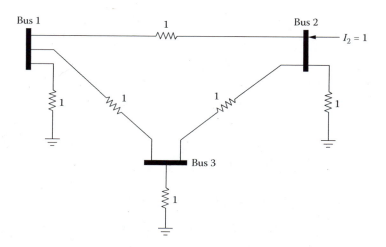

FIGURE E4.5.1

Solution

The bus admittance equations for the network are

$$\mathbf{Y}_{BUS}\mathbf{E}_{BUS} = \begin{bmatrix} 3 & -1 & -1 \\ -1 & 3 & -1 \\ -1 & -1 & 3 \end{bmatrix} \begin{bmatrix} E_1 \\ E_2 \\ E_3 \end{bmatrix} = \mathbf{I}_{BUS} = \begin{bmatrix} 0 \\ 1 \\ 0 \end{bmatrix}$$

The table of factors for forward and backward operations on the right-hand-side vector is

$$[\bullet D\, U] = \begin{bmatrix} \frac{1}{3} & -\frac{1}{3} & -\frac{1}{3} \\ & \frac{3}{8} & -\frac{1}{2} \\ & \bullet & \frac{1}{2} \end{bmatrix}$$

Calculate the bus voltages using the injection current vector:

$$\begin{bmatrix} E_1 \\ E_2 \\ E_3 \end{bmatrix} = U_1 U_2 D U_2^t U_1^t \mathbf{I}_{BUS} = \begin{bmatrix} \frac{1}{4} \\ \frac{1}{2} \\ \frac{1}{4} \end{bmatrix}$$

$$= \begin{bmatrix} 1 & \frac{1}{3} & \frac{1}{3} \\ 0 & 1 & 0 \\ 0 & 0 & 1 \end{bmatrix} \begin{bmatrix} 1 & 0 & 0 \\ 0 & 1 & \frac{1}{2} \\ 0 & 0 & 1 \end{bmatrix} \begin{bmatrix} \frac{1}{3} & 0 & 0 \\ 0 & \frac{3}{8} & 0 \\ 0 & 0 & \frac{1}{2} \end{bmatrix} \begin{bmatrix} 1 & 0 & 0 \\ 0 & 1 & 0 \\ 0 & \frac{1}{2} & 1 \end{bmatrix} \begin{bmatrix} 1 & 0 & 0 \\ \frac{1}{3} & 1 & 0 \\ \frac{1}{3} & 0 & 1 \end{bmatrix} \mathbf{I}_{BUS}$$

Between buses 1 and 2 there is a linear or Thévenin equivalent circuit for the total circuit. The open-circuit voltage (bus 1 positive) between these buses is

$$E_T = E_1(0) - E_2(0) = \tfrac{1}{4} - \tfrac{1}{2} = -\tfrac{1}{4}$$

The Thévenin impedance between these two buses may be found by injecting a unit current into bus 1 and extracting it from bus 2 to obtain the impedance vector

$$\overline{Z}' = \begin{bmatrix} Z_1' \\ Z_2' \\ Z_3' \end{bmatrix} = U_1 U_2 D U_2^t U_1^t \begin{bmatrix} 1 \\ -1 \\ 0 \end{bmatrix} = \begin{bmatrix} \frac{1}{4} \\ -\frac{1}{4} \\ 0 \end{bmatrix}$$

and computing the Thévenin impedance as

$$Z_T = Z_1' - Z_2' = \frac{1}{4} - \left(-\frac{1}{4} \right) = \tfrac{1}{2}\,\Omega$$

The equivalent circuit is shown in Figure E4.5.2. Since the circuit is linear, an impedance Z_F may be attached from bus 1 to bus 2 and the resulting currents and voltages calculated. If $Z_F = -1$, this removes the transmission line. To carry Z_F symbolically until the last step, the terminal voltage in the circuit of Figure E4.5.2 is

$$E_{12}' = -Z_F I'(F) = E_T + Z_T I'(F)$$

$$I'(F) = \frac{-E_T}{Z_T + Z_F}$$

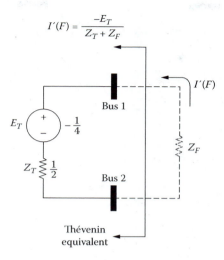

FIGURE E4.5.2

This is solved for the current

$$I'(F) = \frac{-E_T}{Z_T + Z_F}$$

The voltage on the contingency or faulted network is found by superposition of the initial voltages and voltages proportional to the fault current times the impedance vector:

$$\mathbf{E}(F) = \mathbf{E}(0) + I'(F)\bar{\mathbf{Z}}'$$

$$= \begin{bmatrix} \frac{1}{4} \\ \frac{1}{2} \\ \frac{1}{4} \end{bmatrix} + \frac{-E_T}{Z_T + Z_F} \begin{bmatrix} \frac{1}{4} \\ -\frac{1}{4} \\ 0 \end{bmatrix}$$

If the line from bus 1 to bus 2 is taken out of service, $Z_F = -1$ and the system voltage is

$$\mathbf{E}(F) = \begin{bmatrix} \frac{1}{4} \\ \frac{1}{2} \\ \frac{1}{4} \end{bmatrix} + \frac{1/4}{1/2 - 1} \begin{bmatrix} \frac{1}{4} \\ -\frac{1}{4} \\ 0 \end{bmatrix} = \begin{bmatrix} \frac{1}{8} \\ \frac{5}{8} \\ \frac{1}{4} \end{bmatrix}$$

The voltage rises on bus 2 and decreases on bus 1 because the open-circuit line increases the impedance between the injection current and load at bus 1.

The example shows that only *one* application of the table of factors, and not a retriangularization, is necessary for a contingency calculation. Example 4.5 treated a simple network but introduced several fundamental concepts. The first of these is the Thévenin equivalent impedance between buses p and q. This impedance is the difference between injecting a

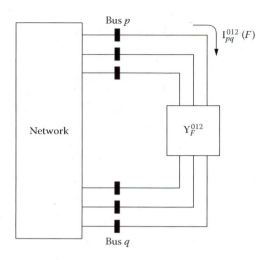

FIGURE 4.9
Fault connected between buses p and q.

unit current into bus p and extracting it from bus q. The mutual impedances are implicitly taken into account by the subtraction. Performing this calculation one step at a time for the network, the Thévenin impedance between buses p and q is

$$\mathbf{Z}_T^{012} = \mathbf{Z}_{pp}^{012} - 2\mathbf{Z}_{pq}^{012} + \mathbf{Z}_{qq}^{012} \text{ (buses } p \text{ and } q) \tag{4.24}$$

which includes all symmetrical component impedances (see denominator of Equation 3.17).

When an external admittance $\mathbf{Y}_F^{012}$ is connected between buses p and q as shown in Figure 4.9, the current

$$\mathbf{I}_{pq}^{012}(F) = \mathbf{Y}_F^{012}\mathbf{E}_{pq}^{012}(F) = \mathbf{Y}_F^{012}\left[\mathbf{E}_p^{012}(F) - \mathbf{E}_q^{012}(F) \right] \tag{4.25}$$

flows from bus p to bus q. The initial or open-circuit voltages $\mathbf{E}_p^{012}(0)$ and $\mathbf{E}_q^{012}(0)$ are known values, so the Thévenin impedance and fault admittance determine the final voltage:

$$\mathbf{E}_{pq}^{012}(F) = \mathbf{E}_p^{012}(F) - \mathbf{E}_q^{012}(F) = \left(\mathbf{u} + \mathbf{Z}_T^{012}\mathbf{Y}_F^{012}\right)^{-1}\mathbf{E}_{pq}(0) \tag{4.26}$$

where $\mathbf{u}$ is the identity matrix. Substituting Equation 4.26 into 4.25, the fault current is

$$\mathbf{I}_{pq}^{012}(F) = \mathbf{Y}_F^{012}(\mathbf{u} + \mathbf{Z}_T^{012}\mathbf{Y}_F^{012})^{-1}\mathbf{E}_{pq}(0)$$
$$= -\mathbf{I}_p^{012}(F) = \mathbf{I}_q^{012}(F) \tag{4.27}$$

which is the same as a simultaneous minus current injection at bus p and a positive injection at bus q. With the fault current references as defined previously, the voltage at any bus *i* is given by

$$
\boxed{
\begin{array}{c}
\text{Voltage at Bus } i \text{ during Fault } \mathbf{Y}_F^{012} \text{ from Bus } p \text{ to Bus } q \\[4pt]
\mathbf{E}_i^{012}(F) = \mathbf{E}_i^{012}(0) - (\mathbf{Z}_{ip}^{012} - \mathbf{Z}_{iq}^{012})\mathbf{Y}_F^{012}(\mathbf{u} + \mathbf{Z}_T^{012}\mathbf{Y}_F^{012})^{-1}[\mathbf{E}_p^{012}(0) - \mathbf{E}_q^{012}(0)] \\[8pt]
= \mathbf{E}_i^{012}(0) - (\mathbf{Z}_{ip}^{012} - \mathbf{Z}_{iq}^{012})\mathbf{I}_{pq}^{012}(F)
\end{array}
}
\tag{4.28}
$$

Equation 4.28 is clearly a generalization of the method used in Example 4.5. The final voltages are calculated from initial voltages and a *weighting factor* times the impedance difference vector. The entries in Equation 4.28 may be calculated by means of the building algorithm for $\mathbf{Z}_{BUS}^{012}$ or the LDU table of factors with appropriate current injections. Two cases of Equation 4.25 to Equation 4.28 that are very useful are summarized in Table 4.4. Observe that if q is ground, Table 4.4 reduces to Table 4.1 with $Z_g = 0$.

Observe that the element $\mathbf{Y}_F^{012}$ between buses p and q can remove the series impedance of a π-section transmission line only by using negative elements. The shunt capacitances of the transmission line must be separately treated or neglected. A contingency calculation usually neglects the shunt elements.

The method of fault calculation by superposition generalizes to simultaneous (multiple) contingencies. In practice, energy management contingency programs rarely treat more than two simultaneous contingencies because their probability of occurrence is extremely small. For two simultaneous, balanced three-phase contingencies, the method [3] is summarized below.

4.4.1 Double Contingencies Using $\mathbf{Y}_{BUS}$ Table of Factors (Balanced Case)

1. Initial voltage magnitudes and phase angles are specified at all buses of the system.

2. Let the double-line contingency occur on lines from bus k to m and bus i to j. Compute to open-circuit Thévenin equivalent voltage between these buses.

$$
\mathbf{E}_{Tkm}(0) = \mathbf{E}_k(0) - \mathbf{E}_m(0)
\tag{4.29a}
$$

$$
\mathbf{E}_{Tij}(0) = \mathbf{E}_i(0) - \mathbf{E}_j(0)
\tag{4.29b}
$$

3. Simultaneously inject and remove unit currents first from buses k and m, then buses i and j, to obtain two impedance vectors,

$$
\bar{Z}_{km} \quad \text{and} \quad \bar{Z}_{ij}
$$

4. Using driving-point and mutual impedances from the vectors of step 3, form the 2×2 Thévenin impedance matrix:

$$
\mathbf{Z}_T =
\begin{bmatrix}
z_{km,km} & z_{km,ij} \\
z_{ij,km} & z_{ij,ij}
\end{bmatrix}
\tag{4.30}
$$

where $z_{km,km} = E_k^{km} - E_m^{km}$ (difference of impedances, unit injection in k, and out of m); $z_{km,ij} = E_i^{km} - E_j^{km}$ (mutual due to unit injection at k, out of m); $z_{ij,ij} = E_i^{ij} - E_j^{ij}$ (difference of impedances, unit injection in i, out of j); and $z_{ij,km} = E_k^{ij} - E_m^{ij}$ (mutual due to unit injection at i, out of j).

TABLE 4.4

Bus-to-Bus Faults

Fault Schematic	Fault Impedance Symmetrical Components	Fault Current, Bus p to Bus q $Z_T^{012} = Z_{pp}^{012} - 2Z_{pq}^{012} + Z_{qq}^{012}$	Bus Voltage at Bus i Due to Fault
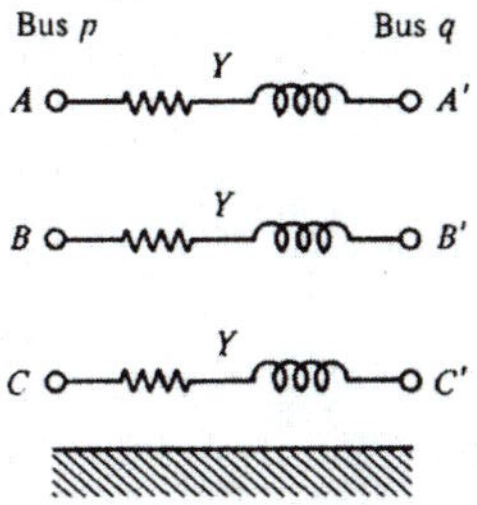 Bus p Bus q A —Y—〇〇〇— A' B —Y—〇〇〇— B' C —Y—〇〇〇— C' ▨	$Y_F^{012} = Y \begin{bmatrix} 1 & 0 & 0 \\ 0 & 1 & 0 \\ 0 & 0 & 1 \end{bmatrix}$	$$I_{pq}^{012}(F) = \begin{bmatrix} \dfrac{Y}{1+Z_T^0 Y} & 0 & 0 \\[2mm] 0 & \dfrac{Y}{1+Z_T^1 Y} & 0 \\[2mm] 0 & 0 & \dfrac{Y}{1+Z_T^2 Y} \end{bmatrix} \begin{bmatrix} E_p^0(0) - E_q^0(0) \\[1mm] E_p^1(0) - E_q^1(0) \\[1mm] E_q^2(0) - E_q^2(0) \end{bmatrix}$$	$E_i^{012}(F) = E_i^{012}(0) - (Z_{ip}^{012} - Z_{iq}^{012})I_{pq}^{012}(F)$
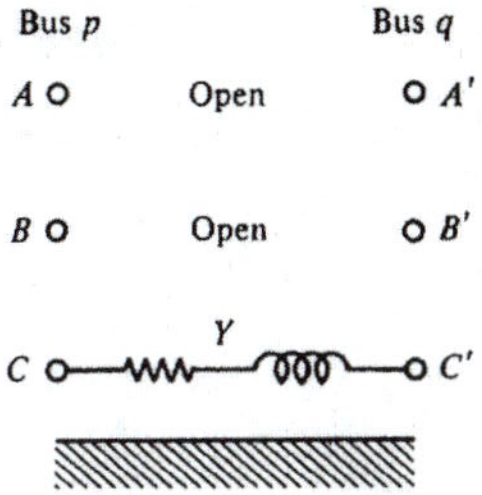 Bus p Bus q A 〇 Open 〇 A' B 〇 Open 〇 B' C —Y—〇〇〇— C' ▨	$Y_F^{012} = \dfrac{Y}{3}\begin{bmatrix} 1 & 1 & 1 \\ 1 & 1 & 1 \\ 1 & 1 & 1 \end{bmatrix}$	$$I_{pq}^{012}(F) = \frac{[E_p^0(0) - E_q^0(0) + E_p^1(0) - E_q^1(0) + E_p^2(0) - E_q^2(0)]}{Z_T^0 + 2Z_T^1 + 3/Y}\begin{bmatrix} 1 \\ 1 \\ 1 \end{bmatrix}$$ $$Z_T^1 = Z_T^2$$	$E_i^{012}(F) = E_i^{012}(0) - (Z_{ip}^{012} - Z_{iq}^{012})I_{pq}^{012}(F)$

5. Connect the external fault impedances,

$$\mathbf{Z}_F = \begin{bmatrix} z_{km}(F) & 0 \\ 0 & z_{ij}(F) \end{bmatrix} \tag{4.31}$$

between the buses and calculate the current that flows:

$$\begin{bmatrix} E_{km}(F) \\ E_{ij}(F) \end{bmatrix} = -\begin{bmatrix} z_{km}(F) & 0 \\ 0 & z_{ij}(F) \end{bmatrix}\begin{bmatrix} I_{km}(F) \\ I_{ij}(F) \end{bmatrix}$$

$$= -\mathbf{Z}_F \begin{bmatrix} I_{km}(F) \\ I_{ij}(F) \end{bmatrix} = \begin{bmatrix} E_{Tkm}(0) \\ E_{Tij}(0) \end{bmatrix} + \mathbf{Z}_T \begin{bmatrix} I_{km}(F) \\ I_{ij}(F) \end{bmatrix} \tag{4.32}$$

Using the impedances and the initial voltages, the fault current is calculated as

$$\begin{bmatrix} I_{km}(F) \\ I_{ij}(F) \end{bmatrix} = -(\mathbf{Z}_T + \mathbf{Z}_F)^{-1}\begin{bmatrix} E_{Tkm}(0) \\ E_{Tij}(0) \end{bmatrix} \tag{4.33}$$

6. Use the fault currents to "weight" the superposition of initial and fault solutions.

$$\mathbf{E}(F) = \mathbf{E}(0) + (I_{km}(F))\overline{Z}_{km} + (I_{ij}(F))\overline{Z}_{ij} \tag{4.34}$$

The method may be generalized to more than two simultaneous contingencies. Example 4.6 demonstrates the method.

Example 4.6

Use the network of Example 4.4 under reactive flow conditions and the corresponding $\mathbf{Y}_{BUS}$ table of factors to calculate the effect of a double-line contingency open-circuiting lines from bus 1 to 2 and bus 5 to 2. As shown in Figure E4.6. The LDU table of factors

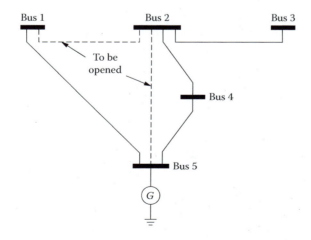

FIGURE E4.6

and initial voltages are

$$[L\,D\,U] = j\begin{bmatrix} \frac{1}{5} & 0 & 0 & 0 & -1 \\ & \frac{1}{3} & 0 & -\frac{2}{3} & -\frac{1}{3} \\ & & \frac{1}{10} & -\frac{3}{5} & -\frac{2}{5} \\ & L & & \frac{15}{121} & -\frac{91}{121} \\ & & & & \frac{121}{182} \end{bmatrix}\begin{matrix} Bus3 \\ Bus1 \\ Bus4 \\ Bus5 \\ Bus2 \end{matrix} \qquad E(0) = \begin{bmatrix} 1.03\angle 0° \\ 1.01\angle 0° \\ 1.04\angle 0° \\ 1.05\angle 0° \\ 1.02\angle 0° \end{bmatrix}$$

Solution

The multiple-contingency method is applied step by step. For the two open-circuit lines, the Thévenin voltages are

$$E_{T12} = E_1(0) - E_2(0) = -0.01$$

$$E_{T52} = E_5(0) - E_2(0) = 0.03$$

Unit currents are injected in 1, –2 and 5, –2 in the reordered sequence so that

$$I_{12} = [0\ 1\ 0\ 0\ -1]^t \qquad I_{52} = [0\ 0\ 0\ 1\ -1]^t$$

The resulting impedance vectors from injecting these currents (ordered 1, 2, 3, 4, 5) are

$$\bar{Z}_{12} = j\begin{bmatrix} 0.296703 \\ -0.109890 \\ -0.109890 \\ -0.043956 \\ 0 \end{bmatrix} \qquad \bar{Z}_{52} = j\begin{bmatrix} -0.054945 \\ -0.164835 \\ -0.164835 \\ -0.065934 \\ 0 \end{bmatrix}$$

According to step 4, the Thévenin impedances are

$$z_{12,12} = j0.296703 - (-j0.109890) = j0.406503$$

$$z_{12,52} = \quad 0 \qquad -(-j0.109890) = j0.109890$$

$$z_{52,52} = \quad 0 \qquad -(-j0.164835) = j0.164835$$

$$z_{52,12} = -j0.054945 - (-j0.164835) = j0.109890$$

In matrix form these are

$$\mathbf{Z}_T = j\begin{bmatrix} 0.406503 & 0.109890 \\ 0.109890 & 0.164835 \end{bmatrix}$$

The fault impedance is the negative of the lines taken out of service:

$$\mathbf{Z}_F = j\begin{bmatrix} -1.0000 & 0 \\ 0 & -0.333333 \end{bmatrix}$$

The resulting fault currents are calculated as

$$\begin{bmatrix} I_{12}(F) \\ I_{52}(F) \end{bmatrix} = -(\mathbf{Z}_T + \mathbf{Z}_F)^{-1} \begin{bmatrix} E_{T12}(0) \\ E_{T52}(0) \end{bmatrix}$$

$$= -(\mathbf{Z}_T + \mathbf{Z}_F)^{-1} \begin{bmatrix} -0.01 \\ 0.03 \end{bmatrix} = \begin{bmatrix} j0.018296 \\ j0.189966 \end{bmatrix}$$

In step 6, the fault currents are used to calculate the contingency voltages.

$$\mathbf{E}(F) = \mathbf{E}(0) + j0.018296\bar{Z}_{12} + j0.189966\bar{Z}_{52}$$

$$= \begin{bmatrix} 1.0150 \\ 1.0533 \\ 1.0633 \\ 1.0533 \\ 1.0500 \end{bmatrix}$$

Examples 4.5 and 4.6 have shown that contingencies are calculated with the $\mathbf{Y}_{BUS}$ table of factors without retriangularization. Contingencies are calculated as fast as a repeated solution of an inverse using a new right-hand-side vector. For a large network this is a considerable savings in execution time for the on-line digital computer.

Problems

4.1. A network power flow condition is shown on the positive-sequence diagram, Figure P4.1. The transmission line positive-sequence impedances are

$$Z_{12} = 0.01(1 + j)$$

$$Z_{13} = 0.02(1 + j)$$

$$Z_{23} = 0.03(1 + j)$$

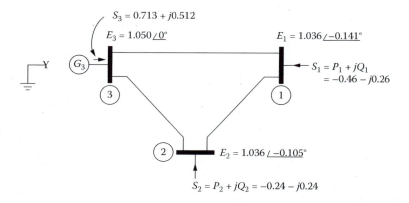

FIGURE P4.1

The line-charging reactances are negligible. The generator transient reactances are

$$X_1' = X_2' = 0.30$$

(a) Write $\mathbf{Z}_{\text{BUS}}$ for the system with *ground* as a reference, utilizing the generator transient reactance. Consider the loads to be open circuits compared to transmission line impedances. © Check the result with **Zbus.exe**.

(b) Find the fault currents for a balanced, three-phase fault occurring at bus 1. Approximate all bus voltages to be 1.05 p.u. before the fault. © Modify Vbase in **Zbus.exe**.

(c) During the steady-state power flow condition given above, line 1–2, open circuits, use the $\mathbf{Z}_{\text{BUS}}$ line contingency method of Section 4.3.1 to estimate the bus voltages during the open-circuit condition.

4.2. A one-line diagram of the power system is shown in Figure P4.2.1. Assuming the resistances to be small in comparison, the reactances for the system are given below in p.u. values.

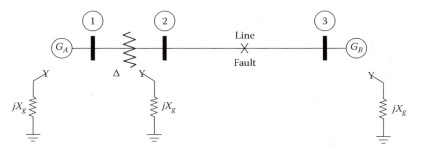

FIGURE P4.2.1

Generators *A* and *B*:

$$X_1 = X_2 = 0.1$$

$$X_0 = 0.04$$

$$X_g = 0.02$$

Transformer:

$$X_1 = X_2 = 0.1$$

$$X_0 = 0.1$$

$$X_g = 0.05$$

The transformer connects generator *A* to a transmission line. The three-phase transmission line has negligible resistance and line charging:

$$\mathbf{Z}_{\text{line}}^{abc} = j \begin{bmatrix} 0.3 & 0.1 & 0.1 \\ 0.1 & 0.3 & 0.1 \\ 0.1 & 0.1 & 0.3 \end{bmatrix}$$

The system is initially in balanced operation. It is desired to study the system short-circuit characteristics for a fault occurring at the *center* of the transmission line. Using the notation of the graph in Figure P4.2.2, do the following:

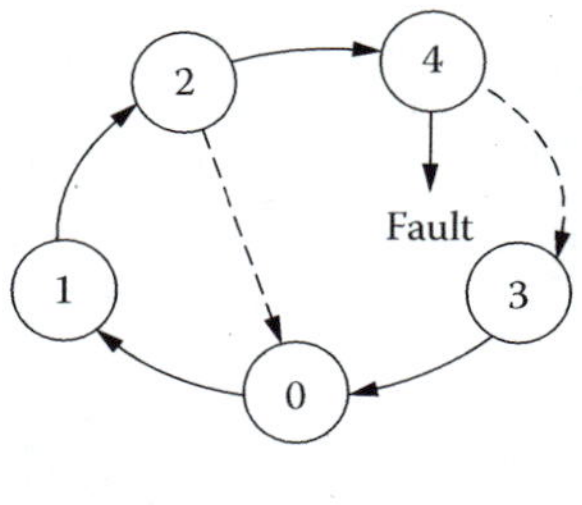

(a) Derive the bus impedance matrix, $\mathbf{Z}_{BUS}^{012}$, for the system using the positive-sequence elements and zero-sequence elements. © Check both matrices using *Zbus.exe*.

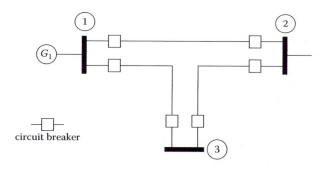

FIGURE P4.2.2

(b) Let the fault be a three-phase short circuit to ground. Find the value of the fault current.

(c) Find the value of the bus voltages during the three-phase fault.

(d) Find the fault current in a single line-to-ground fault.

4.3. Consider the network with positive-sequence impedances shown in Figure P4.3. The data are

Generator transient reactances:

FIGURE P4.3

Gen 1: $Z_d' = j0.30$

Gen 2: $Z_d' = j0.15$

Transmission lines:

$$Z_{12} = j0.04$$

$$Z_{13} = j0.03$$

$$Z_{32} = j0.02$$

(a) Obtain Tinney's table of factors (i.e., his LDU decomposition) using the bus ordering 1, 2, 3.

(b) Using the table of factors, find the total current for a three-phase-to-ground fault occurring at bus 3. © Verify this result with *zbus.exe*.

(c) Find the bus voltages during the fault using the table of factors.

(d) If the circuit breaker near bus 3 on line 2–3 were to open during the fault condition, what is the new value of fault current when only line 1–3 is connected? (*Hint*: Use contingency methods.)

4.4. For a four-bus network including two generators, the positive- and zero-sequence bus impedance matrices are as follows:

$$
\begin{array}{cccc}
1 & 2 & 3 & 4
\end{array}
$$

$$
\mathbf{Z}^1_{\text{BUS}} = \mathbf{Z}^2_{\text{BUS}} = j
\begin{bmatrix}
0.080 & 0.066 & 0.038 & 0.056 \\
0.066 & 0.132 & 0.054 & 0.065 \\
0.038 & 0.054 & 0.118 & 0.084 \\
0.056 & 0.065 & 0.084 & 0.243
\end{bmatrix}
$$

$$
\begin{array}{ccc}
1 & 2 & 3
\end{array}
$$

$$
\mathbf{Z}^0_{\text{BUS}} = j
\begin{bmatrix}
0.038 & 0.023 & 0.003 \\
0.023 & 0.196 & 0.026 \\
0.003 & 0.026 & 0.056
\end{bmatrix}
$$

The zero-sequence circuit is open to bus 4, so it does not appear in the matrix. The buses are all at 1.0 p.u. voltage, balanced, before a line-to-line fault occurs at bus 2.

(a) Find the fault currents for the line-to-line fault at bus 2. (*Hint*: Use the singular line-to-line $\mathbf{Y}^{012}$ matrix from Table 2.1 in Equation 4.10 and Equation 4.11 for this fault condition.)

(b) Find the symmetrical component voltages at all buses during the fault condition. (*Hint*: Table 2.2 shows the symmetrical component equivalent for a line-to-line 2Z connection. It is easier to use an admittance formulation and let $Z = 0$ in the last step to obtain numerical results.)

4.5. Derive analytical expressions for the symmetrical component fault current and voltage as any bus i. In other words, construct a tabulation for a line-to-line fault, 2Z, which could be included in Table 4.1. Let the power system be balanced so that the driving-point impedance at bus k, $\mathbf{Z}^{012}_{kk}$, and mutual impedances, $\mathbf{Z}^{012}_{ik}$, are diagonal 3×3 submatrices of $\mathbf{Z}^{012}_{\text{BUS}}$. Let a line-to-line fault $2Z_f$ occur between phases b and c as shown in Table 2.2.

4.6. A short-circuit program is to determine the four cycle operating requirements of circuit breakers a, b, c, and d on the two-ring bus configuration shown in the positive-sequence network in Figure P4.6. The ring buses are simple points in a less detailed circuit. The resistances in the rings are negligible compared to the reactances for all elements in network. Except for generator 5, which is a system equivalent, the generators are slow-speed, multipole machines driven by hydro-turbines. The positive-sequence reactances of the elements on a 100 MVA base are as follows:

For the transmission lines:

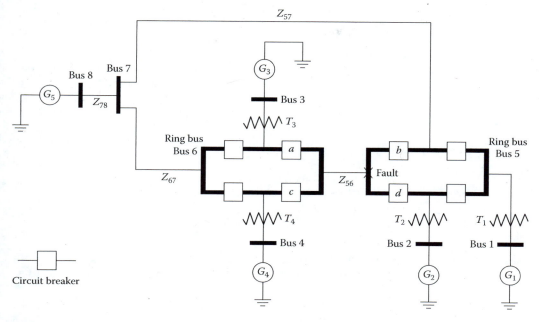

FIGURE P4.6

Bus to Bus	Reactance
7–8	0.02
5–7	0.04
6–7	0.03
5–6	0.02

For the transformers:

T_i	X
T_1	0.082
T_2	0.082
T_3	0.310
T_4	0.120

For the generators:

Gen.	X_d	X'_d
G_1	0.57	0.30
G_2	0.57	0.30
G_3	0.93	0.35
G_4	0.84	0.30
G_5	1.25	0.70

Observe that by combining radial generators and transformers, only three buses are required in the $\mathbf{Z}_{BUS}$ matrix. Determine the effects of a three-phase fault to ground occurring at bus 5 (indicated by an × on Figure P4.6) and do the following:

(a) Use a minimum number of buses. Define a tree and cotree with Z_{56} as a link. Since faults are to be studied, use generator *transient* reactances. Determine the positive-sequence bus impedance matrix $\mathbf{Z}_{BUS}$, adding link Z_{56} as the final element (see part (d)).

(b) Assume that all buses of part (a) are initially at 1.0 p.u. voltage and transformers are on-nominal. Let a three-phase fault to ground occur at the point indicated by an × on the one-line diagram. Determine the voltage at bus 7 and bus 6 during the fault.

(c) Let circuit breakers c and d open so that fault currents are transferred to breakers a and b. Determine the current through circuit breakers a and b.

(d) Let circuit breakers, a, c, and d clear so that only circuit breaker b must open while carrying fault current. Find the current that circuit breaker b must interrupt.

4.7. Problem 4.4 specifies the positive- and negative-sequence impedances as well as the zero-sequence impedance for a four-bus network. The initial balanced bus voltages are:

$$(\mathbf{E}^1_{BUS})^t = [1.01 \quad 1.02 \quad 1.03 \quad 1.04] \angle 0°$$

(a) What is the Thévenin impedance between buses 2 and 3? A transformer whose leakage reactance is $Z = j.05$ is suddenly connected between buses 2 and 3. Do parts (b), (c), and (d) with this modification.

(b) What is the current through the transformer after it is connected?

(c) What are all the bus voltages after the transformer is connected?

(d) If only one phase of the transformer is connected between buses 2 and 3 because two poles of the circuit breaker fail to close, what are the resulting voltages at all four buses? Express the results as phase a, b, and c voltage.

4.8 A 25 MVA cylindrical rotor synchronous generator ($jx'_1 = j0.3$, $jx_0 = j0.20$, $jx_g = 0.10$ p.u.) supplies 4 synchronous motors, each 6 MVA, through a 25 MVA, 13.8/6.9 kV transformer as shown in Figure P4.8. The voltage at bus 2 is 1.0 p.u. before a fault occurs near motor 4. The reactances of the system components as converted to a 6 MVA base are:

Component	jx'_1	jx_0	jx_g
Generator	j0.072	j0.048	j0.024
Transformer	j0.048	Open	Open
Motors	j0.25	j0.15	Solid ground

(a) If a 3-phase fault occurs between motor 4 and its circuit breaker at bus 2, find the fault current in p.u. on the 6MVA base and in amperes.

(b) How much of the fault current comes from the generator and how much comes from the 4 motors?

(c) If the fault is a single phase bolted fault to ground, what is the fault current?

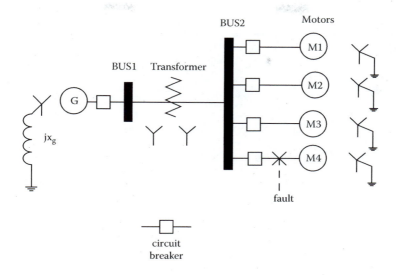

FIGURE P4.8

References

1. Brown, H. E. 1975. *Solution of large networks by matrix methods.* New York: John Wiley & Sons.
2. El-Abiad, A. H., and Stagg, G. W. 1963. Automatic evaluation of power system performance—Effects of line and transformer outages. *Trans. AIEE* 712–16.
3. Tinney, W. A. 1972. Compensation Methods for network solutions by optimally ordered triangular factorization. *IEEE Trans. Power Appar. Syst.* 91, Jan./Feb., 123–127.

5

Power Flow on Transmission Networks

Power flow on transmission networks is calculated using linear mathematical models for transmission lines, transformers, and shunt or series reactances, but nonlinear electrical descriptions for generation and load at the buses. The nonlinear electrical specifications result in nonlinear forms of Kirchhoff's laws for power flow—called the *load-flow problem*. There are two main sources of the nonlinearity. First, the power demand on the transmission network from distribution and subtransmission connections is closely modeled by constant real and reactive power, so if terminal voltages increase, the current demand decreases, and vice versa. This type of load is suitably described as a fixed power demand at a bus (called a P, Q bus),

$$P_i + jQ_i = E_i I_i^* = \text{constant} \tag{5.1}$$

where I_i is the per-unit positive-sequence injection current into the bus, and E_i the per-unit positive-sequence bus voltage. A second reason for the nonlinear problem is that generating plants normally operate at a regulated voltage level and fixed real power injection. The voltage phase angles between generators on the system are not known. If a generator's reactive power output is within acceptable limits, it is permitted to vary to match the system voltage. Prime-mover power to the generator represents a direct operating cost of the plant. There are many mechanical constraints associated with the power, so it is closely monitored and controlled. Not only for a generating plant, but for an entire power system, real power is the basic control signal. For example, control of interconnected power systems is based on real power flow across the system boundaries, part of area control error (ACE), as shown schematically in Figure 5.1.

The ACE signal, as discussed in Chapter 1, is

$$ACE = \sum_{\text{TIES}} P_t - P_S + 10B\,(f - f_0) \tag{5.2}$$

where P_t is the MW tie-line flow defined positive out of the system's boundary, and P_S is the MW desired or scheduled exchange with neighbors. The actual system frequency measured is f, and the reference frequency (say, 50 or 60 Hz) is f_0. The multiplier B is called the *bias* and is specified in megawatts per 0.1 Hz.

Assume for the present that there is no frequency error in Equation 5.2. The real power flow represents real direct cost, so unscheduled flow across boundaries as measured by ACE is undesirable. Total generation in the area is raised or lowered to force ACE to zero without actually measuring the power consumed in the area (total load). After ACE is forced to zero, or the interconnection balanced, the area must supply its load, its transmission line losses, and the P_S exchange. To do this, each generator is usually assigned a power output ratio relative to other generators in order to operate the entire power system at its

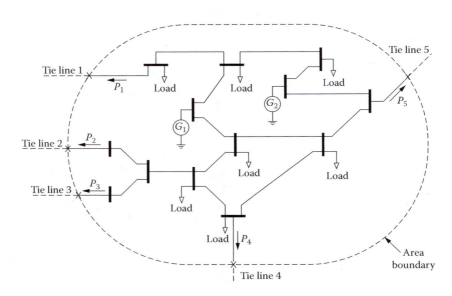

FIGURE 5.1
Power system interconnected to neighbors.

most efficient point with respect to the cost of energy. The generator voltage remains fixed, or in other words, regulated.

Reactive power flow on the transmission network results in higher current flows, and hence higher transmission line losses. Electric utilities have line compensation banks, that is, buses with shunt capacitors or inductors to regulate reactive power. These regulators may be discrete levels such as switched banks of capacitors, or continuously adjustable devices such as synchronous condensers. In recent years, solid-state continuously adjustable reactive power control devices called static var compensators (SVCs) have been introduced. They may be used to establish reactive power consumed at a bus, so that Equation 5.1 holds, or alternatively, their control may be used to maintain constant voltage at a bus remote to the SVC. Thus, regulated buses as well as generator buses are of the type where

$$P_i = \text{fixed, dependent on demand} \tag{5.3a}$$

$$V_i = \text{constant} \tag{5.3b}$$

which is called a *P, V bus* in power flow calculations. If some current or reactive power constraint is violated, such a bus could convert to another type (e.g., a *P, Q* type). Additional combinations of bus voltage and real or reactive power are described later, but the principle worth emphasizing is that a power flow calculation satisfies *power constraints* at the buses. This results in products of current and voltage, and hence nonlinear equations to describe current flow or voltages on the network.

Power flow calculations are mandatory in planning new transmission lines and all studies of generation, load demand, power factor correction, exchange of energy between utilities, or transfer of power between nonadjacent power areas (wheeling power), voltage profiling, and so on. Some of these applications are considered to be off-line or batch executions of a power flow calculation that are performed months or years ahead of the actual situation.

An on-line power flow is a periodically executed program in the digital computers that are monitoring and controlling the power system. They take averaged or processed real-time measured data for the P, Q or P, V conditions at buses in order to calculate bus voltages and phase angles for the entire network. The linear contingency calculations of Chapter 4 are often subprograms of a power flow program. The real-time results of power flow calculation may be used to determine the reactive compensation to establish network voltages or incremental transmission line losses associated with changing the output power of a generator. If total network generation is to change, which generators should be perturbed depends on line losses and the cost of energy for each generating plant, but a power flow calculation is needed to predict the results of this change.

In recent years, distributed computer systems used to monitor power systems have increased the computation capability to the point that contingency cases are now run as power flow cases.

5.1 Slack Bus

Consider two buses connected by an impedance as shown in Figure 5.2. If the complex power injection, S_i, is specified at each bus, this results in two complex (constraint) equations in three complex unknowns:

$$S_1 = P_1 + jQ_1 = E_1 I^* \tag{5.4}$$

$$S_2 = P_2 + Q_2 = E_2 I^* \tag{5.5}$$

The unknown voltages, and current E_1, E_2, and I, are related by Kirchhoff's voltage law:

$$E_1 - ZI = E_2 \quad \text{or} \quad 1 - \frac{ZI}{E_1} = \frac{E_2}{E_1} \tag{5.6}$$

Therefore, the two-bus system has four unknowns—the real and imaginary parts of the voltages when specified in this way. If Equation 5.4 is used to eliminate the current from Equation 5.5 and Equation 5.6, this results in the equations

$$\frac{P_2 + jQ_2}{P_1 + jQ_1} = -\frac{E_2}{E_1} \tag{5.7}$$

$$-1 + \frac{Z(P_1 - jQ_1)}{E_1 E_1^*} = -\frac{E_2}{E_1} \tag{5.8}$$

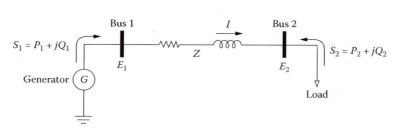

FIGURE 5.2
Two buses with specified injections.

Equating the last two equations yields

$$-1 + \frac{Z(P_1 - jQ_1)}{E_1 E_1^*} = \frac{P_2 + jQ_2}{P_1 + jQ_1} = \frac{S_2}{S_1} \tag{5.9}$$

For arbitrary S_1 and S_2, this equation cannot be satisfied because $E_1 E_1^*$ is real, and hence the solution is overdetermined. As a result, the problem is properly posed if S_2 is specified, while S_1 is permitted to be a *slack complex variable*, or dependent variable. Bus 1 is accordingly called a *slack bus*, and its complex voltage is fixed at E_1. Although it may be any bus on the system, the slack bus for power flow calculations is almost always a generator that operates at regulated terminal voltage.

For a power flow calculation, one bus is designated as the slack bus and has a fixed voltage. Its phase angle is usually used as the reference for other buses on the system. Equation 5.4 to Equation 5.9 are derived for an $n = 2$ bus system, but by inductive arguments, it can be shown that $n - 1$ complex bus power constraints may be imposed in an n-bus load-flow calculation. Stated in another way, $n - 1$ power injections may be specified for an n-bus system. Special cases such as fixed voltage magnitudes at some buses or multiple slack buses are also treated by designating one bus as the slack bus for power flow calculations.

5.2 $\mathbf{Z}_{\text{BUS}}$ Formulation for Load-Flow Equations

Let bus n of an n-bus system be selected as the slack bus for a power flow calculation. Let $P_i + jQ_i$, the load, be specified at $n - 1$ buses. Furthermore, incorporate bus loads with current sources driving the network, so that there are no shunt elements connected between the buses and ground. Then each bus of the power system has a driving-point impedance with respect to the slack bus, and mutual impedances as shown in Figure 5.3. Because there are no elements to ground, the bus impedance matrix is formulated using the slack bus as reference. The bus equations are then

$$\mathbf{E}_{\text{BUS}} = \mathbf{Z}_{\text{BUS}} \, \mathbf{I}_{\text{BUS}} + \bar{E}_r \tag{5.10}$$

where $\bar{E}_r$ is an $n - 1$ column vector of slack bus voltage values and $\mathbf{E}_{\text{BUS}}$ is an $n - 1$ bus voltage vector. In Equation 5.10 and in subsequent balanced three-phase power flow calculations, $\mathbf{Z}_{\text{BUS}}$ is an $(n - 1) \times (n - 1)$ matrix comprised of positive-sequence elements, and $\mathbf{I}_{\text{BUS}}$ and $\mathbf{E}_{\text{BUS}}$ are per-unit values on a specified voltage and MVA base. Alternatively, the slack bus voltage may be used in a difference voltage vector, $\mathbf{E}'_{\text{BUS}}$.

$$\mathbf{E}'_{\text{BUS}} = \begin{bmatrix} E_1 - E_r \\ E_2 - E_r \\ \vdots \\ E_{n-1} - E_r \end{bmatrix} = \mathbf{Z}_{\text{BUS}} \mathbf{I}_{\text{BUS}} \tag{5.11}$$

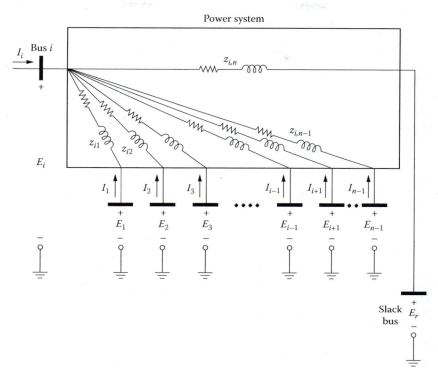

FIGURE 5.3
Power system in $\mathbf{Z}_{BUS}$ with respect to slack bus.

Observe that because it is the reference, the slack bus current is the negative of all injections at the other buses.

$$I_r = -\sum_{i=1}^{n-1} I_i \qquad (5.12)$$

Equation 5.11 may be solved for the bus voltages *if* the bus injection currents are known. However, the bus currents could be calculated from specified power injections if voltage were known. This leads to an iterative scheme of an initial "guess" for the voltage, and then calculating bus currents from specified power injections,

---

Iterative Current Injection (without shunt elements)

$$I_i^k = \frac{(P_i + jQ_i)^*}{\left(E_i^k\right)^*} \qquad (5.13)$$

$$i = 1,\ 2,\ldots,\ n-1$$

---

where k is the iteration number, and then returning to use Equation 5.11. This is called a *Gaussian iterative scheme* to solve for new bus voltages in terms of previous current injections according to

$$
\boxed{
\begin{array}{c}
\text{Gaussian Iterative Method} \\
\mathbf{E}^{k+1}_{\text{BUS}} = \mathbf{Z}_{\text{BUS}} \mathbf{I}^{k}_{\text{BUS}} + \bar{E}_r
\end{array}
}
\tag{5.14}
$$

If bus voltages are calculated sequentially and the updated voltage values are used to calculate subsequent injection currents row by row, a *Gauss-Seidel iterative scheme* to solve the power problem is

$$
\boxed{
\begin{array}{c}
\text{Gauss-Seidel Iterative Method} \\[2mm]
E^{k+1}_i = \displaystyle\sum_{l=1}^{i-1} z_{li} I^{k+1}_l + \sum_{m=i}^{n-1} z_{mi} I^{k}_m + E_r \\[4mm]
i = 1, 2, \ldots, n-1
\end{array}
}
\tag{5.15}
$$

If starting values for all elements of the voltage vector, $\mathbf{E}^{0}_{\text{BUS}}$, are set equal to the slack bus voltage, this is called a *flat start*. Convergence to a solution is faster if the initial values are numerically close to the true solution. However, a flat start is used if previous estimates are not available, or as a common basis for comparing convergence of different iterative methods.

It is also necessary to specify a measure of convergence in order to terminate the iterations. For the schemes of Equation 5.14 and Equation 5.15, a convenient termination is when the error in calculated bus voltage between two successive iterations is less than an acceptable error:

$$
\| E^{k+1}_i - E^{k}_i \| < \epsilon \quad i = 1, 2, \ldots, n-1
\tag{5.16}
$$

All buses must be checked for convergence. Large-magnitude generation or load buses may converge faster than comparatively small load buses at electrically remote locations, and a result based predominantly on the large units may be acceptable. Therefore, judgment is necessary in selecting a convergence tolerance.

Using $\mathbf{Z}_{\text{BUS}}$ and the power injections specified, the computer algorithm for power flow calculation with the Gauss scheme is:

1. Guess an initial $\mathbf{E}^{0}_{\text{BUS}}$, use old solutions, or a flat start for $n-1$ bus voltages. This is the $k=0$ iteration. Save these voltages in an array $\mathbf{E}^{k}_{\text{BUS}}$.
2. Calculate all bus current injections using Equation 5.13 with the voltages stored in $\mathbf{E}^{k}_{\text{BUS}}$.
3. Calculate a new bus voltage vector (bus by bus), $\mathbf{E}^{k+1}_{\text{BUS}}$, using Equation 5.14. Compare each new voltage with old values: $\| E^{k+1}_i - E^{k}_i \| < \epsilon$.
4. If the change in all bus voltages is less than ϵ, terminate. Otherwise, replace $\mathbf{E}^{k}_{\text{BUS}}$ with $\mathbf{E}^{k+1}_{\text{BUS}}$ and return to step 2.

The computer algorithm for power flow calculations using the Gauss-Seidel iterative scheme is a modification of the above method:

1. With an initial $\mathbf{E}^{0}_{\text{BUS}}$, calculate $\mathbf{I}^{0}_{\text{BUS}}$ using Equation 5.13.
2. On a bus-by-bus basis, calculate a new bus voltage E^{k+1}_i using Equation 5.15. Immediately after calculating E^{k+1}_i, find the new value I^{k+1}_i for use in subsequent

$i + 1, i + 2, ...,$ bus calculations. Check each E_i^{k+1} voltage change for convergence according to Equation 5.16.

3. If the change in all bus voltages is less than ϵ, terminate. Otherwise, return to step 2.

An example is next used to demonstrate the Gauss and Gauss-Seidel iterative schemes. An acceleration factor, α, as explained in Section 3.4.2, could be used with both schemes.

Example 5.1

Use the Gauss and Gauss-Seidel methods with a flat start to find the voltage at buses 1 and 2 in the power flow network shown in Figure E5.1.

The power injections at the buses in p.u. are

$$S_1 = -0.4 + j0.2$$

$$S_2 = -0.3 + j0.3$$

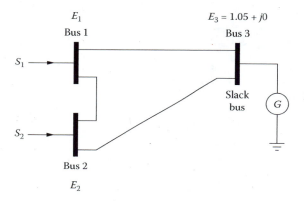

FIGURE E5.1

The line impedances (positive sequence) are

$$\text{Line 1--2: } 0.01 + j0.01$$

$$\text{Line 1--3: } 0.02 + j0.02$$

$$\text{Line 2--3: } 0.03 + j0.03$$

Line charging is negligible. All power injections and impedances are specified using a 50 MW, 13.2 kV base. Calculate voltage convergence within an error tolerance of 10^{-6}. All injections are defined into the bus with respect to ground.

Solution

The bus impedance matrix using the slack bus as a reference is

$$\mathbf{Z}_{\text{BUS}} = (0.01 + j0.01)\begin{matrix} & \begin{matrix} 1 & \quad 2 \end{matrix} \\ & \begin{bmatrix} 1.3333 & 1.0000 \\ 1.0000 & 1.5000 \end{bmatrix} \end{matrix}$$

In the Gaussian iteration, the current injections are calculated by

$$I_1^k = \frac{P_1 - jQ_1}{\left(E_1^k\right)^*} = \frac{-0.4 - j0.2}{\left(E_1^k\right)^*}$$

$$I_2^k = \frac{P_2 - jQ_2}{\left(E_2^k\right)^*} = \frac{-0.3 - j0.3}{\left(E_2^k\right)^*}$$

The Gaussian iterative load flow for this example is

$$\begin{bmatrix} E_1^{k+1} \\ E_2^{k+1} \end{bmatrix} = (0.01 + j0.01)\begin{bmatrix} 1.3333 & 1.0000 \\ 1.0000 & 1.5000 \end{bmatrix}\begin{bmatrix} I_1^k \\ I_2^k \end{bmatrix} + \begin{bmatrix} 1.05 \\ 1.05 \end{bmatrix}$$

The Gauss-Seidel iterative scheme is

$$E_1^{k+1} = (0.01 + j0.01)\left(1.3333I_1^k + I_2^k\right) + 1.05$$

$$E_2^{k+1} = (0.01 + j0.01)\left(I_1^{k+1} + 1.5I_2^k\right) + 1.05$$

where the current, I_1^{k+1}, is calculated immediately after the bus voltage, E_1^{k+1}, has been calculated.

Using a flat start, the numerical results for the two iterative schemes are as given below.

	Gauss Scheme		Gauss–Seidel	
k	E_1	E_2	E_1	E_2
0	1.05 + j0	1.05 + j0	1.05 + j0	1.05 + j0
1	1.047461 – j0.0133314	1.048095 – j0.0142857	1.047461 – j 0.0133314	1.04802 – j0.0142743
2	1.047280 – j0.0133256	1.047901 – j0.0142883	1.047280 – j0.01133277	1.047901 – j0.0142899
3	1.047280 – j0.0133280	1.047901 – j0.0142909	1.047280 – j0.0133297	1.047901 – j0.0142909
4	1.047280 – j0.0133280	1.047901 – j0.0142909	1.047280 – j0.0133297	1.047901 – j0.0142909
5	Converged		Converged	

Using the final values of the voltages at buses 1 and 2, the following line flows are calculated for the network:

Line (Bus to Bus)	P (p.u.)	Q (p.u.)
1–2	0.01671	–0.08311
1–3	–0.4167	0.2831
2–1	–0.01665	0.08317
2–3	–0.2834	0.2168
3–1	0.4213	–0.2785
3–2	0.2868	–0.2134

The number of iterations is the same for both methods. Observe that the out-of-phase part of the voltage is negative, implying negative angles and consequently real power flow to them from the generator at bus 1. Also observe that the voltage magnitudes at the load buses are approximately the same as the generator because the loads inject leading reactive power (leading power factor), which compensates for the inductive transmission lines.

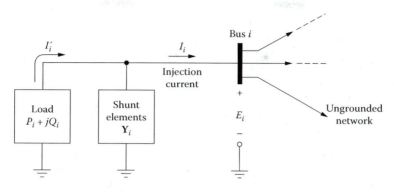

FIGURE 5.4
Loads and shunt element as a bus.

To include shunt elements such as line-charging capacitance, transformer π-equivalent impedances, and so on, from bus to ground, these elements modify the current injection as shown in Figure 5.4. The current injection at each bus with shunt element Y_i is, accordingly, calculated as

$$\boxed{\begin{array}{c}\text{Iterative Current Injection (with shunt elements)}\\[4pt] I_i^k = \dfrac{(P_i + jQ_i)^*}{(E_i^k)^*} - Y_i E_i^k\end{array}} \qquad (5.17)$$

The bus impedance matrix is with respect to the slack bus and does not have any elements to ground.

An interesting aspect of $\mathbf{Z}_{BUS}$ formulated with respect to the slack bus is that the vector $\mathbf{E}'_{BUS}$ in Equation 5.11 represents a voltage difference across the transmission network, so the losses associated with transmitting the power can be calculated. Assume that the network shunt terms from bus to neutral are reactive, so they have no power loss. The power injected into the network at any bus is

$$P_i + jQ_i = I_i^* E_i = I_i^* \left(\sum_{m=1}^{n-1} z_{im} I_m + E_r \right) \qquad (5.18)$$

where z_{im} is the ith row of $\mathbf{Z}_{BUS}$. If the power injections of all buses plus the slack bus are added, and Equation 5.12 is used for the slack bus current, then

$$\begin{bmatrix} \text{transmission line} \\ \text{losses with} \\ \text{reactive shunt} \\ \text{elements} \end{bmatrix} = P_L = \mathbf{Re} \left\{ \sum_{i=1}^{n-1} (P_i + jQ_i) + E_r I_r^* \right\}$$

$$= \mathbf{Re} \left\{ \sum_{i=1}^{n-1} I_i^* \left(\sum_{m=1}^{n-1} z_{im} I_m + E_r \right) - E_r \sum_{i=1}^{n-1} I_i^* \right\} \qquad (5.19)$$

$$= \mathbf{Re} \left\{ (\mathbf{I}_{BUS}^*)^t \, \mathbf{Z}_{BUS} \mathbf{I}_{BUS} \right\}$$

This is a convenient form for transmission line losses. After injection currents at the generators are replaced by $(P_g - jQ_g)/E_g^*$ and V_g is held constant, the incremental transmission losses associated with each generator $\partial P_L/\partial P_g$ are calculated. Even though $\mathbf{Y}_{BUS}$ may be used in Equation 5.19, the incremental losses are more accurately found using $\mathbf{Z}_{BUS}$. Chapter 6 covers this subject.

5.3   Gauss or Gauss-Seidel Iteration Using $\mathbf{Y}_{BUS}$

The advantages of using the bus admittance matrix compared to $\mathbf{Z}_{BUS}$ are reduced computer storage and the fact that there is no need for a bus impedance matrix algorithm or inverting $\mathbf{Y}_{BUS}$. Given an initial guess for the voltage, the current injection at each bus is calculated using Equation 5.17 to obtain $\mathbf{I}_{BUS}^k$, then the bus admittance equations *with respect to ground*,

$$\mathbf{I}_{BUS} = \mathbf{Y}_{BUS}\,\mathbf{E}_{BUS} = [L + D + U]\,\mathbf{E}_{BUS} \tag{5.20}$$

are rearranged to solve iteratively for the diagonal voltage vector in terms of previous currents and voltages.

$$\mathbf{E}_{BUS}^{k+1} = [D]^{-1}\left\{\mathbf{I}_{BUS}^k - (L + U)\,\mathbf{E}_{BUS}^k\right\} \tag{5.21}$$

In this equation, L is the lower-triangular portion of $\mathbf{Y}_{BUS}$, U the upper portion, and D the diagonal or driving-point admittance matrix. In terms of a bus-by-bus calculation, the *Gaussian* iterative scheme is

$$
\boxed{
\begin{array}{l}
\text{Gaussian Iterative Method} \\[2ex]
E_i^{k+1} = \dfrac{1}{y_{ii}}\left\{I_i^k - \displaystyle\sum_{\ell \neq i}^{n} y_{i\ell}E_\ell^k\right\} \\[3ex]
\hfill i = 1, 2, \ldots, n-1
\end{array}
} \tag{5.22}
$$

The slack bus voltage E_r is constant. An acceleration factor, α, may be used, such that the *accelerated Gaussian iterative method* is

$$
\boxed{
\begin{array}{l}
\text{Accelerated Gaussian Iterative Method} \\[2ex]
E_i^{k+1} = \dfrac{\alpha}{y_{ii}}\left\{I_i^k - \displaystyle\sum_{\ell \neq i}^{n} y_{i\ell}E_\ell^k\right\} + (1-\alpha)E_i^k \\[3ex]
\hfill i = 1, 2, \ldots, n-1
\end{array}
} \tag{5.23}
$$

If the *i*th bus voltage calculated is used in subsequent rows $i+1, i+2, \ldots$, this becomes a Gauss-Seidel iterative load-flow calculation. With an acceleration factor this is

Accelerated Gauss-Seidel Method

$$E_i^{k+1} = \frac{\alpha}{y_{ii}} \left\{ I_i^k - \sum_{\ell<i}^{n} y_{i\ell} E_\ell^{k+1} - \sum_{\ell>i}^{n} y_{i\ell} E_\ell^k \right\} + (1-\alpha) E_i^k \tag{5.24}$$

$$i = 1, 2, \ldots, n-1$$

Equation 5.18 to Equation 5.21 assume that shunt elements to ground are taken into account by the current injection calculated for each iteration, Equation 5.17. If the shunt elements are included in $\mathbf{Y}_{\text{BUS}}$ (thus in y_{ii}), they are omitted from Equation 5.17. The computer algorithm for methods of Equation 5.23 and Equation 5.24 are similar to those of Section 5.2.

Thus far only buses where the power is specified have been treated. There are cases where the bus power injection and the magnitude of the bus voltage are constant, such as a generating station or regulated bus.

$$P, V_{\text{bus}} \equiv \begin{array}{l} \text{specified real power injection at bus} \\ \text{at constant voltage magnitude} \end{array}$$

For virtually all power system elements, the reactive flow at the bus is limited to a range, so until the limits are reached, the bus acts as a P, V bus. Beyond the range, the bus converts to a P, Q type. Consequently, the reactive power should be calculated during the iterative process to check against limits.

In the Gauss and Gauss-Seidel schemes, if bus p is a fixed voltage, V_{fp}, the initial reactive injection at bus p is calculated using the initial guess of the bus phase angle and the voltage magnitude:

$$Q_p^0 = -\mathbf{Im}\left\{ \left(E_p^0\right)^* I_P^0 \right\} = -\mathbf{Im}\left\{ \left(E_p^0\right)^* \sum_{\ell \neq p}^{n} y_{p\ell} \left(E_p^0 - E_\ell^0\right) \right\} \tag{5.25}$$

where the bus injection current is the sum of the line flows to other buses. The previous Gaussian iteration schemes calculate a new voltage E_p^k, magnitude $|E_p^k|$, which does not agree with the controlled value, V_{fp}. Therefore, its *calculated phase angle*, δ_p^k,

$$E_p^k = a^k + jb^k = \left|E_p^k\right| \angle \tan^{-1}(b^k/a^k) = \left|E_p^k\right| \angle \delta_p^k \tag{5.26}$$

is used to update the bus reactive power for the $(k+1)$th iteration:

$$Q_p^{k+1} = -\mathbf{Im}\left\{ V_{fp} \angle -\delta_p^k \sum_{\ell \neq p} y_{p\ell} \left(V_{fp} \angle \delta_p^k - E_\ell^k\right) \right\} \tag{5.27}$$

which enhances convergence. The phase-angle changes can be incorporated into either the Gauss or Gauss-Seidel iterative schemes as well as the Newton-Raphson methods to be described in Section 5.4.

There are power systems operated with bus parameters adjusted to satisfy a condition occurring at another location in the power system. For example, a remote-controlled P, V bus has its bus voltage regulated by a nonadjacent generator with variable reactive power injection. Another example is a swing generator whose P, Q output is adjusted to balance an ACE signal measured in the vicinity of the generator, or even in response to an ACE signal from a remote area.

Switched reactance buses have a number of discrete steps of shunt Mvar that are connected to the power system when the calculated voltage reaches predetermined limits. The reactive power varies with the calculated voltage for the next iteration of the load flow according to the connected reactance:

$$Q_p^{k+1} = -\frac{\left|E_p^k\right|^2}{X_p^k} \tag{5.28}$$

where X_p is positive for inductive reactance line to ground.

Variable transformers that tap change under load (TCUL) regulate a voltage at a bus either adjacent to or remote to the transformer. Between iterations, for p on the fixed side of an $\alpha:1$ transformer, the tap ratio is varied to regulate the voltage:

$$\alpha^{k+1} - \alpha^k = \left|E_p^k\right| - \left|V_{fp}\right| \tag{5.29}$$

where V_{fp} is the desired voltage at bus p and E_p^k is the calculated value. Usually, discrete steps or taps are used for α. Example 5.4 demonstrates adjustments of α in a power flow calculation.

An automatic phase-shifting transformer alters taps according to a local or remote difference in power calculated compared to a desired or fixed power (see Section 2.4.1). Consider bus p to be the output side of the transformer. Its p.u. positive-sequence voltage is phase a plus a positive or negative integer, m, of 1% taps added to it from phase b:

$$E_p = \sqrt{3}(E^a + 0.01\, mE^b)$$
$$= \sqrt{3}E^a(1 + 0.01\, me^{\frac{-j2\pi}{3}}) \tag{5.30}$$

This is one method to obtain the phase shift. In many power flow programs, it is assumed that the phase shift varies without changing the magnitude of the voltage output of the transformer. The minimum and maximum of phase-shift range are specified and the tap change between iteration is

$$\alpha^{k+1} - \alpha^k = -\sigma\left\{P_p^k - P_{fp}\right\} \tag{5.31}$$

where σ is a phase-shift constant dependent on the transformer construction. The principle involved in a phase-shifting transformer is that to a first approximation, real power flow in an ac system is proportional to phase-angle differences. If more power flows than specified through a line of the network, then a decrease in phase angle will correct the difference.

The purpose of bus-to-neutral incrementally switched capacitors is that for inductive transmission lines and inductive loads, an increase in capacitance (line to neutral) raises the voltage. The capacitor injects or supplies reactive power to the network. Thus, supplying Mvar to a network raises the voltage level, whether the source is a capacitor bank or a generator increasing its Mvar injection. Mvars from the generator are obtained by setting its voltage regulator to a higher reference level.

Problems at the end of the chapter and examples are used to demonstrate the various types of buses.

5.4   Newton-Raphson Iterative Scheme Using Y_{BUS}

When the simultaneous equations to be iteratively solved are nonlinear,

$$\mathbf{f}(\mathbf{X}) = \mathbf{b} \tag{5.32}$$

where $\mathbf{f}(\mathbf{x})$ is a nonlinear analytic function of the dependent n vector $\mathbf{X}$, and $\mathbf{b}$ is a constant n vector, then the preceding Gauss and Gauss-Seidel iterative schemes can be used. However, the increasing storage capability and computational speed of digital computers, together with sparse matrix methods to invert large matrices, has resulted in the predominance of the Newton-Raphson method to solve Equation 5.32 for the power system load-flow problem. A Taylor series expansion of this equation about the true solution vector, $\mathbf{T}$, is:

$$
\begin{aligned}
\mathbf{f}(\mathbf{X}-\mathbf{T}) &= \mathbf{f}(\mathbf{T}) + \frac{\partial \mathbf{f}}{\partial \mathbf{X}}(\mathbf{X}-\mathbf{T}) + \frac{1}{2}(\mathbf{X}-\mathbf{T})^t \frac{\partial^2 \mathbf{f}}{\partial \mathbf{X}^2}(\mathbf{X}-\mathbf{T})\ldots \\
&= \mathbf{f}(\mathbf{T}) + \mathbf{J}(\mathbf{T})(\mathbf{X}-\mathbf{T}) + \frac{1}{2}(\mathbf{X}-\mathbf{T})^t \,\mathbf{H}(\mathbf{T})(\mathbf{X}-\mathbf{T})\ldots
\end{aligned}
\tag{5.33}
$$

Because they appear in many applications, the first derivative, $\mathbf{J}$, is called the *Jacobian*, and the second derivative, $\mathbf{H}$, is called the *Hessian*. When $\mathbf{f}(\mathbf{X})$ is an analytic function such as in power flow, both the Jacobian and the Hessian may easily be computed.

The Newton-Raphson iterative scheme attempts to find the true solution, $\mathbf{T}$, to Equation 5.32 using successive evaluations of the Jacobian based on the last estimate of $\mathbf{X}$:

$$\mathbf{g}(\mathbf{T}) = 0 = \mathbf{f}(\mathbf{T}) - \mathbf{b} \cong \mathbf{f}(\mathbf{X}^k) - \mathbf{b} + \frac{\partial \mathbf{f}}{\partial \mathbf{X}}(\mathbf{X}^{k+1} - \mathbf{X}^k) = 0 \tag{5.34}$$

so that $\mathbf{X}^k \rightarrow \mathbf{T}$ as $k \rightarrow \infty$. Usually, k is a small number of iterations.

When the two successive estimates for $\mathbf{X}$ are within a small tolerance limit, the $(\mathbf{X}^{k+1} - \mathbf{X}^k)$ goes to zero and $\mathbf{f}(\mathbf{X}^k) - \mathbf{b}$ is also zero at the solution. The elementary iterative scheme solves for the next value of X based on previous quantities:

$$\mathbf{X}^{k+1} = \mathbf{X}^k + \left[\frac{\partial \mathbf{f}}{\partial \mathbf{X}}(\mathbf{X}^k) \right]^{-1} [\mathbf{b} - \mathbf{f}(\mathbf{X}^k)] = \mathbf{X}^k + [\mathbf{J}(\mathbf{X}^k)]^{-1}[\mathbf{b} - \mathbf{f}(\mathbf{X}^k)] \tag{5.35}$$

The Jacobian is updated with the latest $\mathbf{X}^k$ and its inverse computed for each iteration. The following example demonstrates the application of the Newton-Raphson algorithm [1].

Example 5.2

Apply the Newton-Raphson algorithm to find $V_2 \angle \theta$, the per unit voltage and phase angle (rad) at load bus 2, if its real and reactive power are specified as shown in Figure E5.2. All values are in per unit on the generator base. The network has balanced three-phase flow. Iterate until V_2 and θ change less than 0.0001 per iteration. This is a polar form of the Newton-Raphson algorithm.

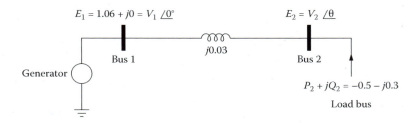

FIGURE E5.2

Solution

At the load bus, the power flow in polar coordinates with bus 1 as the phase reference is written as

$$P_2 + jQ_2 = E_2 I_2^* = V_2(\cos \theta + j\sin \theta) \, I_2^* = V_2 e^{j\theta} I_2^*$$

$$= V_2 e^{j\theta} \left(\frac{E_2 - E_1}{j0.03} \right)^*$$

$$= V_2 e^{j\theta} \left(\frac{V_2 e^{-j\theta} - V_1}{-j0.03} \right)$$

$$= -j \frac{V_1 V_2 e^{j\theta}}{0.03} + j \frac{(V_2)^2}{0.03}$$

$$= -\frac{j}{0.03} \left(V_1 V_2 \, \cos\theta - V_2^2 \right) + \frac{V_1 V_2 \, \sin \theta}{0.03}$$

The power at bus 2 is specified, so the real and reactive parts may be separated into two nonlinear equations and written as the vector:

$$[\mathbf{f(X)} - \mathbf{b}] = 0 = \begin{bmatrix} \dfrac{+V_1 V_2 \sin\theta}{0.03} - P_2 \\ \hline \dfrac{V_2^2 - V_1 V_2 \cos\theta}{0.03} - Q_2 \end{bmatrix} = \begin{bmatrix} f_1(\mathbf{X}) - b_1 \\ f_2(\mathbf{X}) - b_2 \end{bmatrix}$$

In the vector $\mathbf{X}$, the unknowns are $x_1 = \theta$ and $x_2 = V_2$. Since P_2 and Q_2 are specified constants, the partial derivatives in the Jacobian, Equation 5.33, are:

$$\mathbf{J(X)} = \begin{bmatrix} \dfrac{\partial f_1}{\partial x_1} & \dfrac{\partial f_1}{\partial x_2} \\ \hline \dfrac{\partial f_2}{\partial x_1} & \dfrac{\partial f_2}{\partial x_2} \end{bmatrix} = \begin{bmatrix} \dfrac{V_1 V_2 \cos\theta}{.03} & \dfrac{V_1 \cos\theta}{.03} \\ \hline \dfrac{V_1 V_2 \sin\theta}{.03} & \dfrac{2V_2 - V_1 \cos\theta}{.03} \end{bmatrix}$$

The Newton-Raphson iterative scheme in the form of Equation 5.35 is

$$\mathbf{X}^{k+1} = \begin{bmatrix} \theta^{k+1} \\ V_2^{k+1} \end{bmatrix} = \begin{bmatrix} \theta^k \\ V_2^k \end{bmatrix} + 0.03 \begin{bmatrix} V_1 V_2^k \cos \theta^k & V_1 \sin \theta^k \\ V_1 V_2^k \sin \theta^k & 2V_2^k - V_1 \cos \theta^k \end{bmatrix}^{-1} [\mathbf{b} - \mathbf{f}(\mathbf{X}^k)]$$

If the starting voltage for bus 2 is set equal to the reference generator, this is a flat start. With $\theta^0 = 0.0$ and $V_2^0 = 1.06$, the equation above becomes at $k = 0$

$$\begin{bmatrix} \theta^1 \\ V_2^1 \end{bmatrix} = \begin{bmatrix} 0.0 \\ 1.06 \end{bmatrix} + 0.03 \begin{bmatrix} (1.06)^2 & 0 \\ 0 & 1.06 \end{bmatrix}^{-1} \begin{bmatrix} -0.5 & +0 \\ -0.3 & +0 \end{bmatrix}$$

$$= \begin{bmatrix} 0.0 \\ 1.06 \end{bmatrix} + 0.03 \begin{bmatrix} \dfrac{1}{(1.06)^2} & 0 \\ 0 & \dfrac{1}{1.06} \end{bmatrix} \begin{bmatrix} 0.5 \\ 1.3 \end{bmatrix} = \begin{bmatrix} -0.013350 \\ 1.0515 \end{bmatrix}$$

The second and third iterations yield the values

$$\begin{bmatrix} \theta^2 \\ V_2^2 \end{bmatrix} = \begin{bmatrix} -0.013460 \\ 1.0513 \end{bmatrix} \qquad \begin{bmatrix} \theta^3 \\ V_2^3 \end{bmatrix} = \begin{bmatrix} -0.013460 \\ 1.0513 \end{bmatrix}$$

Further iterations do not change the fifth significant decimal place, so that

$$\| \mathbf{X}^3 - \mathbf{X}^2 \| < \begin{bmatrix} 0.0001 \\ 0.0001 \end{bmatrix} = \in$$

And the solution has converged to within this tolerance in three iterations. The solution is $E_2 = 1.0513 \angle -0.771°$. The injection current is $I_2 = -0.471719 + j0.291735$

To apply this method to a load-flow calculation with power injections specified at $n - 1$ buses of the n-bus system, the $2(n - 1)$ vector $\mathbf{X}$ is the *state* of the power system (elements of E_{BUS}):

$$\mathbf{X} = \begin{bmatrix} \delta_1 \\ \delta_2 \\ \vdots \\ \delta_{n-1} \\ V_1 \\ V_2 \\ \vdots \\ V_{n-1} \end{bmatrix} = \begin{bmatrix} \boldsymbol{\delta} \\ \mathbf{V} \end{bmatrix} \tag{5.36}$$

where the slack bus voltage is a known, fixed reference. In Equation 5.36, δ_i is the phase angle of the ith bus with respect to the slack bus and V_i the magnitude of the ith bus voltage. The $2n - 2$ nonlinear equations $\mathbf{f}(\mathbf{X})$ are the calculated real and reactive power injections, while $\mathbf{b}$ is the specified power at the bus. The power flow problem is to find the power

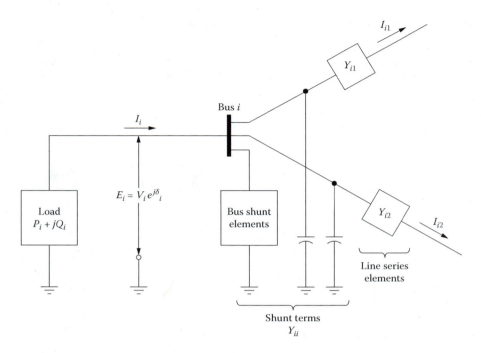

FIGURE 5.5
Bus *I* injection.

system state vector $\mathbf{X}^k$ that satisfies the desired or constrained power at the buses.

$$f(\mathbf{X}^k) = \mathbf{b} \tag{5.37}$$

At convergence for the Newton-Raphson method, two successive iterations are within an error tolerance, ϵ, of each other at every bus:

$$|X_i^{k+1} - X_i^k| = |\Delta X_i^k| \le \epsilon \tag{5.38}$$

At convergence, Equation 5.37 is also satisfied, so the Jacobian, $\partial f/\partial \mathbf{X}$, does not affect the accuracy of the solution. As a consequence, it is possible to use many approximations to the Jacobian matrix. These are discussed in the next section.

The power injection at any bus *i* using an admittance formulation as shown in Figure 5.5 is

$$S_i = P_i + jQ_i = E_i I_1^* = E_i \sum_{\substack{m \ne i}}^{n} Y_{im}^* (E_i^* - E_m^*) + E_i E_i^* y_{ii}^*$$

$$= V_i e^{j\delta_i} \sum_{\substack{m \ne i}}^{n} |Y_{im}| e^{-j\delta_{im}} (V_i e^{-j\delta_i} - V_m e^{-j\delta_m}) + V_i^2 y_{ii}^* \tag{5.39}$$

In this equation, Y_{im} is the series element of a π model for each line connected to bus *i*. The polar form is used to express these series and shunt admittances. If each bus power injection is divided by the voltage at the bus, Equation 5.39 may be written in matrix form as

$$
\begin{bmatrix}
\dfrac{S_1}{E_1} \\
\dfrac{S_2}{E_2} \\
\vdots \\
\dfrac{S_{n-1}}{E_{n-1}}
\end{bmatrix}
= \mathbf{I}^*_{BUS} = \mathbf{Y}^*_{BUS}\mathbf{E}^*_{BUS}
\tag{5.40}
$$

Of course, $\mathbf{E}_{BUS}$ is unknown until the iterative process has been completed. The important point is that the matrix form of Equation 5.39 with the ith row multiplied by voltage E_i retains the same nonzero matrix entries as Equation 5.40, thus the *sparsity* of $\mathbf{Y}_{BUS}$. In Equation 5.39, each self-admittance, y_{ii}, includes shunt admittances due to line-charging capacitance, off-nominal transformers, shunt reactors or capacitor banks, and so on. The right-hand side of Equation 5.39 is separated into real and reactive terms as

$$
P_i = V_i \sum_{\substack{m \neq i}}^{n} |Y_{im}|\{V_i \cos \delta_{im} - V_m \cos (\delta_i - \delta_m - \delta_{im})\} + V_i^2 \ \mathbf{Re} \ (y_{ii}^*)
\tag{5.41}
$$

$$
Q_i = V_i \sum_{\substack{m \neq i}}^{n} |Y_{im}|\{-V_i \sin \delta_{im} - V_m \sin (\delta_i - \delta_m - \delta_{im})\} + V_i^2 \ \mathbf{Im} \ (y_{ii}^*)
\tag{5.42}
$$

which are, respectively, the ith and $(n - 1 + i)$th elements of the nonlinear $\mathbf{f}$ vector. Vector $\mathbf{b}$ is comprised of the specified $2(n - 1)$ bus injections:

$$
\mathbf{b} =
\begin{bmatrix}
P_{f1} \\
P_{f2} \\
\vdots \\
Q_{f1} \\
Q_{f2} \\
\vdots \\
Q_{fn-1}
\end{bmatrix}
\tag{5.43}
$$

The difference between the specified injection and the kth iteration calculated power using Equation 5.40 and Equation 5.41,

$$
\begin{bmatrix} \Delta\mathbf{P} \\ \Delta\mathbf{Q} \end{bmatrix}^k =
\begin{bmatrix}
P_{f1} - P_1 \\
P_{f2} - P_2 \\
\vdots \\
P_{fn-1} - P_{n-1} \\
Q_{f1} - Q_1 \\
\vdots \\
Q_{fn-1} - Q_{n-1}
\end{bmatrix}^k
= \mathbf{b} - \mathbf{f}(\mathbf{X}^k)
\tag{5.44}
$$

is called the bus power *mismatch*. It is another measure of the convergence of an iterative process for power flows, often replacing the voltage criterion for Newton-Raphson based methods.

The Jacobian or gradient, $\partial \mathbf{f}/\partial \mathbf{X}$, is comprised of partial derivatives of P_i and Q_i with respect to the bus phase angles and voltages. Substituting Equation 5.36, Equation 5.41, Equation 5.42, and Equation 5.44 into Equation 5.35, the Newton-Raphson scheme for power flows is

$$
\begin{bmatrix} \delta \\ \hline \mathbf{V} \end{bmatrix}^{k+1} = \begin{bmatrix} \delta \\ \hline \mathbf{V} \end{bmatrix}^{k} + \begin{bmatrix} \frac{\partial \mathbf{P}}{\partial \delta}(\mathbf{X}^k) & \vdots & \frac{\partial \mathbf{P}}{\partial \mathbf{V}}(\mathbf{X}^k) \\ \hline \frac{\partial \mathbf{Q}}{\partial \delta}(\mathbf{X}^k) & \vdots & \frac{\partial \mathbf{Q}}{\partial \mathbf{V}}(\mathbf{X}^k) \end{bmatrix}^{-1} \begin{bmatrix} \Delta \mathbf{P} \\ \hline \Delta \mathbf{Q} \end{bmatrix}^{k}
$$

$$
= \begin{bmatrix} \delta \\ \mathbf{V} \end{bmatrix}^{k} + \begin{bmatrix} \mathbf{J}_1 & \mathbf{J}_2 \\ \mathbf{J}_3 & \mathbf{J}_4 \end{bmatrix}^{-1} \begin{bmatrix} \Delta \mathbf{P} \\ \Delta \mathbf{Q} \end{bmatrix}^{k}
$$

(5.45)

The Jacobian is $2(n-1) \times 2(n-1)$ for power injections specified at $n-1$ buses. Each $(n-1) \times (n-1)$ quadrant of the Jacobian retains the *sparsity* of $\mathbf{Y}_{BUS}$; hence, sparse matrix techniques are suitable to compute the inverse. The Jacobian is usually nonsingular because of diagonal dominance in each quadrant.

The partial derivative elements of the Jacobian in polar form are as follows:

$\mathbf{J}_1$:

$$
\frac{\partial P_i}{\partial \delta_m} = -V_i \mid Y_{im} \mid V_m \sin(\delta_i - \delta_m - \delta_{im})
$$

(5.46a)

$$
= -V_i V_m \{G_{im} \sin(\delta_i - \delta_m) - B_{im} \cos(\delta_i - \delta_m)\} \qquad m \neq i
$$

$$
\frac{\partial P_i}{\partial \delta_i} = V_i \sum_{m \neq i}^{n} \mid Y_{im} \mid V_m \sin(\delta_i - \delta_m - \delta_{im})
$$

(5.46b)

$$
= V_i \sum_{m \neq i}^{n} V_m \{G_{im} \sin(\delta_i - \delta_m) - B_{im} \cos(\delta_i - \delta_m)\}
$$

$\mathbf{J}_2$:

$$
\frac{\partial P_i}{\partial V_m} = -V_i \mid Y_{im} \mid \cos(\delta_i - \delta_m - \delta_{im})
$$

(5.47a)

$$
= -V_i \{G_{im} \cos(\delta_i - \delta_m) + B_{im} \sin(\delta_i - \delta_m)\} \qquad m \neq i
$$

$$
\frac{\partial P_i}{\partial V_i} = 2V_i \ \mathbf{Re} \ (y_{ii}^*) + \sum_{m \neq i}^{n} \mid Y_{im} \mid \{2V_i \cos \delta_{im} - V_m \cos(\delta_i - \delta_m - \delta_{im})\}
$$

(5.47b)

$$
= 2V_i \ \mathbf{Re} \ (y_{ii}^*) + \sum_{m \neq i}^{n} 2V_i G_{im} - V_m \{G_{im} \cos(\delta_i - \delta_m) + B_{im} \sin(\delta_i - \delta_m)\}
$$

J_3:

$$\frac{\partial Q_i}{\partial \delta_i} = V_i \mid Y_{im} \mid V_m \cos (\delta_i - \delta_m - \delta_{im})$$

(5.48a)

$$= V_i V_m \{ G_{im} \cos (\delta_i - \delta_m) + B_{im} \sin(\delta_i - \delta_m) \} \quad m \neq i$$

$$\frac{\partial Q_i}{\partial \delta_i} = -V_i \sum_{m \neq i}^{n} \mid Y_{im} \mid V_m \cos (\delta_i - \delta_m - \delta_{im})$$

(5.48b)

$$= -V_i \sum_{m \neq i}^{n} V_m \{ G_{im} \cos (\delta_i - \delta_m) + B_{im} \sin (\delta_i - \delta_m) \}$$

J_4:

$$\frac{\partial Q_i}{\partial V_m} = -V_i \mid Y_{im} \mid \sin (\delta_i - \delta_m - \delta_{im})$$

(5.49a)

$$= -V_i \{ G_{im} \sin (\delta_i - \delta_m) - B_{im} \cos (\delta_i - \delta_m) \} \quad m \neq i$$

$$\frac{\partial Q_i}{\partial V_i} = 2V_i \ \mathbf{Im} \ (y_{ii}^*) - \sum_{m \neq i}^{n} \mid Y_{im} \mid \{ 2V_i \sin \delta_{im} + V_m \sin (\delta_i - \delta_m - \delta_{im}) \}$$

(5.49b)

$$= 2V_i \ \mathbf{Im} \ (y_{ii}^*) - \sum_{m \neq i}^{n} 2V_i B_{im} + V_m \{ G_{im} \sin(\delta_i - \delta_m) - B_{im} \cos (\delta_i - \delta_m) \}$$

The forms for the Jacobian are for evaluation in the numerical examples and comparison of magnitudes. For example, there are often a few degrees of phase-angle difference between adjacent buses, so $\delta_i - \delta_m$ is small or zero while δ_{im} tends to approach $\pi/2$ for highly reactive lines. As a consequence, in Equation 5.46 to Equation 5.49, approximations can be made to simplify the Jacobian.

If M buses of the system are voltage controlled, then at each of these buses successive iterated voltage changes must be zero:

$$\left. \begin{aligned} \Delta V_m &= V_m^{k+1} - V_m^k = 0 \\ \frac{\partial Q_m}{\partial V_m} &= 0 \end{aligned} \right\} m = 1, 2, \ldots, M$$

(5.50)

and M equations are deleted from the voltage part of Equation 5.45. The Jacobian is then of order $(2n - 2 - M) \times (2n - 2 - M)$, and there are $n - 1$ real power constraints and $n - 1 - M$ reactive power constraints. An alternative method to reducing the order of the equations and Jacobian is to utilize the error between the specified and calculated voltages at the $m = 1, 2, \ldots, M$ controlled buses:

$$\Delta V_m^2 = V_{fm}^2 - \mid E_m \mid^2 = V_{fm}^2 - \left(e_m^2 + f_m^2 \right)$$

(5.51)

where the calculated voltage, E_m, is expressed in rectangular coordinates:

$$e_m = V_m \cos \delta_m \tag{5.52a}$$

$$f_m = V_m \sin \delta_m \tag{5.52b}$$

The Jacobian elements for this error measure are

$$\frac{\partial \Delta V_m^2}{\partial e_i} = \begin{cases} 0 & \text{if } i \neq m \\ -2e_m & \text{if } i = m \end{cases} \tag{5.53}$$

$$\frac{\partial \Delta V_m^2}{\partial f_i} = \begin{cases} 0 & \text{if } i \neq m \\ -2f_m & \text{if } i = m \end{cases} \tag{5.54}$$

By retaining all $(2n - 2)$ equations a bus may be switched from a $P - V$ type to a $P - Q$ type during iterations if Q exceeds its limits.

The complete formulation for the Newton-Raphson iterative method for load flows in rectangular coordinates including voltage-controlled buses is

$$\begin{bmatrix} e \\ -- \\ f \end{bmatrix}^{k+1} = \begin{bmatrix} e \\ -- \\ f \end{bmatrix}^{k} + \begin{bmatrix} \frac{\partial \mathbf{P}}{\partial e}\left(\mathbf{E}_{\text{BUS}}^k\right) & \vdots & \frac{\partial \mathbf{P}}{\partial f}\left(\mathbf{E}_{\text{BUS}}^k\right) \\ \frac{\partial \mathbf{Q}}{\partial e}\left(\mathbf{E}_{\text{BUS}}^k\right) & \vdots & \frac{\partial \mathbf{Q}}{\partial f}\left(\mathbf{E}_{\text{BUS}}^k\right) \\ -\frac{\partial \Delta \mathbf{V}_m^2}{\partial e_m} & \vdots & -\frac{\partial \Delta \mathbf{V}_m^2}{\partial f_m} \end{bmatrix}^{-1} \begin{bmatrix} \Delta \mathbf{P} \\ ---- \\ \Delta \mathbf{Q} \\ \Delta \mathbf{V}_M^2 \end{bmatrix} \tag{5.55}$$

The voltage-controlled buses are ordered *last* (even using optimal ordering) to avoid a possible singularity condition due to the zeros of Equation 5.53 and Equation 5.54. In Equation 5.55, the bus voltages are in rectangular components:

$$E_i = e_i + jf_i = V_i \cos \delta_i + jV_i \sin \delta_i \tag{5.56}$$

and the power system admittances are also in rectangular coordinates:

$$Y_{im} = G_{im} + jB_{im} = |Y_{im}| (\cos \delta_{im} + j \sin \delta_{im})$$

The elements of the Jacobian may also be derived in rectangular components from Equation 5.39. The Jacobian entries of Equation 5.55 in rectangular form are as follows:

$$\frac{\partial P_i}{\partial e_m} = -e_i G_{im} - f_i B_{im} \qquad i \neq m \tag{5.57a}$$

$$\frac{\partial P_i}{\partial e_i} = -2e_i \, \mathbf{Re}(y_{ii}^*) + \sum_{m \neq i}^{n} \{2e_i G_{im} - e_m G_{im} + f_m B_{im}\} \tag{5.57b}$$

$$\frac{\partial P_i}{\partial f_m} = -e_i B_{im} - f_i G_{im} \qquad i \neq m \tag{5.57c}$$

$$\frac{\partial P_i}{\partial f_i} = -2f_i \ \mathbf{Re} \ (y_{ii}^*) + \sum_{m \neq i}^{n} \{2f_iG_{im} - e_mB_{im} - f_mG_{im}\} \tag{5.57d}$$

$$\frac{\partial Q_i}{\partial e_m} = -f_iG_{im} + e_iB_{im} \qquad i \neq m \tag{5.57e}$$

$$\frac{\partial Q_i}{\partial e_i} = 2e_i \ \mathbf{Im} \ (y_{ii}^*) + \sum_{m \neq i}^{n} \{-2e_i \ B_{im} + e_m \ B_{im} + f_m \ G_{im}\} \tag{5.57f}$$

$$\frac{\partial Q_i}{\partial f_m} = f_iB_{im} + e_iG_{im} \qquad i \neq m \tag{5.57g}$$

$$\frac{\partial Q_i}{\partial f_i} = -2f_i \ \mathbf{Im} \ (y_{ii}^*) + \sum_{m \neq i}^{n} \{-2f_i \ B_{im} - e_m \ G_{im} + f_m \ B_{im}\} \tag{5.57h}$$

In either polar form or rectangular component form, the Newton-Raphson method exhibits strong convergence. When sparse matrix techniques are used to compute the inverse of the Jacobian, the computational disadvantage is overcome, and the Newton-Raphson method can solve power flow problems for the widest range of networks. Table 5.1 compares the Newton-Raphson method with Gauss-Seidel techniques.

TABLE 5.1

Comparison of Two Load-Flow Methods

Type of Problem	Gauss-Seidel Method	Newton-Raphson Method
1. Heavily loaded systems	Usually cannot solve systems with more than 70° phase shift	Solves systems with shifts up to 90°
2. Systems containing negative reactance such as three-winding transformers or series line capacitors	Unable to solve	Solves with ease
3. Systems with slack bus at a desired location	Often requires trial and error to find a slack bus location that will yield a solution	More tolerant of slack bus location
4. Long and short lines terminating on the same bus	Usually cannot solve if long-to-short ratio is above 1,000:1	Can solve a system with a long-to-short ratio at any bus of 1,000,000:1
5. Long radial type of system	Difficulty in solving	Solves a wider range of such problems
6. Acceleration factors	Number of iterations depends on choice of factor	None required

Source: Harold Wood in reviewing Tinney and Hart's paper in "Power Flow by Newton's Method," *IEEE Trans. Power Appar. Syst.*, 86, 1967.

Example 5.3

A six-bus, seven-line power system is shown in Figure E5.3.1. The bus load schedule (p.u.) is as follows:

Bus	P	Q
1	−0.577	− 0.100
2	−0.100	− 0.050
3	+0.800	+ 0.100
4	−0.250	− 0.150
5	−0.200	− 0.150
6	Slack at 1.06 p.u.	

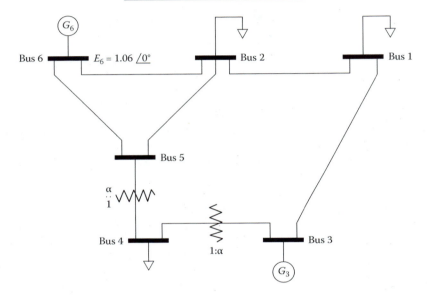

FIGURE E5.3.1
System diagram.

The line parameters for a 100 MW base are:

Line	R	X	ωC/2 Charging or Turns Ratio
1–2	0.02	0.035	0.025
1–3	0.03	0.080	0.030
3–4	0	0.060	$\alpha = 1.02$
5–4	0	0.080	$\alpha = 1.01$
2–5	0.04	0.150	0.035
2–6	0.01	0.030	0.030
5–6	0.05	0.180	0.010

Apply the Newton-Raphson load-flow method to calculate the system state. Use the generator at bus 6 with voltage held at 1.06 p.u. as the slack bus. Iterate until both the real and reactive power mismatch at all buses is less than 0.0001 p.u. on a 100 MW floating voltage base.

Solution

The Jacobian for this example is a 10×10 matrix. The initial Jacobian for $k = 0$ is shown in Figure E5.3.2.

	$\partial P/\partial \delta$					$\partial P/\partial V$			
1	2	3	4	5	1	2	3	4	5
36.51	−24.20	−12.31	0	0	17.4	−13.05	−4.356	0	0
−24.20	64.90	0	0	−6.993	−13.05	25.41	0	0	−1.759
−12.31	0	30.67	−18.36	0	−4.356	0	4.356	0	0
0	0	−18.36	32.27	−13.91	0	0	0	0	0
0	−6.993	0	−13.91	26.69	0	−1.759	0	0	3.278
−18.45	13.83	4.618	0	0	34.33	−22.83	−11.62	0	0
13.83	−26.93	0	0	1.865	−22.83	61.04	0	0	−6.598
4.618	0	−4.618	0	0	−11.62	0	28.19	−17.32	0
0	0	0	0	0	0	0	−17.32	31.39	−13.120
0	1.865	0	0	−3.475	0	−6.598	0	−13.12	24.830

| | $\dfrac{\partial Q}{\partial \delta}$ | | | $\dfrac{\partial Q}{\partial V}$ | |

FIGURE E5.3.2
Jacobian for iteration $k = 0$.

Observe the nonzero pattern in each quadrant, which is the same as $\mathbf{Y}_{\text{BUS}}$. Except for the initial flat start iteration, the quadrants are not symmetrical. The power mismatch at the buses for iterations $k = 0, 1, 2$ is given in Figure E5.3.3.

Iter.	$\Delta\delta$, Change in Bus Phase Angle				
k	Bus 1	Bus 2	Bus 3	Bus 4	Bus 5
1	−0.01052	−0.00671	0.02034	0.00049	−0.00774
2	0.00033	0.00010	0.00084	0.00045	0.00022

Iter.	ΔV, Change in Bus Voltage				
k	Bus 1	Bus 2	Bus 3	Bus 4	Bus 5
1	−0.00074	−0.00036	−0.00054	−0.00230	−0.00122
2	−0.00018	−0.00010	−0.00029	−0.00030	−0.00019

Iter.	ΔP, Real Power Mismatch				
k	Bus 1	Bus 2	Bus 3	Bus 4	Bus 5
0	−0.57700	−0.10000	0.80000	−0.25000	−0.20000
1	−0.00471	0.00003	0.00435	−0.00233	−0.00481
2	−0.00002	−0.00001	0.00005	−0.00002	−0.00002

Iter.	ΔQ, Reactive Power Mismatch				
k	Bus 1	Bus 2	Bus 3	Bus 4	Bus 5
0	−0.03820	−0.05112	0.49370	−0.65625	0.03824
1	−0.00682	−0.00115	−0.01296	−0.01799	−0.00021
2	0	0	−0.00004	−0.00005	0

FIGURE E5.3.3
Incremental changes during solution.

The mismatch for the second iteration is less than 10^{-4} at any bus for both real and reactive power; the convergence tolerance is satisfied and the iterations are halted. Observe that convergence is very rapid, with ΔP and ΔQ decreasing by ratios $1/20$ to $1/100$ per iteration. The real power, for this example, converges faster than reactive power. Bus 4 is the slowest to converge. If a smaller convergence tolerance is used, it accordingly takes more iterations to reach the desired solution.

For $k = 0$ the bus voltage magnitudes and phase angles are set to the slack bus values for a flat start. The changes in bus phase angles and voltage magnitudes are also shown in Figure E5.3.3. These changes are calculated at each iteration of the Newton-Raphson formula, Equation 5.45.

In the computer program, the calculated bus power injection at each bus is obtained on a line-by-line basis, summing the line flows to obtain the total injection, Equation 5.41 and Equation 5.42. For the converged solution, the line flows are shown in Figure E5.3.4 from the bus into the line. Notice that real power into a line and out of the line differ because of resistive losses. Only lines 4–5 and 3–4 do not have resistive losses, as their resistance is negligible compared to the transformer reactance. The reactive power into a line and out of the line differ due to line-charging capacitance and inductance in the series element. The power injected by the slack generator is calculated from the system state and line parameters because the load flow does not perform this calculation.

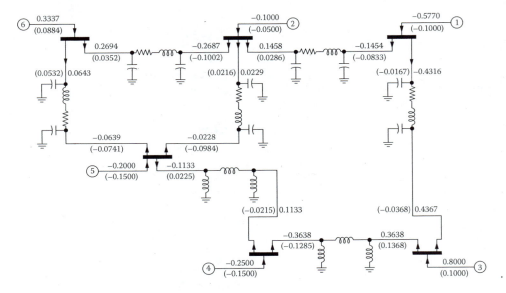

FIGURE E5.3.4
Injections and line for six-bus test case. (Real power flow is given above the line, and reactive power flow is in parentheses below the line; 100 MVA base.)

The state of the power system after the iterations is as follows:

Bus	V (p.u.)	δ (deg)
1	1.0509	−0.584
2	1.0555	−0.379
3	1.0625	1.213
4	1.0340	0.0538
5	1.0461	−0.4312
6	1.0600	0

© The program *p_flow*, included in the software, computes the power flow for any network with properly formatted data. Bus injection and line parameter data for Example 5.3 are formatted in the file *example53.sss*. Copy the *example53.sss* into *p_flow.sss*, then execute *p_flow*. The computed results match the above results within round off. The file *p_flow.sss* can be edited to examine other cases. Note the transmission lines in *example53.sss* are pi lines, Ltype = 1, and transformers are modeled by off-nominal pi's.

Example 5.3 demonstrates the strong convergence of the Newton-Raphson method. For this example and in general, the rate of convergence depends on the number of buses and lines, as well as the line parameters, load distribution, and topology of the network. Because power flow calculations are nonlinear, it is not possible to predict the number of iterations necessary to converge to within a specified tolerance. Prior solutions for the network state are often used for better starting values in order to reduce the number of iterations to convergence.

Adjustments are often made on bus injections or network parameters during the iterations of a load flow. Examples of these adjustments are the tap position of a TCUL transformer that regulates a bus voltage, selecting an increment of a capacitor bank to correct a line power factor, and so on. The adjustment is usually made in response to an error detected after an iteration. The new tap setting or parameter is entered for the next iteration, with the result that another source must conversely change.

Let a group of load and generation buses constitute an area of the power system. If this area is scheduled to have zero ACE, the real power generation is modified for the next iteration. As a consequence, the network current flow pattern is changed, the transmission line losses change, and the slack bus supplies incremental power equal and opposite to the area and losses. The shifting of real power in this case, or reactive power in other situations, decreases the rate of convergence of a load-flow calculation.

The following example demonstrates the application of Equation 5.29 to regulate the voltage of a load bus by means of varying a local transformer tap ratio.

Example 5.4

A four-bus, five-line power system is shown in Figure E5.4.1. The bus load schedules (p.u.) are:

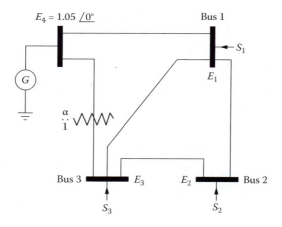

FIGURE E5.4.1

Bus	P	Q
1	−0.3000	−2.9300
2	−0.4000	−0.3000
3	0.4000	0.4000

The line parameters (100 MW base) are:

Line	R	X	1/2 Charging/Turns
1–2	0.02	0.06	0
2–3	0.01	0.03	0
3–1	0.02959	0.071	0
4–3	0	0.05	$\alpha = 0.9625$
4–1	0.00588	0.02353	0

The transformer in the line from bus 4 to bus 3 is of the TCUL type with automatic tap changing, which regulates the voltage at bus 3 at 1.0 ± 0.01 for all load conditions. The transformer tap ratio is adjustable between the limits $0.955 \le \alpha \le 1.045$ in increments of 0.00375. The generator at bus 4 is the slack bus at voltage 1.05 p.u. Use the Newton-Raphson method for load flows until the real power mismatch at any bus is less than 0.001 p.u. and the reactive power mismatch at any bus is less than 0.01 p.u.

Solution

After each new voltage vector is calculated during the iterations, the new tap setting is obtained using Equation 5.29 by means of a computer integer function that truncates to the lowest integer value:

$$\alpha^{k+1} = \alpha^k + 0.003758 * \text{INT} \left[(|E_3^k| - V_{fp})/0.00375 + 0.5000 \right]$$

The function name INT is different for various programming languages, but the operation is identical.

The results of a Newton-Raphson iterative scheme to solve for the system state from a flat start are given in Figure E5.4.2.

Power Mismatch

Iteration k	ΔP_1	ΔQ_1	ΔP_2	ΔQ_2	ΔP_3	ΔQ_3	α
0	−0.30000	−2.9300	−0.40000	−0.30000	0.40000	1.25909	0.9625
1	−0.00946	−0.12869	−0.01699	−0.00538	−0.00470	−1.02008	1.00750
2	−0.00137	−0.00178	−0.0015	−0.00002	0.00165	−0.33950	1.0225
3	−0.00021	0.00001	−0.00013	0.00001	0.00049	−0.15430	1.03000
4	−0.00004	0.00000	−0.00003	0.00000	0.00023	−0.07508	1.03375
5	−0.00001	0.00000	−0.00001	0.00000	0.00001	−0.00016	1.03375

FIGURE E5.4.2
Power mismatches and off-nominal tap ratio *a*. Transformer tap ratio varied to regulate $|V_3| = 1.0$.

Observe the reactive power at bus 3 is the mismatch that delays the convergence. The voltages and phase angles after five iterations are as follows:

Bus	Voltage	Phase Angle (deg.)
1	0.985628	0.375676
2	0.989008	−0.115006
3	1.00387	0.163526
4	1.05	0

After the first iteration the tap changer moves +12 positions, then subsequently +4, +2, +1, and 0 positions during the iterations..

Two additional load-flow solutions, with the tap changer fixed at $\alpha = 1.000$ and $\alpha = 0.9625$, converge within the same power mismatch tolerance in three iterations and result in the following bus voltages:

Bus	$\alpha = 1.0000$ Voltage	$\alpha = 1.0000$ Phase Angle (deg.)	$\alpha = 0.9625$ Voltage	$\alpha = 0.9625$ Phase Angle (deg.)
1	0.993270	0.345105	1.00237	0.309835
2	1.00488	−0.220623	1.02380	−0.342480
3	1.02365	0.0060436	1.04725	−0.173784
4	1.05	0	1.05	0

© The program *p_flow*, included in the software, computes the power flow for any network with properly formatted data. Bus injection and line parameter data for Example 5.4 are formatted in the file *example54.sss*. Copy the file *example54.sss* into the file *p_flow.sss* and execute *p_flow* to obtain the results for a fixed tap bus 4 to 3 of $\alpha = 0.9625$. Observe that the α side of the transformer is always the "from bus." Also note a transformer is Ltype = 2 in the data file.

Example 5.4 shows how adjusting transformer taps during solution requires more iterations for convergence to the same bus power mismatch tolerance. This is a computational burden to the Newton-Raphson method that requires only a few iterations for convergence. Each additional iteration represents a proportionately large increase in execution time. As a result, some operational load-flow programs alternate a Newton-Raphson iteration with a relatively fast Gauss-Seidel iteration and perform the tap adjustment in the Gauss-Seidel portion.

In general, adjusted solutions require more iterations than a straightforward load flow for convergence to a solution of the same tolerance. The type of adjustment—TCUL transformers, phase shifters to regulate a line real power flow, generator reactive power to regulate voltage at a remote bus, and so on—as well as the basic load-flow method determine the adjustment scheme. There does not appear to be a universally applicable technique to treat adjustments.

5.4.1 Approximations to the Jacobian in the Newton-Raphson Method

The power system *state* vector in polar coordinates,

$$\Delta X^k = \begin{bmatrix} \Delta \delta^k \\ \Delta V^k \end{bmatrix}$$

(5.58)

is calculated using the Newton-Raphson iterative scheme:

$$\mathbf{X}^{k+1} = \mathbf{X}^k + [\mathbf{J}(\mathbf{X}^k)]^{-1} \begin{bmatrix} \mathbf{P}_f - \mathbf{P} \\ \mathbf{Q}_f - \mathbf{Q} \end{bmatrix}^k = \mathbf{X}^k + \begin{bmatrix} \mathbf{J}_1 & \mathbf{J}_2 \\ \mathbf{J}_3 & \mathbf{J}_4 \end{bmatrix}^{-1} \begin{bmatrix} \Delta \mathbf{P} \\ \Delta \mathbf{Q} \end{bmatrix}^k \qquad (5.59)$$

where the Jacobian determines convergence, but to a great extent does not affect final accuracy because $\Delta \mathbf{P}^k \to 0$, $\Delta \mathbf{Q}^k \to 0$, and $\Delta \mathbf{X}^k \to 0$ as $k \to \infty$. Therefore, a number of iterative methods have approximated the Jacobian in order to reduce computer storage requirements or arithmetic computations. The most significant of these approximations is to *decouple* the real power part of the Jacobian from the reactive part:

$$\mathbf{J}(\mathbf{X}^k) = \begin{bmatrix} \dfrac{\partial \mathbf{P}}{\partial \delta}(\mathbf{X}^k) & 0 \\ 0 & \dfrac{\partial \mathbf{Q}}{\partial \mathbf{V}}(\mathbf{X}^k) \end{bmatrix} \qquad (5.60)$$

This is justified because to a first approximation, the real power flow in an ac system depends on phase-angle difference, and the reactive flow depends on voltage magnitude differences. In addition, when transmission lines are highly reactive, the ratio of their inductive to resistance parts is high, so that δ_{im},

$$\delta_{im} = \tan^{-1}\left\{ \frac{\mathbf{Im}(Y_{im})}{\mathbf{Re}(Y_{im})} \right\} = \tan^{-1}\left\{ \frac{B_{im}}{G_{im}} \right\} \qquad (5.61)$$

approaches $-70°$ for B/G on the order of 3. As a consequence, the incremental power terms, Equation 5.46 to Equation 5.49, are such that

$$\frac{\partial P_i}{\partial \delta_m} = -V_i \,|\, Y_{im} \,|\, V_m \sin(\delta_i - \delta_m - \delta_{im}) >> -V_i \,|\, Y_{im} \,|\, \cos(\delta_i - \delta_m - \delta_{im}) = \frac{\partial P_i}{\partial V_m} \qquad (5.62a)$$

and also

$$\frac{\partial Q_i}{\partial \delta_m} = -V_i \,|\, Y_{im} \,|\, V_m \cos(\delta_i - \delta_m - \delta_{im}) << -V_i \,|\, Y_{im} \,|\, \sin(\delta_i - \delta_m - \delta_{im}) = \frac{\partial Q_i}{\partial V_m} \qquad (5.62b)$$

so the antidiagonal quadrants of Equation 5.60 are approximated as zero. This decouples the Jacobian to an $(n - 1) \times (n - 1)$ matrix and an $(n - 1 - M) \times (n - 1 - M)$ matrix when M voltage-controlled buses are in the network.

Investigators [2] using what is termed a *fast-decoupled load flow* (FDLF) have made further approximations to the Jacobian. Because transmission lines are predominantly reactive and the phase difference between connected buses is small, the incremental power through the series element of the π transmission line is

$$\frac{\partial P_i}{\partial \delta_m} = V_i \,|\, Y_{im} \,|\, V_m \sin(\delta_i - \delta_m - \delta_{im})$$

$$\simeq +V_i \,|\, Y_{im} \,|\, V_m \sin \delta_{im} = V_i V_m B_{im} \qquad (5.63)$$

and incremental reactive power is

$$\frac{\partial Q_i}{\partial V_m} \simeq V_i \mid Y_{im} \mid \sin \delta_{im} = V_i B_{im} \tag{5.64}$$

Since V_i appears in each element of the ith row of the Jacobian, the power mismatch may be divided by V_i (see Equation 5.41 and Equation 5.42). The voltage terms V_m remaining in the Jacobian, Equation 5.63 and Equation 5.64, are set to unity because they are nominally close to this value. The result is a Jacobian comprised of only constant *susceptance* terms:

$$\begin{bmatrix} \dfrac{\Delta P_1}{V_1} \\ \vdots \\ \dfrac{\Delta P_{n-1}}{V_{n-1}} \\ \hline \dfrac{\Delta Q_1}{V_1} \\ \vdots \\ \dfrac{\Delta Q_{n-1-M}}{V_{n-1-M}} \end{bmatrix} = \begin{bmatrix} -\mathbf{B}' & 0 \\ \hline 0 & -\mathbf{B}'' \end{bmatrix} \begin{bmatrix} \Delta \delta \\ \hline \Delta \mathbf{V} \end{bmatrix} \tag{5.65}$$

where $\mathbf{B}'$ is of order $n-1$ and $\mathbf{B}''$ is of order $n-1-M$.

Finally, the FDLF assumes for the Jacobian that shunt reactances from buses to ground are negligible and that all transformers have unity for the off-nominal tap ratio, without any phase shift. The resulting iterative method for the FDLF is

$$\begin{bmatrix} \delta \\ \hline \mathbf{V} \end{bmatrix}^{k+1} = \begin{bmatrix} \delta \\ \hline \mathbf{V} \end{bmatrix}^{k} + \begin{bmatrix} -\mathbf{B}' & 0 \\ \hline 0 & -\mathbf{B}'' \end{bmatrix}^{-1} \begin{bmatrix} \dfrac{\Delta P_1}{V_1} \\ \vdots \\ \dfrac{\Delta P_{n-1}}{V_{n-1}} \\ \hline \dfrac{\Delta Q_1}{V_1} \\ \vdots \\ \dfrac{\Delta Q_{n-1-M}}{V_{n-1-M}} \end{bmatrix} \tag{5.66}$$

where $\mathbf{B}'$ and $\mathbf{B}''$ are line susceptances from the $\mathbf{Y}_{BUS}$ matrix. Observe that $\mathbf{B}'$ has the non-zero pattern for elements corresponding to $\mathbf{Y}_{BUS}$, so sparse matrix methods are applicable. The matrix $\mathbf{B}''$ is also sparse for large networks, and is decoupled, so it is separately triangularized. The advantage of the FDLF is that $\mathbf{B}'$ and $\mathbf{B}''$ do not vary with iterations, even if transformer taps vary, so they are constant matrices that are inverted only one time and repeat solutions are calculated for different right-hand-side vectors. More iterations are

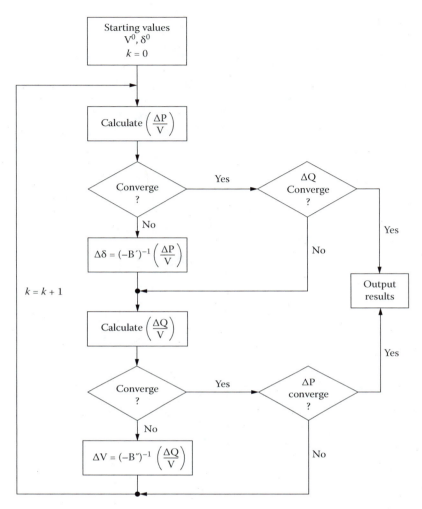

FIGURE 5.6
Iterative loop for the fast-decoupled load-flow algorithm.

required for the FDLF to converge compared to a Newton-Raphson method, but there is a net computational advantage because of the once-only triangularization on smaller matrices, less storage, and half the multiplications of inverse times the right-hand-side vector (power mismatch). The iterative flowchart is shown in Figure 5.6. The FDLF exhibits best convergence when restricted to sequentially calculating δ^{k+1}, then V^{k-1} as opposed to several iterations on phase angle, then one voltage iteration, for example. The FDLF is not as strongly convergent as the Newton-Raphson method, and furthermore, the FDLF may diverge for power systems without reactance-dominated transmission lines. In fact, dummy lines or dummy buses may be introduced in order to achieve the reactance-dominated condition. Development continues on methods to approximate the Jacobian.

Example 5.5

Apply the fast-decoupled load-flow algorithm using the **B'** matrix to find the state of the power system shown in Figure E5.5.1. The data are shown in the figure.

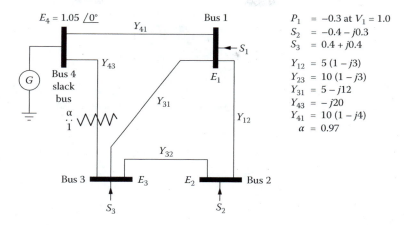

FIGURE E5.5.1

Consider the solution converged when the real power mismatch at any bus is less than 0.001 p.u. and the reactive power is less than 0.01 p.u. on a 100 MW base. Bus 1 is voltage-controlled to $|E_1| = 1.0$. The transformer tap ratio is 0.97.

Solution

The power balance at buses 1, 2, and 3 is

$$S_1 = -0.3 + jQ_1 = E_1 I_1^* = V_1 e^{j\delta_1} [Y_{41}^*(E_1^* - E_4^*) + Y_{21}^*(E_1^* - E_2^*) + Y_{31}^*(E_1^* - E_3^*)]$$

$$S_2 = -0.4 - j0.3 = V_2 e^{j\delta_2} [Y_{12}^*(E_2^* - E_1^*) - Y_{23}^*(E_2^* - E_3^*)]$$

$$S_3 = 0.4 + j0.4 = V_3 e^{j\delta_3} [Y_{32}^*(E_3^* - E_2^*) + Y_{31}^*(E_3^* - E_1^*) + Y_{34}^*(E_3^* - E_4^*)]$$

The reactive power mismatch at bus 1 is set to zero because the bus is voltage controlled to be 1.0 p.u. The matrix $-\mathbf{B}'$ may be written by inspection from the network as

$$-\mathbf{B}' = \begin{bmatrix} 40+12+15 & -15 & -12 \\ -15 & 15+30 & -30 \\ -12 & -30 & 30+12+20 \end{bmatrix}$$

The same matrix is used for voltage iterations with the reactive power mismatch vector

$$\frac{\Delta \mathbf{Q}}{\mathbf{V}} = \left[0.0, \frac{\Delta Q_2^k}{V_2^k}, \frac{\Delta Q_3^k}{V_3^k} \right]'$$

Using a flat start for bus 2 and bus 3, the real power mismatch vector needed to start the iterations is

$$\left[\frac{\Delta P_1^0}{V_1^0}, \frac{\Delta P_2^0}{V_2^0}, \frac{\Delta P_3^0}{V_3^0} \right] = [-0.3000, -0.3809, +0.3809]$$

The results of the iterations are given below in Figure E5.5.2.

Iteration k	$\Delta P_1/V_1$	$\Delta P_2/V_2$	$\Delta P_3/V_3$	δ_1 (rad)	δ_2 (rad)	δ_3 (rad)	$\Delta Q_2/V_2$	$\Delta Q_3/V_3$	V_2	V_3
1	−0.30000	−0.3809	+0.3809	−0.00686	−0.01113	−0.0005692	−1.2114	0.57691	1.0139	1.0389
2	0.76512	0.17802	−0.44404	−0.00610	−0.00350	−0.001528	0.17326	0.02297	1.0218	1.0435
3	0.02736	−0.07784	0.01239	0.00572	−0.00611	−0.02669	−0.06503	−0.01984	1.0186	1.0415
4	−0.02598	−0.02730	0.00122	0.00558	−0.00529	−0.00227	0.02470	0.01226	1.0199	1.0423
5	0.01181	−0.01044	−0.00151	0.00566	−0.00560	−0.00243	−0.01022	−0.00548	1.0194	1.0420
6	−0.00505	0.00430	0.00073	0.00563	−0.00547	−0.00237				
7		Converged								

FIGURE E5.5.2
Power mismatch and calculated state for a four-bus test case using the FDLF algorithm. Bus 1 is a regulated *P-V* bus.

The reactive power converged at the fifth iteration, but another real power iteration was required, so the fast-decoupled load flow for this example converged in 5½ iterations.

5.5 Adjustment of Network Operating Conditions

The steady-state methods presented thus far have not provided insight as to how network real power flow or voltage levels are adjusted for seasonal and daily variations in load reconfigurations in the network, or power exchange with select neighbors. The contingency methods of Chapter 4 are "fast" methods to determine the approximate effects of the wide variety of planned and abnormal events that occur continuously on bulk power transmission networks. However, an accurate load flow is usually required to determine real power flow or voltage levels both near and remote from the location of the event.

The system control devices used to redistribute real power flow on the network are fewer than those for reactive flow simply because there are far fewer real power sources. Among the system devices that can be remotely controlled by an operator (dispatcher) to adjust real power flows directly are:

1. Generator real power set point
2. Phase-shift transformer positions
3. Pumped storage input or output
4. Load shedding
5. Switched series capacitance
6. Solid-state electronically controlled series flexible alternating currents transmission system (FACTS) devices

Of these alternatives, method 4 is undesirable because service to consumers is interrupted, so this is mainly an emergency control to be used when demand exceeds generation capability. Item 3 is not used as a normal control, but more as a periodic control. For example, it may be used to reallocate the daily load cycle for economic reasons by means of pumping water up to a higher reservoir during minimum load periods, then using the same apparatus as a hydrogeneration plant during peak-load periods, as shown in Figure 5.7.

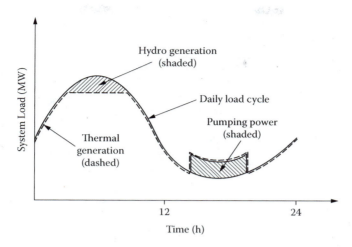

FIGURE 5.7
Daily load cycle showing recovery of pumped storage energy.

The disadvantages of pumped storage facilities are they are more expensive than conventional hydrogeneration, and they generally are located at remote sites, from where they cannot redistribute real power near the load center of the network.

Power system planners and dispatchers rely primarily on shifting generator outputs to alter real power flow patterns, and in a few systems, upon phase-shift transformers. Reactive power flow is controlled by voltage regulation devices (compensation) that planners strategically locate on the network. These compensators are used to obtain desired reactive flows for power factor correction, and desired voltage. The most common devices for reactive compensation are as follows:

1. Generator reactive power set points
2. Switched shunt capacitor banks
3. Switched inductor banks
4. Changes in transformer taps
5. Switching of long extra-high-voltage (EHV) lines that have substantial charging capacitance
6. Synchronous condenser set points
7. Solid-state electronically controlled shunt reactances (FACTS devices)

Observe that devices 2, 3, and 5 are incremental, while the others may be adjusted almost continuously over a range of values.

The voltage regulators used for generator field excitation have fast dynamic response, so they rapidly change the generator's output reactive power to regulate high-voltage transmission lines to which they are connected. All methods that employ high-power electronic devices have comparatively faster response, so they are applied in critical locations to regulate a voltage that fluctuates due to a random load, or provide dynamic stability to the system.

In general, the type of fixed or variable compensation that is applied satisfies power system dynamic considerations, which are not treated in this book, and is cost-effective for steady-state power flow. Table 5.2 presents a general overview of FACTS devices, closely following reference [3]. All the FACTS devices may be remotely controlled from the SCADA

TABLE 5.2

Characteristics of Power Electronic Devices (FACTS Devices) for Power Systems

Principle	Devices	Power Flow	Stability	Voltages	Scheme
Vary series impedance for power flow	TCSC (Thyristor Controlled Series Compensation)	Medium	Strong	Small	
	TCPAR (Thyristor Controlled Phase Angle Compensation)	Strong	Strong	Small	
Shunt control of voltages	SVC (Static Var Compensator)	Low	Medium	Strong	
	STATCOM (Static synchronous Compensator)	Low	Medium	Strong	
Power flow control	HVDC (High Voltage Direct Current)	Strong	Strong	Medium	
	UPFC (Unified Power Flow Controller) For real and reactive power control	Strong	Strong	Strong	

(supervisory control and data acquisition) system, and have fast response, on the order of one line frequency cycle, such that they may be applied for dynamic response as well as to adjust steady-state power flow and voltages. The first FACTS devices were mercury arc controlled converters at both ends of a HVDC (high voltage direct current) line. The mercury arc devices were replaced solid-state SCR valves [4].

The static var generator (SVG), also called static var compensator (SVC) [5], is one of the most employed FACTS devices. Its typical installation would be near large time-varying industrial loads such as arc furnaces used in steelmaking. A schematic diagram of a solid-state SVC is shown in Figure 5.8. The capacitor bank in Figure 5.8 may be one of several that are switched in to provide a range of shunt compensation.

The arc furnace shown in Figure 5.8 is an excellent application for the fast response of the SVC because the load is time varying with power fluctuations distributed from steady state to 10 or 20 Hz. By sensing the critical bus voltage disturbances and controlling the thyristors shown in Figure 5.8 (response time two cycles or less of the 60 Hz waveform), the arc furnace fluctuations are compensated by reactive power flow into the L element.

The silicon-controlled rectifiers/thyristors (SCRs) shown in Figure 5.8 have the capability of steady-state partial "turn-on" or firing angle control to adjust reactor 60 Hz current from zero to maximum value. A thyristor "string" is shown in Figure 5.9a. The thyristor strings are for indoor use only. The reactors and capacitors are generally outdoors, as shown in Figure 5.9b.

Depending on the relative size of C or L, the static var compensator is capable of supplying or removing var from the network. For steady-state power flow calculations, the reactive power of the SVC is scheduled between the lower limit of full current in the

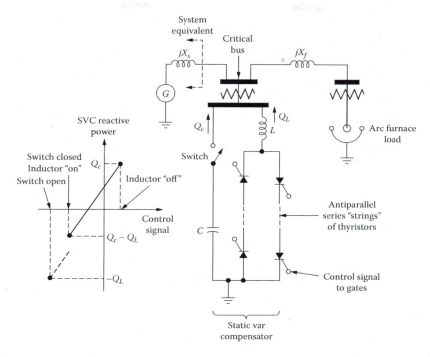

FIGURE 5.8
Static var compensator (SVC) used to control voltages at a critical bus. The inductors (L) and thyristors are shown as line to neutral for equivalence. They are usually connected line to line to suppress the third harmonic generated.

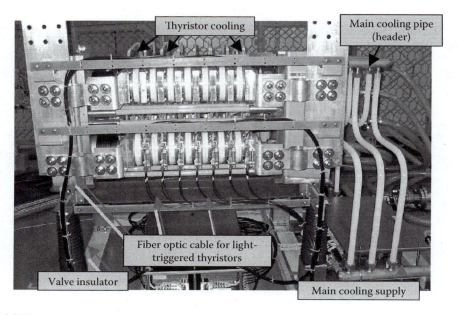

FIGURE 5.9(a)
Antiparallel light-triggered thyristor strings rated at 16.5 kV, 2.6 kA rms and cooled by demineralized water through heat sinks. Thyristors are white-colored elements alternated with heat sinks. Used to control one line-to-line inductor of a static var compensator. Indoor use only. (Photograph used with permission of, and copyrighted by, Mitsubishi Electric and its affiliates).

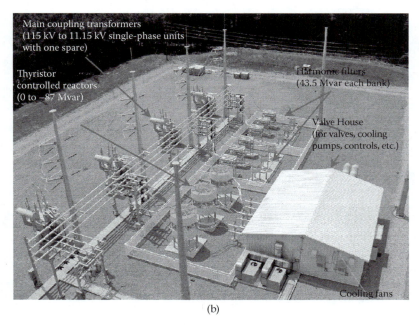

(b)

FIGURE 5.9 (b)
Photograph of a static var compensator installation with transformer coupling into the utility 115 kV network. The harmonic filters are capacitive reactance at 60 Hz and supply +87 Mvar for this installation. When the inductors (such as L in Figure 5.8) are fully conducting, net reactive power injected into the network is 0.0 Mvar. (Photograph used with permission of, and copyrighted by, Mitsubishi Electric and its affiliates)

inductance (−var when the capacitor switch is open) and upper limit with zero inductor current (+var when the capacitor switch is closed).

The older, more traditional synchronous condenser is also capable of supplying +var or −var, within its design limits, to the system, depending on whether it is over- or under-excited, respectively. There are many situations where both positive and negative VAR compensations are required at the same bus, as demonstrated by Example 5.6.

Example 5.6

In the network shown in Figure E5.6, determine the compensation necessary from an SC, TSC, or SVC to regulate bus voltage V_2 to 1.0 ± 0.05 p.u. as the passive load is switched in or out of the circuit. The line-charging capacitance is large, as could result from underground coaxial cables.

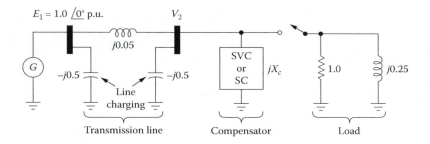

FIGURE E5.6

Solution

When the load is disconnected (switch open) and there is no compensation, the voltage V_2 is higher than desired because of reactive cancellation between the line-charging and line series inductances (Ferranti effect).

$$\frac{V_2}{E_1} = \frac{-j0.5}{-j0.5 + j0.05} = 1.111 \qquad \text{no load, uncompensated}$$

If the compensator is to limit the voltage to ±5%, then at no load the compensation is selected to be half the line-charging capacitance. In other words, the reactance of the compensator must be inductive, $jX_c = j0.5$ p.u. impedance, or −var. Thus, the no-load compensated voltage is

$$\frac{V_2}{E_1} = \frac{\frac{-j0.5(jXc)}{jXc - j0.5}}{j0.05 - \frac{-j0.5(jXc)}{jXc - j0.5}} = 1.0 \qquad \text{no load, compensated}$$

At full-load conditions and no compensation, the load inductive component causes the voltage to drop to

$$\left|\frac{V_2}{E_1}\right| = \left|\frac{\frac{j0.5}{1.0 + j0.5}}{j0.05 + \frac{j0.5}{1.0 + j0.5}}\right| = 0.908 \qquad \text{full load, uncompensated}$$

This is an excessive voltage drop, such that +var must be added to bus 2 in order to raise the voltage to an acceptable level. If the compensator impedance is capacitive, $-j0.5$, the inductive component of the load is canceled. The combination of line charging and compensator regulate the output voltage to

$$\left|\frac{V_2}{E_1}\right| = \left|\frac{1.0}{j0.05 + 1.0}\right| = 0.9988 \qquad \text{full load, compensated}$$

Thus, ±var is necessary to compensate adequately the voltage at bus 2 depending on the load condition.

In Figure 5.10, the shunt reactive injection at bus 2 raises the voltage at bus 2 and contributes to a larger power transfer. In almost every case, a power flow calculation is necessary to determine the compensation necessary for power flow and voltage profiles on the entire

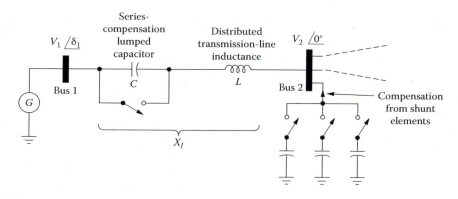

FIGURE 5.10
Shunt and series compensated transmission line.

network. Example 5.7 demonstrates using adjustable shunt compensation at a bus as part of a load-flow solution.

Example 5.7

A radial extension of a power system is very lightly loaded and its line charging is negligible, so that all buses on the radial lines were maintained at 1.048 p.u., which was the supply voltage. The load on the 13.8 kV output bus was subsequently increased by six synchronous motors operated at almost unity power factor. Their load total is 32.8 MW and 4.4 Mvar. The system is shown in Figure E5.7.1.

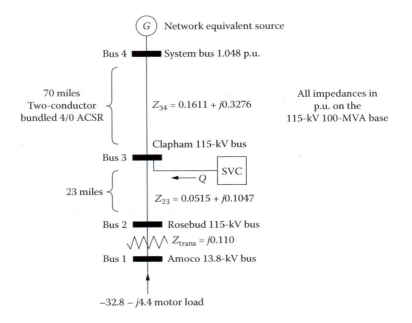

FIGURE E5.7.1

Use a Newton-Raphson load flow to determine the system bus voltages after the motor load is added. Also, calculate the bus voltages with the motor load plus 12.5 Mvar of capacitive compensation added at the 115 kV Clapham bus. Consider the load flow to converge when the real and reactive power mismatch at all buses is less than 0.01 p.u.

Solution

A Newton-Raphson load-flow method is used to calculate the voltages at buses 1, 2, and 3. The state vector is

$$\mathbf{X} = \begin{bmatrix} \delta_1 \\ \delta_2 \\ \delta_3 \\ V_1 \\ V_2 \\ V_3 \end{bmatrix}$$

The loads are zero at all buses except bus 1. The ΔP and ΔQ mismatches are defined in Equation 5.44.

Using a flat start, the Jacobian for the first Newton-Raphson iteration computed from Equation 5.46 to Equation 5.49 is

$$
J = \begin{bmatrix} J_1 & J_2 \\ J_3 & J_4 \end{bmatrix} = \begin{bmatrix}
9.984 & -9.984 & 0 & 0 & 0 & 0 \\
-9.984 & 18.431 & -8.446 & 0 & 3.964 & -3.964 \\
0 & -8.446 & 11.533 & 0 & -3.964 & 5.225 \\
0 & 0 & 0 & 9.527 & -9.527 & 0 \\
0 & 4.155 & -4.155 & -9.527 & 17.587 & -8.060 \\
0 & -4.155 & 5.477 & 0 & -8.060 & 10.462
\end{bmatrix}
$$

The results of the iterations for voltages, phase angles, and mismatches are given in Figure E5.7.2 for both cases—without compensation and with compensation.

Iteration	V_1	δ_1	ΔP_1	ΔQ_1	V_2	δ_2	ΔP_2	ΔQ_2	V_3	δ_3	ΔP_3	ΔQ_3
					Computed results without compensation							
0	1.048	0	−0.3200	−0.0440	1.048	0	0	0	1.048	0	0	0
1	0.9606	−0.1495	−0.0499	−0.0080	0.9653	−0.1175	−0.0099	−0.0008	0.9854	−0.0890	0.0112	−0.217
2	0.9364	−0.1675	−0.0029	−0.0004	0.9422	−0.1280	−0.0011	0.0001	0.9664	−0.0946	−0.0010	−0.0009
					Computed results with compensation							
0	1.048	0	−0.3200	−0.0440	1.048	0	0	0	1.048	0	0	0.1250
1	0.9997	−0.1677	−0.0275	−0.0067	1.004	−0.1357	−0.0074	−0.0053	1.024	−0.1072	−0.0027	−0.0201
2	0.9805	−0.1768	−0.0013	−0.0002	0.9861	−0.1405	−0.0006	0	1.009	−0.1098	−0.0003	0

FIGURE E5.7.2
Voltage magnitudes, phase angles, and bus power mismatches computed with N–R load flow versus iteration. Uncompensated and compensated networks.

For the case without compensation, the voltage at bus 1 drops to 0.9364 p.u. When +12.5 Mvar compensation is added at bus 3, it raises the voltage to 0.9805 at bus 1. For customers supplied by bus 1, this compensated voltage level is more acceptable.

Two tables with the resulting line flows are:

Line	P	Q
Line flows without compensation		
1–2	−0.3171	−0.0436
2–1	0.3171	0.0564
2–3	−0.3182	−0.0565
3–2	0.3242	0.0688
3–4	−0.3252	−0.0679
4–3	0.3442	0.1066
Line flows with compensation		
1–2	−0.3187	−0.0438
2–1	0.3187	0.0556
2–3	−0.3193	−0.0556
3–2	0.3248	0.0669
3–4	−0.3251	0.0585
4–3	0.3423	−0.0234

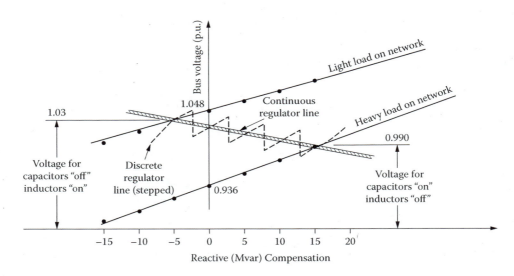

FIGURE 5.11
Regulation line for shunt reactive compensation.

Shunt reactive compensation may be introduced in a network to correct power factor for a line. For a bus voltage, the control error signal is usually obtained by comparing the actual bus voltage with a reference voltage level. Let the site of the bus be chosen because of its desirable geographic location or sensitive connection to the network. The computed bus voltage (using a power flow calculation) for variable amounts of reactive injection, $-Q_4$, $-Q_3$, 0, $+Q_2$, $+Q_1$, is examined for the two worst load cases to determine the necessary compensation and the voltage regulation line for control. Figure 5.11 shows bus voltages calculated for injections of −15, −10, −5, 0, +5, +10, and +15 Mvar. A regulation line for a continuously variable shunt compensator and a discrete step regulator response are also shown. Both methods control the voltage at 1.0 p.u., with +0.03, −0.01 tolerance for light to heavily loaded networks.

In Figure 5.11, when the bus voltage is 0.990 p.u., the control has turned on +15 Mvar capacitive reactance (i.e., shunt capacitors). When the voltage is 1.03, the control has turned on −5 Mvar of compensation. The compensation may be added in discrete steps (some overlap) or by a continuous element. A continuously controlled reactive source such as a static var compensator (SVC) or a synchronous condenser has a regulation line of −0.04/20 = −0.002 p.u. V/Mvar, while the discrete regulator shown has five levels.

When the p.u. V/Mvar slope of the regulator is small compared to the other transmission bus voltage variations, the slope may be neglected and a compensated bus is treated as a constant-voltage bus for power flow. In other words, it is treated as a regulated (P, V) bus in power flow computations. There are a number of methods to introduce shunt compensation to a network; some of these are shown in Figure 5.12.

For regulation of real power flow, one of the simplest methods is a switched series capacitor, as shown schematically in Figure 5.10. Continuously variable series capacitive FACTS devices of Table 5.2 are increasingly used. For negligible resistive losses, the series capacitor application shown in Figure 5.10 has real power flow between buses 1 and 2 determined by the series reactance:

$$P_{12} = \frac{V_1 V_2 \sin \delta_1}{x_\ell} = \frac{V_1 V_2 \sin \delta_1}{\omega L - 1/\omega C} \tag{5.67}$$

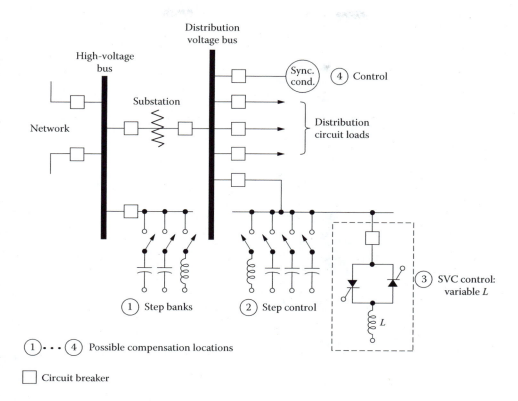

FIGURE 5.12
Typical distribution point showing four various shunt reactive compensation equipment and possible locations.

A series capacitor is often applied to compensate very long transmission lines. The capacitive reactance is in the range 25 to 70% of the reactance of *L*. Among advantages of this method are that a relatively small capacitor is required and the steady-state voltage across it is much less than the line-to-line voltage of the transmission line. One of its disadvantages is that it may cause a zero input impedance in the frequency range 5 to 55 Hz "looking into" the transmission line. When this line resonance point is near a critical shaft torsional frequency of the generator, electromechanical energy may be exchanged between the shaft and the line. If the energy exchange builds up in amplitude, shaft problems arise. This is called a *subsynchronous resonance*. Another disadvantage is that line and capacitor switching transients may be severe, and require special equipment to limit the voltage overshoot.

Often the necessary compensation for a transmission system is adjusted as the load distribution changes hourly or seasonally. Sometimes it is desired to adjust primarily the real power flow on a line of the network. Since real power flow is to a first approximation due to phase-angle differences, real power flow can be controlled by phase-shift transformers and phase-shifting FACTS devices called thyristor-controlled phase-angle regulators (TCPARs).

In order to accommodate a phase-shift branch in a power flow calculation, either a transformer or TCPAR, the Hermetian transfer functions of Section 2.4.1 must be employed for power flow and partial derivatives in the Jacobian terms, Equation 5.46 to Equation 5.57. Figure 5.13 shows the phase-shift device and defines the branch variables. The p bus angle is shifted as $e^{j\delta} = a + jb = \cos(\delta) + j\sin(\delta)$ with respect to the q bus. In the following terms,

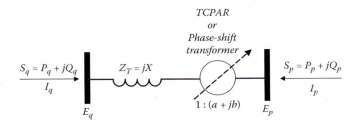

FIGURE 5.13
Definitions for a TCPAR device or a phase-shift transformer.

the jb term is conjugated between buses, and the currents, power flows, and some partial derivatives are derived as

Phase-shift branch equations:

$$I_p = \left(\frac{1}{Z_T}\right)[E_p - (a + jb)E_q] = \left[\frac{E_p}{Z_T} + \left(j\frac{a}{X} + \frac{b}{X}\right)E_q\right] = \left[\frac{E_p}{Z_T}e^{j\theta_p} + \left(j\frac{a}{X} + \frac{b}{X}\right)V_q e^{j\theta_q}\right] \quad (5.68a)$$

$$I_q = \left(\frac{1}{Z_T}\right)[E_q - (a - jb)E_p] = \left[\frac{E_q}{Z_T} + \left(j\frac{a}{X} - \frac{b}{X}\right)E_p\right] \quad (5.68b)$$

The voltage phase angles θ_p and θ_q are used to calculate the line power flows as

$$S_p = E_p I_p^* = -\left(\frac{b}{X}\right)V_p V_q \cos(\theta_p - \theta_q) + \left(\frac{a}{X}\right)V_p V_q \sin(\theta_p - \theta_q)$$
$$- j\left[V_p^2/X + \left(\frac{b}{X}\right)V_p V_q \sin(\theta_p - \theta_q) + \left(\frac{a}{X}\right)V_p V_q \cos(\theta_p - \theta_q)\right] \quad (5.69a)$$

$$S_q = E_q I_q^* = \left(\frac{b}{X}\right)V_p V_q \cos(\theta_q - \theta_p) + \left(\frac{a}{X}\right)V_p V_q \sin(\theta_q - \theta_p)$$
$$- j\left[V_q^2/X - \left(\frac{b}{X}\right)V_p V_q \sin(\theta_q - \theta_p) + \left(\frac{a}{X}\right)V_p V_q \cos(\theta_q - \theta_p)\right] \quad (5.69b)$$

The Hermetian property of the phase angle jb is demonstrated by the following partial derivatives:

$$\frac{\partial P_p}{\partial \theta_q} = -\left(\frac{b}{X}\right)V_p V_q \sin(\theta_p - \theta_q) - \left(\frac{a}{X}\right)V_p V_q \cos(\theta_p - \theta_q) \quad (5.70a)$$

$$\frac{\partial P_q}{\partial \theta_p} = \left(\frac{b}{X}\right)V_p V_q \sin(\theta_q - \theta_p) - \left(\frac{a}{X}\right)V_p V_q \cos(\theta_q - \theta_p) \quad (5.70b)$$

The remaining partial derivatives for $\frac{\delta P}{\delta V}$, $\frac{\delta Q}{\delta \theta}$, and $\frac{\delta Q}{\delta V}$ for the Newton-Raphson Jacobian terms may be derived from 5.69a and 5.69b.

The following example demonstrates the power flow impact of a phase-shift transformer or a TCPAR in a network.

Example 5.8

A TCPAR is installed in the transmission line from bus 2 to bus 4 as shown in Figure E5.8.1 in order to increase the real power delivered on this line. The phase shift is from bus 4

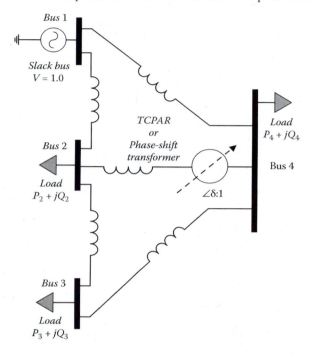

FIGURE E5.8.1

toward bus 2. The load schedule and parameters of the transmission lines are specified below on a 100 Mvar, 345 kV base. The base case for comparison is power flow on the network with a transmission line from bus 2 to 4. Case 2 is to regulate the voltage at bus 2 to 1.01 p.u. by means of a variable tap transformer on line 2–4. Finally, the third case is to insert a TCPAR with +1.156° phase shift in the line to demonstrate impact on line power flow due to phase-angle control.

Bus	P	Q
1	Slack	
2	−.02	−.84
3	−1.98	+.14
4	−7.99	−2.5

Line	R	X	1/2 Charging/Turns
1–2	0.0	0.01	0
1–4	0.0	0.01	0
2–3	0.0	0.01	0
2–4	0.0	0.01	0
3–4	0.0	0.01	0

Solution

A Newton-Raphson power flow for the base case without a transformer of phase shifter has the following solution for convergence to <=.00007 real or reactive power mismatch at any bus:

	Base Case	
Line (Bus to Bus)	**P (p.u.)**	**Q (p.u.)**
1–2	4.00	1.67
2–1	–4.00	–1.48
1–4	5.99	2.21
4–1	–5.99	–1.81
2–3	1.99	0.18
3–2	0.2868	–0.2134
2–4	1.99	0.46
4–2	–1.99	–0.42
3–4	0.01	0.28
4–3	–0.01	–0.28

	Base Case	
Bus	**V (p.u.)**	**Angle (deg.)**
1	1.0000	0.0, slack
2	0.9841	–2.32675
3	0.9825	–3.50428
4	0.9797	–3.50843

When a transformer is used between buses 2 and 4, with tap ratio 0.88:1 the results of the power flow calculation show excessive MVAR flow on the line.

	Transformer $\alpha = 0.88$	
Line (Bus to Bus)	**P (p.u.)**	**Q (p.u.)**
1–2	4.19	–1.22
2–1	–4.19	1.41
1–4	5.80	5.97
4–1	–5.80	–5.28
2–3	1.99	3.55
3–2	–1.99	–3.39
2-4	2.19	–5.80
4-2	–2.19	6.18
3-4	0.01	3.53
4-3	–0.01	–3.40

	Transformer	
Bus	**V (p.u.)**	**Angle (deg.)**
1	1.0000	0.0, slack
2	1.0131	–2.3728
3	0.9782	–3.5225
4	0.9421	–3.5276

When a phase-shift device such as a TCPAR is used between buses 2 and 4, with tap ratio $\angle\delta:1$, with the phase-angle shift of bus 2 to +1.156°, the results of the power flow are:

Line (Bus to Bus)	TCPAR	
	P (*p.u.*)	*Q* (*p.u.*)
1–2	4.49	1.71
2–1	−4.49	−1.48
1–4	5.50	2.19
4–1	−5.50	−1.84
2–3	1.50	0.16
3–2	−1.50	−0.14
2–4	2.96	0.47
4–2	−2.96	−0.38
3–4	−0.47	0.28
4–3	0.47	−0.28

Bus	TCPAR	
	V (*p.u.*)	Angle (deg.)
1	1.0000	0.0, slack
2	0.9840	−2.6140
3	0.9824	−3.5047
4	0.9796	−3.2205

The computed line flow from bus 2 to 4 for the transformer case compared to the base case shows a large change in the reactive flow and a small change in the real power flow. The voltage increase at bus 2 causes increased var flow.

Compared to the base case of a normal transmission line from bus 2 to 4, when the TCPAR phase-shift device is inserted it causes a large increase in the real power flow and almost zero change in the reactive flow. The TCPAR increases the sending end phase angle compared to the receiving end to force more real power flow through the line. All network bus phase angles adjust to allow more real power through the TCPAR line, so the TCPAR affects all lines and buses of the network.

© The program *p_flow*, included in the software, computes the power flow for any network with properly formatted data. Copy the formatted bus injection and line parameter data *example58.sss* into *p_flow.sss*. Execute *p_flow* to duplicate results of the base case with a .01 p.u. reactance transmission line from bus 2 to 4. Next, modify this line data in *p_flow.sss* so that 2–4 is a transformer, Ltype = 2, with tap ratio $\alpha = 0.88$ to duplicate case 2. Finally, modify line 2–4 in *p_flow.sss* to be a phase shifter, Ltype = 0, with $\delta = 1.156°$ to duplicate case 3 results.

5.6 Operational Power Flow Programs

Power flow programs are used by electric utilities and power systems research/development organizations as a planning tool, on-line monitoring mechanism, or simulation of the real system. The size of the programs in terms of number of buses and lines is set at compilation time and varies with the application. The power flow programs of utilities are usually more dedicated in purpose and have fewer diagnostics to assist with difficulties.

Virtually all operational power flow programs have features that facilitate data handling in terms of reading base case information, storing results, and manipulating power system control variables. The capability to delete lines, change a bus type during iterations, check limits, use a different slack bus, and so on, is in computer program manipulations that apply the basic methods of Section 5.1 to Section 5.5.

Some features of more complete power flow programs are as follows:

A. Buses

1. Buses are often identified by a combination name-voltage, for example, CHATHAM 115, where the 115 refers to the line-to-line voltage. Bus numbers referring to this name are internal, changeable, program labels.

2. Generation, load, shunt capacitors, and shunt resistors are represented separately, so ratings and limits are available individually. The internal program may combine quantities such as MW generation and MW load at a bus, but their external identity is maintained.

3. Bus types are tabulated in Table 5.3. The slack bus identity is retained because its phase angle is a reference (slightly different from a swing bus).

4. Shunt admittances are usually represented as:

    (a) Fixed admittance, inductive or capacitive

    (b) Switched capacitor in steps with on and off voltage points

    (c) Switched reactor in steps with on and off voltage points

TABLE 5.3

Types of Buses for Power Flow

Specified Parameters (p.u.)	Definition
P, Q (or unregulated)	Scheduled real and reactive power injections into the network. Power flow calculation determines the voltage magnitude and phase angle. High- and low-voltage limits are possible. In the event that a voltage limit is reached, the Mvar rating is converted to a fixed reactive element.
P, V (or regulated)	Scheduled real power injection into the network at fixed voltage magnitude. The voltage is maintained at a constant level by means of an adjustable internal or remote reactive source, such as a synchronous condenser, generator field excitation, static var generator, saturable reactor, or other control device. Both + and − Mvar limits are entered. In the event that a Mvar limit is reached, the voltage schedule is no longer held, and the reactive source is fixed Mvar.
Remote	Scheduled MW and variable Mvar. The Mvar injection is varied to maintain constant voltage at a remote bus. Mvar limits are enterable, beyond which the bus becomes a P, Q type, with Q at the limit. The remote bus changes to a fixed reactive element and specified P.
Swing (area)	Variable MW and scheduled voltage. The MW injection is varied to maintain the real power part of desired area control error. In the event that an Mvar limit is reached, the voltage schedule is no longer held.
Slack	Scheduled voltage magnitude and fixed-phase-angle reference for the power flow calculation. Necessary because of overdetermined calculation (see Section 5.1), but has variable MW, Mvar output.

B. Branches (lines or elements)

1. The series element of a branch between two buses may be specified in terms of impedance $(R + jX)$ or admittance $(G + jB)$ in per unit on MVA base or bases as specified by the user.

2. There are no restrictions on the magnitude or sign of branch impedances. Zero, low-value, or negative impedances are acceptable, but at least one nonzero value per branch.

3. Pi elements having unequal legs are acceptable, with capacitive or inductive shunt elements.

4. Parallel lines are permitted with identity retained. Mutual coupling is an input quantity.

5. Branches are identified by terminal bus names. Branch numbers are not required.

6. Provision is made for line current ratings and transformer MVA ratings for overload checking.

7. Provision is made for calculating line currents at terminals on selected lines.

8. HVDC lines which may be operated in either of 2 modes: constant current or constant power.

C. Transformers and phase shifters

1. Fixed tap transformer ratios may be entered in terms of rated kV on each terminal. Transformers may be 2 or 3 winding types.

2. TCUL transformer voltage range and step size are specified by the program user. The desired voltage and bus name to be automatically controlled are also inputs.

3. TCUL control on Mvar injection at a remote or adjacent bus specified by user.

4. Phase shifters have through power or angle set by means of the user's schedule. Phase shifters have a 1:1 voltage ratio. Phase-shifter impedance adjustment with step change is automatic. Phase-shifter angle range and step size are inputs by the user.

D. Area control error (power flow only)

1. Areas may be identified within the network by means of user-specified metering points at either end of a transmission line (or branch). The lines are treated as tie lines to the remaining network.

2. Each area has an individual swing machine, whose real power (sum of ties) is adjusted within limits between iterations to satisfy a user-specified real power exchange as part of ACE.

3. It is possible to combine individual interchange areas into larger interchange areas without changing bus or branch specifications.

4. Area interchange convergence tolerance is specifiable by the user and is separate from the overall solution convergence tolerance.

5. A pause is often inserted in the program execution, so a user has a rough check prior to the first voltage balance in order to prevent possible solution failure or abnormal computer time due to initial area interchange errors (due to inaccurate load, generation, or loss estimates).

In addition to subjects A, B, C, and D, specialized power flow programs have the capability to adjust the operating points of FACTS devices. The FACTS devices have problems in control and harmonic generation. Harmonics are generated in direct-current (dc) transmission lines by the accompanying converter or inverter terminals, which change the ac to dc, and vice versa [4]. This subject is beyond our scope here.

The output of the power program is often a detailed bus-by-bus, line-by-line description of power flow, voltages, currents, and transformer tap settings. An additional summary indicates buses out of the specified voltage range, Mvar limits, or converted to another type of bus. Transmission lines or transformers out of desired operating ranges are also tabulated. Each power area performance is summarized in terms of ACE, total generation, total load, and transmission line losses.

Problems

© The program *p_flow*, included in the software, is a polar-form Newton-Raphson algorithm power flow that can be used to verify problem solutions.

5.1. Apply accelerated Gauss and Gauss-Seidel iterative methods to solve Example 5.1. Find the best value of α to improve convergence.

5.2. For the power system shown in Figure P5.2, bus 4 is selected as the slack bus. The incrementally switched capacitor bank at bus 1 is disconnected from the system. The bus load schedule (100 MVA base) is:

Bus	P (p.u.)	Q (p.u.)
1	−0.40	−0.30
2	0.30	0.20
3	−0.20	−0.15
4	Slack bus	

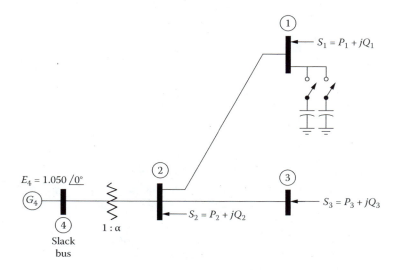

FIGURE P5.2

The parameters for the transmission lines (100 MVA, 100 kV Base) are:

Line	R_1 (p.u.)	X_1 (p.u.)	Shunt Susceptance $\omega C/2$ (p.u.)
1–2	0.01	0.03	0.0264
2–3	0.021	0.14	0.0180

For the transformer,

$$X_1 = 0.06, \qquad \alpha = 0.97$$

(a) From a flat start, use either the Gauss or Gauss-Seidel iterative method to determine the voltages at buses 2, 3, and 4. Terminate the iterative process when the voltage change is less than 1×10^{-6} for every bus on successive iterations.

(b) Let bus 1 be voltage controlled at $|E_1| = 1.04$ by variable Q_1 from switched capacitors. Repeat part (a) for this condition when Q_1 is variable.

5.3. (a) For the power system shown in Figure P5.2, use the rectangular form of the Newton-Raphson iterative method to find voltages at buses 2, 3, and 4.

(b) Let the switched capacitor control the voltage at bus 1 to $|E_1| = 1.02$ by varying Q_1. Calculate the voltage at buses 2 and 3, and the phase angle at bus 1 for this operating condition. From the resulting power system state, determine the total reactive injection Q_1.

5.4. (a) Figure P5.4 shows the network topology of the classic IEEE five-bus network of Stagg and El-Abiad [6]. The line parameters and injections on a 100 MVA, 345 kV base are listed below for the network.

Bus	P (p.u.)	Q (p.u.)
1	Slack bus V = 1.06 ∠0°	
2	0.20	0.20
3	−0.45	−0.15
4	−0.40	−0.05
5	−0.60	−0.10

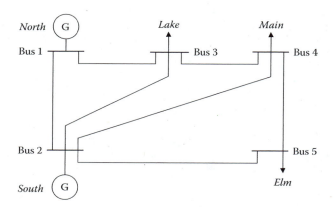

FIGURE P5.4
IEEE five-bus test case.

Line	R_1 (p.u.)	X_1 (p.u.)	Shunt Susceptance $\omega C/2$ (p.u.)
1–2	0.02	0.06	0.030
1–3	0.08	0.24	0.025
2-3	0.06	0.18	0.020
2-4	0.06	0.18	0.020
2-5	0.04	0.12	0.015
3-4	0.01	0.03	0.010

© The program *p_flow*, included in the software, computes the power flow for any network with properly formatted data. Copy the formatted bus injection and line parameter data *Stagg.sss* into *p_flow.sss*, and execute *p_flow* to obtain the power flow solution for this five-bus case. Use the *p_flow* program to:

(b) Modify the real power load at bus 5 to be 100 MW, or 1.0 p.u., and observe the voltage at bus 5.

(c) Increase the reactive injection at bus 5, as if shunt capacitive compensation is present, until the bus voltage increases to 1.0, or rated voltage.

5.5. A four-bus power system is shown in Figure P5.5. The bus scheduled power (p.u.) is

$$P_1 = -0.3$$
$$S_2 = -0.4 - j0.3$$
$$S_3 = +0.4 + j0.4$$

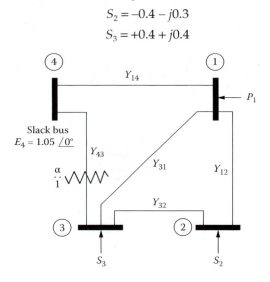

FIGURE P5.5

The line parameters (p.u.) are:

Series Admittance	$\omega C/2$
$Y_{12} = 5(1 - j3)$	0.02
$Y_{23} = 10(1 - j3)$	0.04
$Y_{31} = 5 - j12$	0.03
$Y_{43} = -j20,\ \alpha = 0.97$	0
$Y_{14} = 10(1 - j4)$	0.02

Bus 2 is a tie-line bus. Bus 1 is voltage ccontrolled to $|V_1| = 1.0$ p.u. by means of a variable capacitance at the bus.

(a) Use any computer language or software available to apply the fast-decoupled load-flow algorithm from a flat start to find the state of the power system below. Consider the solution converged when the real power mismatch at any bus is less than 0.0001 p.u., and the reactive power is less than 0.001 p.u. on a 100 MW base.

(b) Using the same convergence tolerance as in part (a), determine the tie-line distribution factors (TLDFs), which are

$$\text{TLDF}_3 = \frac{\Delta P_T}{\Delta P_3} = \frac{\Delta P_2}{\Delta P_3} \qquad \text{TLDF}_4 = \frac{\Delta P_T}{\Delta P_4} = \frac{\Delta P_2}{\Delta P_4}$$

These ratios may be obtained by selecting buses 3 and 4 as the slack bus, then increasing the tie flow by a small amount, say 5%, allowing only the slack bus to provide the incremental change. The TLDFs are a method to direct power flow on a network.

5.6. The positive-sequence equivalent for two hydro generating stations is shown in Figure P5.6. The slack bus represents a large interconnection that accepts all power generated and maintains constant 230 kV (1.0 p.u.). The real and reactive power delivered to the system by the hydro units is measured (monitored) at the slack bus. Generator 1 is designated as a swing machine that must supply transmission line losses and the reactive power required to deliver 438 MV at unity power factor ($4.38 + j0$ p.u.) to the system represented by the slack bus. The transmission lines have negligible line charging. Their parameters on a 100 MVA base are given in the table below. The transformer impedances are also specified

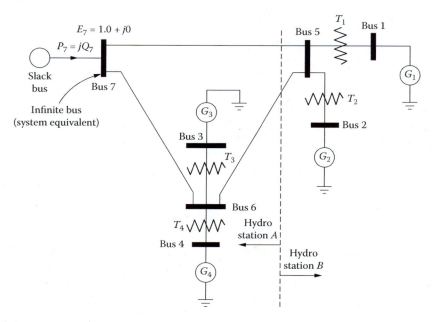

FIGURE P5.6

below. All are on-nominal. The line parameters (p.u.) are:

Bus-to-Bus	$R + jX$
5–7	$0.0041 + j0.0399$
6–7	$0.0052 + j0.0319$
5–6	$0.0015 + j0.0170$

The transformer impedances (p.u.) are:

Trans.	$R + jX$
T1	$0.00163 + j0.08166$
T2	$0.00163 + j0.08166$
T3	$0.0258 + j0.3100$
T4	$0.00241 + j0.1195$

The generation schedule (p.u.) is:

Generator	Power
1	Swing
2	$1.50 - j0.10$
3	$0.40 + j0.10$
4	$1.20 + j0.10$

There are no loads at buses 5 and 6. The power system equivalent is the slack bus at voltage $E_7 = 1.0 + j0$ p.u. Do the following:

(a) Make a simplified flowchart that demonstrates a method to vary the swing generator output between iterations of a power flow solution. What is the initial power schedule for generator 1?

(b) © Use the *p_flow* program in a "cut-and-try" fashion to vary the swing bus output real and reactive power until $4.38 + j0$ is delivered to the slack bus (system).

5.7. Power flow on a multiterminal high-voltage direct-current (HVDC) system is a special case of ac power flow where all the impedances are real and the phase angles are zero. A four-bus system, line resistances, and power injections are shown in Figure P5.7. Do the following:

(a) Write the admittance matrix using bus 4 as reference.

(b) Set up a Newton-Raphson iterative method to determine the state of the system, which in this case is V_1, V_2, V_3, all at zero phase angle.

(c) Use a flat start and iterate until all voltages change less than 10^{-5} between successive iterations.

5.8. © The program *p_flow*, included in the software, computes the power flow for any network with properly formatted data. Copy the formatted bus injection and line parameter data *Stagg.sss* into *p_flow.sss*. Use the *p_flow* program to do the following:

(a) Modify the line parameter and bus injection data in the file *p_flow.sss* to describe the IEEE 14-bus power flow problem given below in Figure P5.8.

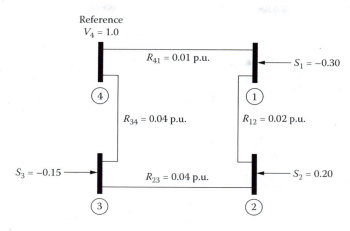

FIGURE P5.7

The topology of the network is shown in Figure P3.7. Note the α:1 connections at the bottom of Figure P5.8. Execute *p_flow* to obtain the 14-bus power flow solution.

(b) Write a fast-decoupled load-flow (FDLF) algorithm using any computational program available, to solve the power flow for the 14-bus test case. Compare the FDLF with part (a) results for convergence to a real or reactive power mismatch at any bus <= 0.00007 p.u. It is possible to converge to the solution with ½ iterations if you test for convergence after the P and after the Q calculations.

		Line Data			Bus Data (100 MVA Base)		
Line	Bus 1	Bus 2	100 MVA Base Admittance	$\omega C/2$ Line Charging	Bus	MW Load	MVAR Load
1	1	2	$5 - j15.263$	0.0264	1	(Slack E = 1.06 + j0)	
2	1	5	$1.026 - j4.235$	0.0246	2	0.183	$-j0.297$
3	2	3	$1.135 - j4.782$	0.0219	3	-0.942	$+j0.044$
4	2	4	$1.686 - j5.116$	0.0187	4	-0.478	$+j0.039$
5	2	5	$1.701 - j5.194$	0.017	5	-0.076	$-j0.016$
6	3	4	$1.986 - j5.069$	0.0173	6	-0.112	$-j0.0474$
7	4	5	$6.841 - j21.58$	0.0064	7	0	0
8[a]	4	7	$-j4.782$	0	8	0	$+j0.174$
9[b]	4	9	$-j1.798$	0	9	-0.295	$+j0.046$
10[c]	5	6	$-j3.968$	0	10	-0.090	$+j0.058$
11	6	11	$1.955 - j4.094$	0	11	-0.035	$-j0.018$
12	6	12	$1.526 - j3.167$	0	12	-0.060	$-j0.016$
13	6	13	$3.099 - j6.103$	0	13	-0.135	$-j0.058$
14	7	8	$-j5.677$	0	14	-0.149	$-j0.050$
15	7	9	$-j9.09$	0			
16	9	10	$3.902 - j10.37$	0			
17	9	14	$1.424 - j3.029$	0			

18	10	11	$1.881 - j4.395$	0
19	12	13	$2.489 - j2.252$	0
20	13	14	$1.137 - j2.315$	0

[a] Trans. ratio 0.978, α:1, bus 4: bus 7.
[b] Trans. ratio 0.969, α:1, bus 4: bus 9.
[c] Trans. ratio 0.932, α:1, bus 5: bus 6.

FIGURE P5.8
IEEE 14-bus standard test case.

References

1. Tinney, W. F., and Hart, C. E. 1967. Power flow solution by Newton's method. *IEEE Trans. Power Appar. Syst.* 86 (November 1967): 1449–1460.
2. Stott, B., and Alsac, O. 1974. Fast decoupled load flow. *IEEE Trans. Power Appar. Syst.* 93 (May 1974): 859–869.
3. Sitnikov, V., et al. 2004. Benefits of power electronics for transmission enhancement. Paper presented at the Conference on Russia Power, Moscow, March 10–11, 2004.
4. Arrillaga, J., Arnold, C. P., and Harker, B. J. 1983. *Computer modelling of electrical power systems.* New York: John Wiley & Sons.
5. Gyugyi, L., and Taylor, E. R., Jr. 1978. Characteristics of static, thyristor-controlled shunt compensators for power transmission system applications. *IEEE Trans. Power Appar. Syst.* 97(5) (September/October 1978): 1795–1804.
6. Stagg, G., and El-Abiad, A. 1968. *Computer methods in power system analysis.* New York: McGraw-Hill.

6

Generator Base Power Setting

Historically, the first applications of on-line computers in power systems originated in the need for a central facility to economically operate several generating plants supplying the load of the system. Present power systems often have a mixture of generating plants, including coal-fired steam turbine generators, oil or natural gas combustion units used for peak periods, hydro installations, pumped storage units, solar conversion, and nuclear generation. Each generation source has specific operating purposes and constraints. A modern fossil-fired generating station or a nuclear unit may represent capital investments of hundreds of millions of dollars and employ the most sophisticated computer controls available. When such a generating station is under the control of central dispatching, a master–slave relation is not implied, but more of an information exchange and coordination, as every station is capable of independent operation in the power grid.

Generators in the system may be in one of the following operating states when on-line:

1. *Base not regulating*: Output power level remains constant.
2. *Base regulating*: Output level changes with system total load.
3. *Automatic noneconomic regulating*: Output level changes around a base setting as area control error (ACE) changes.
4. *Automatic economic regulating*: Unit output is adjusted with the area load and area control error, while tracking an economic setting.

Regardless of the unit's operating mode, it has a contribution to the economic operation, even though its output is changed for different reasons. Economic operation is defined here in terms of fuel cost per power generated even though the true cost could involve capital investment and maintenance, depreciation, and so on. The cost basis is more direct and easily obtained than numerical values for all utility operations, which ultimately determine consumer price in dollars per megawatt-hour.

Economic dispatch is a fundamental consideration for power systems [1, 2], and has probably had more study than any other operating practice. Assume initially that M generating units are participating in economic dispatch and that each unit is within its active regulating region. Start-up and shutdown costs are later considered, and the number of units in operation is called the commitment problem, which is later discussed. For economic dispatch, assume that the cost in dollars per hour of operating each of the M generating units is a known, differentiable function, F_i, over the entire power output range for the generating unit i:

$$F_i(P_i) \equiv \begin{bmatrix} \text{operating} \\ \text{cost} \\ (\$/h) \\ (\text{MBtu}/h) \end{bmatrix} \quad \text{for } P_{i\min} \leq P_i \leq P_{i\max}$$

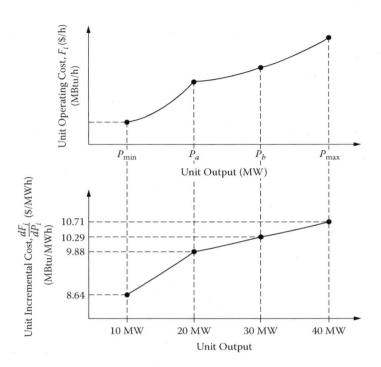

FIGURE 6.1
Typical cost and incremental costs for a generating unit.

F_i is a piecewise continuous function that is valid for ranges of the output power, and whose derivative (called the incremental cost curve or heat rate) takes on the same value as the limit is approached from adjacent ranges. Typical cost and incremental cost curves are shown in Figure 6.1.

The discontinuities in the operating cost curve may be due to the firing of additional boilers, steam condensors, or other discrete equipment necessary to extend the power output of the generator from minimum to maximum. Discontinuities also appear if the operating costs represent an entire station and the paralleling of additional generators changes the cost.

It is often convenient to express the curves in terms of Btu rather than dollars, as fuel costs change monthly or daily in comparison with unit efficiency, which is relatively constant over the lifetime of the unit. The cost and incremental cost curves are fundamental to the economic dispatch of the power system, as shown below.

6.1 Economic Dispatch of Generation without Transmission Line Losses

Consider the special case where M generators are supplying a common load, P_R, without transmission losses, or the M generators are within a single station and are paralleled to an output bus. The units are schematically shown in Figure 6.2 and a cost curve, F_i, is assumed known for each unit. It is desired to minimize the generation cost:

$$J = \sum_{i=1}^{M} F_i(P_i) = \text{ performance index} \tag{6.1}$$

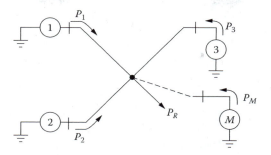

FIGURE 6.2
Special case of supplying a load without power transmission losses.

subject to the conservation of power constraint:

$$\sum_{i=1}^{M} P_i = P_R = \text{load power} \tag{6.2}$$

Assuming that F_i is a continuous function of its argument, the maximum or minimum of Equation 6.1 is at the extremal point [3] when

$$\frac{dJ}{dP_i} = 0 = \frac{dF_i}{dP_i} + \sum_{j \ne i}^{M} \frac{dF_j}{dP_j}\left(\frac{dP_j}{dP_i}\right) = 0 \quad \text{for } i = 1, 2, \ldots, M \tag{6.3}$$

as subject to the power flow constraint. Since the load P_R is a fixed value, an increase in generator i must be accompanied by a decrease in the other machines; thus, the partial derivatives obtained from Equation 6.2 are

$$\frac{dP_R}{dP_i} = 0 = 1 + \sum_{j \ne i}^{M} \frac{dP_j}{dP_i} \quad \text{for } i = 1, 2, \ldots, M \tag{6.4}$$

Equation 6.3 and Equation 6.4 have a nontrivial solution only if each of the incremental costs are equal to the same value:

$$\boxed{\begin{array}{l} \text{Minimum cost of operation without line losses} \\ \lambda = \dfrac{dF_1}{dP_1} = \dfrac{dF_2}{dP_2} = \dfrac{dF_3}{dP_3} = \ldots = \dfrac{dF_M}{dP_M} \end{array}} \tag{6.5}$$

as may be easily verified for the case when $M = 2$. The value of λ is called the *system incremental cost* or simply the *system lambda*, although in the general case when line losses are present, it serves a different purpose. Note that λ is expressed in MBtu/MWh or $/MWh, so it represents a cost of generating electrical energy. According to Equation 6.5, the output power of a generating station is raised or lowered until its incremental cost is equal to the system λ, ensuring the most efficient system operation.

Example 6.1

The fuel input costs for two units supplying a common load without transmission losses are given as

$$F_1 = 8.644P_1 + 0.10707P_1^2 \quad \text{(MBtu/h)} \quad 0 \le P_1 \le 50$$

$$F_2 = 7.552P_2 + 0.014161P_2^2 \quad \text{(MBtu/h)} \quad 0 \le P_2 \le 100$$

The incremental costs for these two generators are

$$\partial F_1 / \partial P_1 = 8.644 + 0.21414 P_1 \quad (\text{MBtu/MWh}) \quad 0 \le P_1 \le 50$$

$$\partial F_2 / \partial P_2 = 7.552 + 0.028322 P_2 \quad (\text{MBtu/MWh}) \quad 0 \le P_2 \le 100$$

Subject to the power flow constraint,

$$P_1 + P_2 = P_R$$

the two units are to be optimally operated. Determine the output of each unit as the common system load, P_R, varies between 0 and 150 MW.

Solution

Find the minimum of the cost functional, Equation 6.1, at the extremal point by means of differentiating the performance index $J = F_1 + F_2$ with respect to P_1:

$$\frac{dJ}{dP_1} = 0 = \frac{d}{dP_1}(F_1 + F_2) = \frac{dF_1}{dP_1} + \frac{dF_2}{dP_2}\frac{dP_2}{dP_1} \tag{1}$$

Then substitute the incremental costs into equation 1 to obtain

$$0 = 8.644 + 0.21414 P_1 + (7.552 + 0.02832 P_2)\frac{dP_2}{dP_1} \tag{2}$$

Next, use the power constraint $P_2 = P_R - P_1$ to obtain the partial derivative of P_2 with respect to P_1:

$$\frac{dP_2}{dP_1} = \frac{d}{dP_1}(P_R - P_1) = -1 \tag{3}$$

Substitute equation 3 and the power constraint in equation 2 to obtain

$$P_1 = 0.1168 P_R - 4.504 = 0.1168(P_R - 38.56) \tag{4}$$

Since only positive values of generation are permitted, unit 1 remains at zero output level until the system load reaches 38.56 MW, as shown in Figure E6.1a.

The output power of unit 2 is $P_2 = P_R - P_1 = 0.8832 P_R + 4.504$, and the system lambda is

$$\lambda = \frac{dF_1}{dP_1} = \frac{dF_2}{dP_2} = 7.6795 + 0.025 P_R \quad (\text{MBtu/MWh}) \tag{5}$$

Only in the region of system load, $38.56 \le P_R \le 108.1$, are the two units economically operated. In the region of high system demand unit 2 reaches its maximum output limit and unit 1 must supply the additional load. The total cost of operating the system in MBtu/h is obtained by introducing P_1 and P_2 values into the sum $J = F_1 + F_2$. Total cost versus P_R is plotted in Figure E6.1c.

© The power allocation in Example 6.1 at the dispatch point $P_R = 108.125$ MW may be verified by using the computer program ***econ.exe***, which uses the Bolzano iterative method (presented later in this chapter) to find the dispatch powers. This program contains line losses on the order of 1.2% that do not appreciably change the results of Example 6.1. Additional cases for dispatch points P_R may be executed by modifying P_R in the file ***econ.sss***.

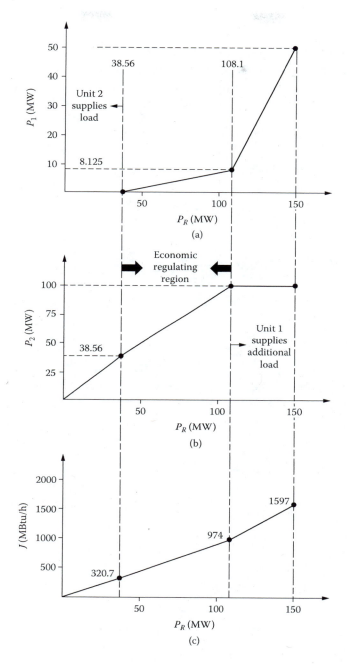

FIGURE E6.1
Two generating units supplying a variable load without transmission losses—limited region of operation.

In general, whenever a number of units are in economic dispatch, maximum or minimum unit operating limits are reached for various system loads, such that the digital computer algorithm must continually test for feasible values. In the event that system demand cannot be met by the total capacity of all available units, additional generation must be provided through either increased tie-line flows or start-up of idle units.

6.2   Economic Dispatch of Generation with Line Losses

In normal operations, a power system has a mixture of units supplying the load-nuclear units, thermal units, hydro units, units operating at fixed output, automatic economic regulating, and tie lines to adjacent utilities. Let M units participate in the economic dispatch while both the K tie lines and N load buses are considered to be operating at constant MW, Mvar values (Figure 6.3). The ties may be receiving or transmitting system power.

The economic dispatch problem is to minimize the total cost of operating the M units.

$$J = \sum_{i=1}^{M} F_i(P_i) \tag{6.6}$$

subject to the power balance constraint:

$$\sum_{i=1}^{M} P_i = P_R + P_L \tag{6.7}$$

where P_R is the total power received by the $N - K$ loads plus K tie lines, and P_L is the total of the transmission line losses. If the received power P_R is constant, and the line loss P_L depends on the source of power, the constraint in the form of a zero value is adjoined to the cost function by means of an undetermined multiplier, λ, called a *Lagrange multiplier* [3]:

$$J' = \sum_{i=1}^{M} F_i(P_i) - \lambda \left\{ \sum_{i=1}^{M} (P_i) - P_R - P_L \right\} \tag{6.8}$$

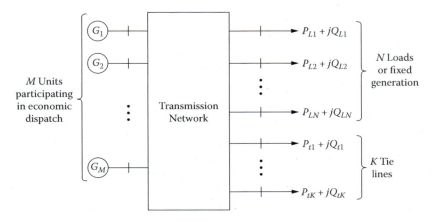

FIGURE 6.3
General system model for economic dispatch.

and minimized with respect to $M + 1$ variables, $P_1, P_2, ..., P_M, \lambda$. The minimum (or maximum) is the extremal point that satisfies the $M + 1$ simultaneous equations:

Coordination equations

$$\frac{dJ'}{dP_1} = 0 = \frac{dF_1}{dP_1} - \lambda \left\{ 1 - \frac{\partial P_L}{\partial P_1} \right\}$$

$$\frac{dJ'}{dP_2} = 0 = \frac{dF_2}{dP_2} - \lambda \left\{ 1 - \frac{\partial P_L}{\partial P_2} \right\}$$

$$\vdots \qquad \vdots \qquad \vdots$$

$$\frac{dJ'}{dP_M} = 0 = \frac{dF_M}{dP_M} - \lambda \left\{ 1 - \frac{\partial P_L}{\partial P_M} \right\}$$

$$= 0 = \sum_{i=1}^{M} (P_i) - P_R - P_L$$

(6.9)

The solution to these equations, called *coordination equations*, is the classical economic dispatch described by Steinberg and Smith [1] and Kirchmayer [2] and has served as a standard method of the electric utility industries. If the numerical value of the partial derivative of the line losses with respect to each generator, $\partial P_L / \partial P_i$, is known, the generator output power may be adjusted to satisfy the equation

Minimum cost of operation with line losses

$$\frac{dF_i}{dP_i} \left\{ \frac{1}{1 - \partial P_L / \partial P_i} \right\} = \lambda = \text{system lambda (MBtu/MWh)}$$

$$i = 1, 2, ... M$$

(6.10)

The name *system lambda* is a carryover from early years of dispatching when Equation 6.5 was the basis for establishing the generator output power level, and the system lambda was telemetered to the stations. In the present form, it is one of the $M + 1$ simultaneous solutions. The term in braces is defined as

$$\left\{ \frac{1}{1 - \partial P_L / \partial P_i} \right\} = \text{penalty factor} = \text{PF}_i \text{ for unit } i$$

(6.11)

and is retained as an entity during the course of the solution for both its physical significance and its use in the iterative process to find λ.

In modern power system control, the individual generating units are controlled from the central dispatch computer by means of raise/lower pulses as shown schematically in Figure 6.4. The generator output is monitored, and the digital control loop around the generator is closed to ensure that desired power is attained.

In order to understand the physical meaning of the penalty factor, assume that a small increment of load power ΔP_R is received from generator i, which increased its output by ΔP_i.

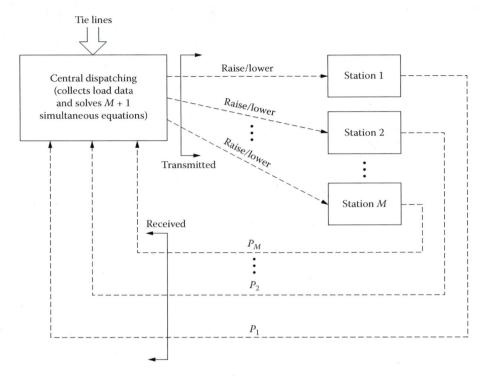

FIGURE 6.4
Control of system generation by central dispatching. (Dashed lines represent telemetry signals or communication lines.)

The power received by the load is decreased by line losses:

$$\Delta P_R = \Delta P_i - \Delta P_L$$

For small increments of power, the penalty factor is approximately

$$\mathrm{PF}_i = \frac{1}{1 - \Delta P_L/\Delta P_i} \cong 1 + \frac{\Delta P_L}{\Delta P_i}$$

Therefore, as the ΔP_i increment has a larger proportion of power dissipated as the loss derivative, $\Delta P_L/\Delta P_i$ increases and the penalty factor increases. This means that if unit i has a large penalty factor, unit i should be operated at low incremental cost, dF_i/dP_i, implying a low output power, or

$$\left\{\begin{array}{l}\text{low output power}\\\text{low incremental cost}\end{array}\right\} \times \left\{\begin{array}{l}\text{large penalty}\\\text{factor}\end{array}\right\} = \left\{\begin{array}{l}\text{system}\\\text{lambda}\end{array}\right\}$$

Generally, the system transmission losses are a small percentage of the total generation and $\partial P_L/\partial P_i \ll 1$, such that using the approximation

$$\frac{1}{1-x} = 1 + x + x^2 + x^3 \cdots. \quad 1 + x$$

Equation 6.10 may be rewritten as

$$\frac{dF_i}{dP_i}\left\{\frac{1}{1-\partial P_L/\partial P_i}\right\} \cong \frac{dF_i}{dP_i}\left\{1+\frac{\partial P_L}{\partial P_i}\right\} \cong \lambda \tag{6.12}$$

This approximation avoids the division, so is a more convenient form in which to solve the coordination equations. Various methods are used to obtain the transmission line partial derivative of transmission line losses for each generator. The following example uses a classical method [1], where P_L is expressed as a quadratic function of the M generator output powers called B coefficients.

Example 6.2

Two generation units are supplying a system load of 156.1 MW (1.561 p.u. on a 100 MW base). The transmission losses are expressed as B coefficients dependent upon the output power of the two generators:

$$P_L = 0.01065P_1^2 + 0.01415P_1P_2 + 0.0181P_2^2 \quad \text{(p.u., 100 MW base)}$$

This formula, which may be in B matrix form, was calculated at the present system operating point:

$$\text{Operating levels (p.u.)} \begin{cases} P_1 = 0.3952 \\ P_2 = 1.2 \\ P_R = 1.561 \\ P_L = 0.0344 \end{cases}$$

where all generators, lines, and loads have real power flow for simplicity.

It is desired to bring these two units into economic dispatch considering transmission losses such that generation costs are minimized for the power system. The cost of fuel for the two units ($/MBtu) is the same, and the thermal conversion curves for the two units are

$$F_1 = 864.4P_1 + 1070.7P_1^2 \quad \text{(MBtu/h)} \quad 0 \le P_1 \le 0.50 \text{ p.u.}$$

$$F_2 = 755.2P_2 + 141.6P_2^2 \quad \text{(MBtu/h)} \quad 0 \le P_2 \le 1.50 \text{ p.u.}$$

Solution

At the present operating point, the cost of supplying the load is $F_1 + F_2 = 1{,}619$ MBtu/h. The generator incremental costs and penalty factors are calculated in order to compare their respective values for lambda. For unit 1, the value of lambda is

$$\frac{dF_1}{dP_1}(PF_1) = \frac{dF_1}{dP_1}\left\{\frac{1}{1-\partial P_L/\partial P_1}\right\} = \frac{864.4+2141.4P_1}{1-0.0213P_1-0.01415P_2}\Bigg|_{\substack{P_1=0.3952 \\ P_2=1.2}} = 1766 \text{ MBtu/h} \tag{1}$$

A similar calculation for unit 2 yields

$$\frac{dF_2}{dP_2}(PF_2) = \frac{755.2+283.2P_2}{1-0.01415P_1-0.0362P_2}\Bigg|_{\substack{P_1=0.3952 \\ P_2=1.2}} = 1152 \text{ MBtu/h} \tag{2}$$

Thus, the cost of delivering power to the load by means of operating unit 1 is much greater than that by unit 2, such that outputs should be adjusted accordingly. Any iterative numerical method that decreases P_1 and increases P_2 until λ is the same for both units while satisfying the power constraint equation is suitable for the calculation. The constraint is the power balance equation:

$$P_1 + P_2 = P_L + P_R = P_T \tag{3}$$

To minimize the constrained cost functional,

$$J' = F_1 + F_2 - \lambda(P_1 + P_2 - P_L - P_R)$$

the extremal point of the coordination equations, Equation 6.9, is found using a Newton-Raphson iterative technique (see Chapter 5).

Define the vector of variables as

$$\mathbf{X} = [P_1 \; P_2 \; \lambda]^t$$

and the nonlinear vector function $\mathbf{F}(P_1, P_2, \lambda)$, which has zero value upon solution to the coordination Equation 6.9 for the case $M = 2$. The function $\mathbf{F}(\mathbf{X})$ is a vector:

$$\mathbf{F}(\mathbf{X}) = \begin{bmatrix} 864.4 + 2141.4P_1 - \lambda(1 - 0.0213P_1 - 0.01415P_2) \\ 755.2 + 283.2P_2 - \lambda(1 - 0.01415P_1 - 0.0362P_2) \\ P_1 + P_2 - 1.561 - P_L \end{bmatrix} = 0$$

The Newton-Raphson algorithm requires the 3×3 Jacobian, $\partial \mathbf{F}/\partial \mathbf{X}$, to be evaluated for values of $\mathbf{X}$ computed in the kth iterative step, then inverted for each iteration. The algorithm to calculate $\mathbf{X}$ for the next step is

$$\mathbf{X}^{k+1} = \mathbf{X}^k - \left[\frac{\partial \mathbf{F}}{\partial \mathbf{X}}(\mathbf{X}^k)\right]^{-1} \mathbf{F}(\mathbf{X}^k)$$

This numerical process is repeated until $\mathbf{X}^{k+1}$ and $\mathbf{X}^k$ agree to a convergence tolerance of four significant decimal places. In this example, the results of the calculation for the new system minimum operating cost point are

$$\begin{array}{l} \text{Minimum} \\ \text{operating} \\ \text{cost point} \\ \text{at } P_R = 1.561 \end{array} \left\{ \begin{array}{l} P_1 = 0.1577 \text{ p.u.} \\ P_2 = 1.445 \text{ p.u.} \\ \lambda = 1231 \text{ MBtu/p.u.-h} \end{array} \right\}$$

The generation is shifted from unit 1 to unit 2 such that the product of incremental cost and penalty factor is the same for both units. Lambda for each of the units is

Generator 1:

$$\frac{dF_1}{dP_1}(\text{PF}_1) = 1202 \times 1.024 = 1231 = \lambda \quad \frac{\text{MBtu}}{\text{h}}$$

Generator 2:

$$\frac{dF_2}{dP_2} \text{PF}_2 = 1164 \times 1.058 = 1231 = \lambda \quad \frac{\text{MBtu}}{\text{h}}$$

The thermal costs for the two units at this new operating point are

$$F_1(P_1 = 0.1577) = 163 \text{ Mbtu/h}$$

$$F_2(P_2 = 1.445) = 1386 \text{ MBtu/h}$$

$$J = \text{total heat rate} = 1549 \text{ MBtu/h}$$

This new operating point is 1619 – 1549 = 70 MBtu/h savings compared to the initial setting, even though the transmission network losses have increased from 3.44 MW to 4.16 MW. In general, the minimum operating cost does not agree with minimum transmission losses except in the special case when generation costs are the same for every unit.

© The computer program *ecnr.exe* may be used to verify these results. Copy the file *example62.sss* into *econ.sss*, then execute *ecnr.exe*

A major problem of economic dispatch is to determine the penalty factors, or equivalently the loss partial derivatives $\partial P_L / \partial P_i$, as dependent on the output power of each of the M generation units. A large number of researchers have contributed to the solution of this problem, usually with assumptions concerning the behavior of loads on the system and the reactive power flow. If fixed numerical values are used, two common assumptions are

1. The variation of each bus load (or substation of the transmission network) is a fixed percentage of the variation in system total load, and this percentage remains fixed irrespective of total system load.

2. The reactive power load or generation is directly proportional to the real load, and this proportionality remains fixed as the load magnitude varies.

It is easy to enumerate special conditions under which these assumptions are violated. As computational capability of computers has increased, the loss partial derivatives are calculated periodically, say every 10 min. Power flow methods and network reduction methods based upon measured state estimation are common to obtain the loss factors. One approach is presented later in this chapter.

6.3   On-Line Execution of the Economic Dispatch

Considering the generating units on-line at any time of the day, the economic dispatch is calculated periodically during the day. The power system usually experiences a daily load cycle where the system load peaks in the afternoon and has a minimum during the night. The ratio of these values is called the daily load factor:

$$\text{load factor} = \frac{\text{minimum daily load}}{\text{maximum daily load}}$$

This ratio varies with the type of load—industrial versus domestic, with weather, workdays and vacation days, and so on. A load factor close to unity is desirable because generating units do not have to be shut down or changed to off-line status during periods of low load. Two typical winter days of load cycle for a utility in the northeastern part of the United States are given in Table 6.1. The Friday represents a workday, in contrast with the Sunday, which is not a workday. Observe that minimum load occurs at 5 A.M. and maximum load occurs at 7 P.M. for both days, which is usually not true. This particular utility serves an area that does not have high-energy industry employing equipment such as arc furnaces or induction furnaces.

As load varies on a daily cycle, it is necessary to raise or lower generation in response to frequency deviation or inadvertent power flows on tie lines to neighbors. The basic signal to perform this adjustment to load changes is the area control error (ACE), as discussed in Chapter 1. ACE is not proportioned to all generators, because some operate at constant, dispatched levels. Even when ACE is zero it is necessary to share changes in the total area

TABLE 6.1

Winter Workday and Weekend Load Demand for a
Utility in Northeastern United States

	Recorded Load (MW)	
Hour	Friday, January 31	Sunday, January 26
01	1134.0	1004.5
02	1121.8	973.7
03	1095.5	966.9
04	1063.7	930.3
05	1029.3	914.4
06	1082.6	925.9
07	1203.1	951.5
08	1323.4	959.8
09	1397.2	1006.8
10	1444.0	1112.3
11	1457.5	1158.5
12	1473.4	1177.2
13	1438.6	1188.1
14	1433.5	1205.9
15	1438.5	1188.6
16	1375.6	1177.9
17	1427.4	1222.9
18	1460.9	1272.2
19	1513.7	1301.5
20	1478.8	1275.7
21	1397.8	1253.9
22	1360.4	1221.0
23	1336.3	1164.5
24	1183.3	1103.4

load. An automatic generation control (AGC) that performs ACE and total load sharing is shown in Figure 6.5. In the block diagram of Figure 6.5, unit i has $K_{ACE,i}$ share of ACE and $K_{ECO,i}$ share of the change in total area load. Their sum is shown as ΔP.

To satisfy the daily variations in load demand at minimum operating cost, the economic dispatch program is periodically executed at 2 to 10 min intervals to calculate the unit's power dispatch set point P_{iD} shown in Figure 6.5. Another method is to execute economic dispatch whenever the system load exceeds the last generation set point by a specified amount. The difference is called base load deviation (BLD). It is shown schematically in Figure 6.5 and defined as

$$\begin{bmatrix} \text{base load} \\ \text{deviation} \\ \text{(BLD)} \end{bmatrix} = \begin{bmatrix} \text{previously} \\ \text{dispatched} \\ \text{total power} \end{bmatrix} - \sum_{i=1}^{M} P_i \qquad (6.13)$$

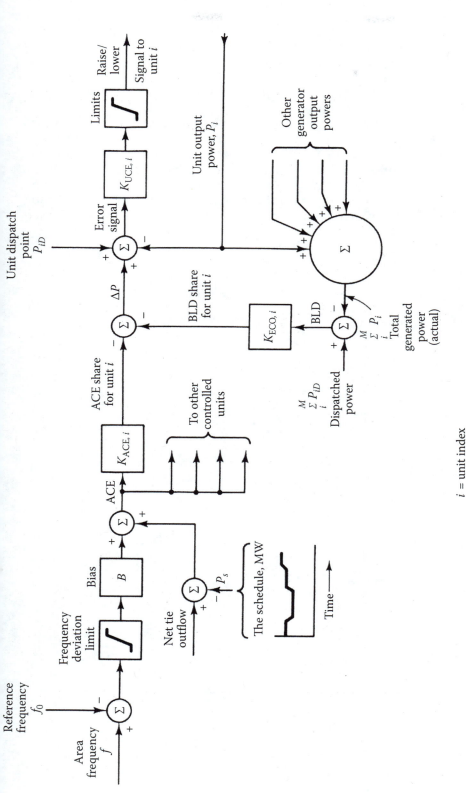

FIGURE 6.5

ACE and BLD sharing by means of an automatic generation control.

where ΣP_i is the presently measured output power of the M generators participating in economic dispatch and ΣP_{iD} is the previously dispatched power.

Other criteria used to trigger the economic dispatch programs are whenever a generating unit changes status from on- to off-line, enters a manual mode of operation, and so on. Operator demand and a periodic basic (e.g., every 5 min) are often used in the same energy management system (EMS).

In most cases the power system load is not measured because there are too many load buses to fully instrument and telemeter data to the central computer where the total load summation can be performed. As a result, P_R, the total load, is unknown, and the economic dispatch program adds generator outputs to obtain P_T, the power to be dispatched. Using P_T, the basic iterative loop (related to the Bolzano or bisection method) to solve $M + 1$ economic dispatch equations appears as shown in Figure 6.6. For a fixed load in this iterative method, it is easily envisioned that a readjustment of all generator output powers

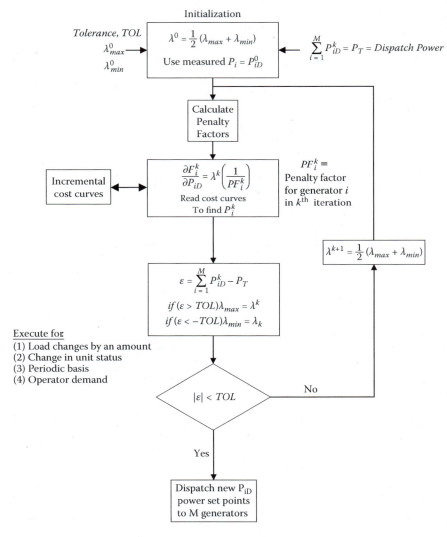

FIGURE 6.6
Economic dispatch iterative loop for the real-line computer of a power system.

also changes the transmission line losses, and hence a newly calculated set point does not satisfy load plus line losses. However, line losses, P_L, are on the order of 3% or less, and the error is a small part of 3%, and hence is negligible in comparison to P_T, which is allocated.

> © The computer program *econ.exe* uses the algorithm of Figure 6.6 to find the optimum dispatch point for as many as nine generators with quadratic cost curves and transmission line loss partial derivatives, $\Delta P_L / \Delta P_i$, as calculated from a B matrix for generator losses. Since the algorithm of Figure 6.6 does not use the Newton-Raphson method, the results of Example 6.2 may be approximated for $P_T \approx 1.6027$ using the data file *example62.sss*. Copy the file *example62.sss* and paste it into *econ.sss* to find the Bolzano solution to Example 6.2. Vary the loss partial derivatives for generators 1 and 2 to examine the change in dispatch as dependent upon transmission line losses.

Significant problems in economic dispatch are to calculate costs and loss coefficients, B coefficients, as dependent upon the variable set of on-line generators, while observing the power transfer capability of transmission lines on the network. When transmission lines limit power transfer, this is called congestion on the network.

6.4 Day-Ahead Economic Dispatch with a Variable Number of Units On-Line

Often the quadratic form, or piecewise linear form of the cost of generation with regard to fuel consumption, is difficult to justify when considerations of capitalization, maintenance, and other costs are considered. Large power systems with many generating plants and pools quote prices in $/MWh, such that the cost of generation and incremental cost are, respectively, a linear expression and constants:

$$F_1(P) = K_{01} + K_1 P \quad \$/h \tag{6.14}$$

$$\partial F_1 / \partial P = K_1 = Incr_1 \quad \$/MWh \tag{6.15}$$

With this cost simplification for each generator, the system minimum cost of operation without line losses, λ in Equation 6.5, is a constant for dispatching the generators. A consequence of employing constant incremental costs is that the system (or pool) power demand is satisfied by using the full capacity of the cheapest generator, then the full capacity of the next cheapest, etc., up to the most expensive generation. Let $Incr_i$ be the constant incremental cost of unit i. Then the economic dispatch becomes an ordered use of generation:

$$\{Generator\ power\ order\} = \{Incr_1 < Incr_2 < Incr_3 < Incr_4, ..., < Incr_{HIGHEST}\}$$

An example of power allocation by increasing $/MWh is graphically depicted in Figure 6.7 for a daily load cycle. The most expensive generation may be off-line when its power is not needed.

Unless a generating unit has a "must run" requirement, there are many restrictions and costs when and if a unit should be taken off-line, or brought back on-line. Nuclear power plants are in the "must run" category for many safety reasons. Other generation plants may have minimum hours of on-line or off-line restrictions. Table 6.2 lists conditions that determine whether or not a generation unit may be off-line when not needed to satisfy daily load cycle demand, or spinning reserve requirements.

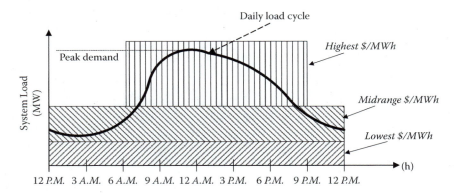

FIGURE 6.7
On-line operation of three constant $/MWh generators to meet daily load cycle at minimum cost.

A power system with many power generating units forecasts the load demand hour by hour for the next day and the units needed to satisfy the load considering the constraints in Table 6.2. The software that calculates the unit operation schedule to satisfy load and spinning reserve requirements is called a *unit commitment program* (UCP). The UCP minimizes the total cost over a daily load cycle to satisfy the load considering the list of constraints. Other UCPs may allocate generation for weekly schedules.

For increased reliability and economics, utilities have merged into power pools with central allocation of the generators within the connection. This has led to a market concept of buyer and seller members within the pool, instead of the single area demand-cost considered thus far. The interconnection network is considered lossless, and member requests to buy power and sell power at constant incremental costs, Incr, are entered a day ahead (as if forecast) by the members.

In pool operation, the pool central coordinator balances the ACE for the pool and each member's area, then calculates the economic dispatch forecast for various loads at times of the forthcoming day. After the day, the difference between requested MWh and actual power MWh demand are balanced, and the costs are adjusted from system power flow measurements. The generator operating constraints plus constraints of the pool network—MVar flows, current flow restrictions, lines out of service, etc.—are taken into account

TABLE 6.2

Generator Economic Constraints

Uhigh = Maximum power output of the generator (MW)
Ulowp = Minimum power output of the generator (MW)
Ofset = A step cost, added to the constant incremental cost, to bring a unit back on-line ($)
Uincr = Incremental cost constant for operating the unit ($/MWh)
Quadt = Quadratic term for unit cost ($/MW²h)
Upmin = Minimum operating hours on-line (h)
Dnmin = Minimum hours downtime when off-line (h)
Colds = A unit off-line for more than **x** hours requires a preheat before it can be on-line (h)
Hstrt = A penalty cost to prepare a unit if it has been off-line less than **x** hours ($)
Cstrt = A penalty cost to preheat a unit, off for more than **x** hours, to come back on-line ($)
Bmatrix = Loss coefficient matrix of transmission line losses of the generators in terms of the real power output of each generator (assumes reactive power is proportional to real power) (1/MW)

during the day. Deviations between predicted and actual MWh are adjusted according to the after-the-fact economic dispatch costs for the day. This is called locational marginal pricing (LMP) [4].

Whether a single area is studied, or a power pool is considered, the constant incremental cost method requires the least expensive power to be used, progressing up to the more expensive power as the load demand increases. Example 6.3 demonstrates power allocation by incremental costs and pricing by this method.

Example 6.3a

The five-bus lossless pool network shown in Figure E6.3 has power demand at buses B, C, and D. There are four suppliers of constant $/MWh power at buses A, C, D, and E with the capacity and costs shown in the figure.

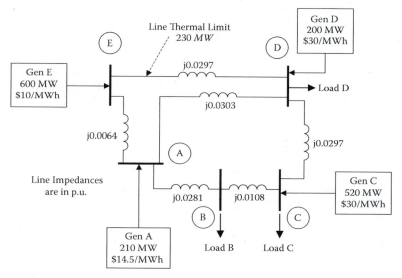

FIGURE E6.3
Five-bus network for constant $/MWh dispatch.

(a) If the power demand is 223 MW at each of bus B, C, and D (total 669 MW), determine the LMP cost to each load. Use the dc power flow method to calculate the power flow on the network.
(b) If the demand increases to 300 MW at each load, determine the LMP cost if the power transfer of transmission line D–E is limited to 230 MW.

Solution

(a) To satisfy the 669 MW demand, the full 600 MW of the lowest incremental cost power $10/MWh at bus E is allocated. The next cheapest power is to use 69 MW of power at $14.5/MWh from generation at bus A. This makes the total cost:

Total cost = 600 MW*$10/MWh + 69 MW*$14.5/MWh = $7,000.5/h

Each load pays $2,333.5 as its share of the total cost. Power flow on the network is approximated by using the *dc_power* program (see Chapter 1) with bus B as the slack bus. The dc power flow equation on a 100 MVA base, with small angle assumptions and 1.0 p.u. bus voltages, is as follows:

$$
P = \begin{bmatrix} P_A \\ P_C \\ P_D \\ P_E \end{bmatrix} = \begin{bmatrix} .69 \\ -2.33 \\ -2.23 \\ 6 \end{bmatrix}
$$

$$
= \begin{bmatrix}
\dfrac{1}{X_{AB}} + \dfrac{1}{X_{AD}} + \dfrac{1}{X_{AE}} & 0 & \dfrac{-1}{X_{AD}} & \dfrac{-1}{X_{AE}} \\
0 & \dfrac{1}{X_{BC}} + \dfrac{1}{X_{CD}} & \dfrac{-1}{X_{CD}} & 0 \\
\dfrac{-1}{X_{AD}} & \dfrac{-1}{X_{CD}} & \dfrac{1}{X_{AD}} + \dfrac{1}{X_{CD}} + \dfrac{1}{X_{DE}} & \dfrac{-1}{X_{DE}} \\
\dfrac{-1}{X_{AE}} & 0 & \dfrac{-1}{X_{DE}} & \dfrac{1}{X_{AE}} + \dfrac{1}{X_{DE}}
\end{bmatrix}
\begin{bmatrix} \delta_A \\ \delta_C \\ \delta_D \\ \delta_E \end{bmatrix}
$$

The slack bus is maintained at 1.0∠0°. The *dc_flow* computer program results, with numerical values for transmission lines and loads in the above equation, are as follows with phase angles (deg.) and line flows (p.u.):

$$
\delta = \begin{bmatrix} \delta_A \\ \delta_C \\ \delta_D \\ \delta_E \end{bmatrix} = \begin{bmatrix} 4.717\,\text{deg} \\ -9.440\,\text{deg} \\ 2.172\,\text{deg} \\ 6.076\,\text{deg} \end{bmatrix}
\qquad
\begin{bmatrix} P_{AB} \\ P_{BC} \\ P_{CD} \\ P_{DE} \\ P_{AE} \\ P_{AD} \end{bmatrix} = \begin{bmatrix} 2.930 \\ 0.700 \\ -1.530 \\ -2.294 \\ -3.706 \\ 1.466 \end{bmatrix}
$$

The power flow in line DE is 229 MW, which is below the rating, so the economic dispatch is feasible.

(b) When the loads at buses B, C, and D increase to 300 MW each, the least expensive power suppliers are buses A and E. These two buses supply 810 MW, so the remaining 90 MW of power must come from generators with the higher incremental cost of $30/MWh. To investigate the impact of using generation at buses C and D, the *dc_power* flow program is first employed with generation at bus D supplying 90 MW of power at $30/MWh. The results are:

$$
\text{Injections } P = \begin{bmatrix} P_A \\ P_C \\ P_D \\ P_E \end{bmatrix} = \begin{bmatrix} 2.10 \\ -3.00 \\ -2.10 \\ 6.00 \end{bmatrix}
\quad \text{force line flows} \quad
\begin{bmatrix} P_{AB} \\ P_{BC} \\ P_{CD} \\ P_{DE} \\ P_{AE} \\ P_{AD} \end{bmatrix} = \begin{bmatrix} 3.838 \\ 0.838 \\ -2.162 \\ -2.523 \\ -3.477 \\ 1.739 \end{bmatrix}
$$

The transmission line DE power constraint is violated in this case. When bus C is supplying 90 MW of power, the results of the *dc_flow* program are:

$$
\text{Injections } P = \begin{bmatrix} P_A \\ P_C \\ P_D \\ P_E \end{bmatrix} = \begin{bmatrix} 2.10 \\ -2.10 \\ -3.00 \\ 6.00 \end{bmatrix}
\quad \text{force line flows} \quad
\begin{bmatrix} P_{AB} \\ P_{BC} \\ P_{CD} \\ P_{DE} \\ P_{AE} \\ P_{AD} \end{bmatrix} = \begin{bmatrix} 3.524 \\ 0.524 \\ -1.576 \\ -2.667 \\ -3.333 \\ 1.910 \end{bmatrix}
$$

The conclusion of these two approximate power flow results is that generators with lower incremental costs of \$10/MWh and \$14.5/MWh cannot be fully utilized and prevent an excessive power flow on transmission line DE. Either the generation at bus A or bus E must be reduced for line DE to remain less than 230 MW. Various methods are available to exactly solve this constrained line dispatch problem [5] at minimum cost.

The incremental cost solution to alleviate the overload is to decrease the next most expensive unit, bus A, by 60 MW and increase the generation at bus D to 150 MW. The results of this calculation are as follows:

$$\text{Injections } P = \begin{bmatrix} P_A \\ P_C \\ P_D \\ P_E \end{bmatrix} = \begin{bmatrix} 1.50 \\ -3.00 \\ -1.50 \\ 6.00 \end{bmatrix} \text{ force line flows } \begin{bmatrix} P_{AB} \\ P_{BC} \\ P_{CD} \\ P_{DE} \\ P_{AE} \\ P_{AD} \end{bmatrix} = \begin{bmatrix} 3.722 \\ 0.722 \\ -2.201 \\ -2.302 \\ -3.698 \\ 1.476 \end{bmatrix}$$

If the lower incremental cost generation at buses A and E could be fully utilized, the system cost would be \$11,745/h. When the transmission line DE constraint forces the use of an additional 60 MW of \$30/MWh power from bus D, the cost increases to \$12,675/h. The additional cost over the minimum possible is called a congestion cost and is passed on to all loads, as each pays 1/3 of the total cost, \$4,225/h.

© The transmission line flows presented in Example 6.3a may be verified by using the computer program *dc_flow* and varying the power injections at buses A, B, C, D, and E to match cases (a) and (b).

In Example 6.3a, it was necessary to use more expensive power because a transmission line is overloaded. In general, the economic dispatch is first calculated to obtain the minimum cost allocation of power, then a power flow calculation is used to check network transmission line constraints. Sometimes adjustment of network parameters—Var adjustments, transformer tap changes, FACTS devices, etc.—permits operation at the minimum cost point.

When operation of generation over a period of time, such as for a daily load cycle (24 h), is considered, the total cost for this period is to be minimized considering constraints such as presented in Table 6.2. The combination of generating units on-line is called a state of operation, which is usually a binary number of 0 for off-line and 1 for on-line operation. The simplest example is: given a power system cyclic load and the initial combination of generating units on-line; find the minimum total cost over a time period using various combinations of generating units to be on-line. This problem was elegantly formulated in 1957 by Richard Bellman as the *principle of optimality*:

> An optimal policy has the property that whatever the initial state and initial decision are, the remaining decisions must constitute an optimal policy with regard to the state resulting from the first decision.

By modern computer techniques [6, 7], this principle is implemented by means of exhaustively computing all possible combinations of generators on-line to satisfy the load demand for a time interval and all possible combinations that can make the transition to the next time interval. Transitions from state to state (paths) are saved in memory and the optimal path is found to the final state and power demand. The program that computes the optimal path is called a unit commitment program (UCP). The following example demonstrates a UCP that considers factors from Table 6.2 for selection of generators to be on-line at any time for economic dispatch. The UCP calculation often is called an economy B power allocation.

Example 6.3b

Four generators, U1, ..., U4, are available to satisfy the system power demand on a load cycle. Initially the system power demand is 669 MW and units U1 and U4 are on-line to supply this load. Generators U1 and U4 have been on-line for 8 h and units U2 and U3 have been off-line for 5 h. Additional data for the generators' power and operation constraints are as follows (as defined in Table 6.2):

TABLE E6.3a

Power and Operating Constraints for Four Generators

Unit	U1	U2	U3	U4
Uhigh	210	500	300	600
Ulowp	30	60	40	80
Ofset	0	500	100	0
Uincr	14.5	30.1	30	10
Quadt	0	0	0	0
Upmin	0	0	0	0
Dnmin	0	0	0	0
Colds	4	5	5	0
Hstrt	150	170	500	0
Cstrt	350	400	1,100	200
Bmatrix	0	0	0	0
(Row/Gen)				

(a) After the first hour, the load demand increases from 669 MW to 900 MW. Minimize the accumulated total cost of operating and transitions to satisfy the 669 load in the first hour and the 900 MW load in the second hour. This is the complete schedule.

(b) It is desired to minimize the total generation cost over an 8 h load cycle by allocating power between four generators, and switching units off-line or on-line as needed. The predicted load over the forthcoming 8 h period is shown in Table E6.3b.

TABLE E6.3b

System Load for an 8 h Cycle

Hour	Load (MW)
1	669
2	900
3	1,000
4	1,200
5	900
6	300
7	400
8	610

To start the analysis, it is first necessary to define the possible operating states (combinations) of generators to satisfy the load. There are $2^4 - 1 = 15$ possible combinations of operating the four generators. Let the operating states be binary 1 for a unit on-line and binary 0 when the unit is off-line, as follows:

Operating State	U1	U2	U3	U4
0	0	0	0	0
1	1	0	0	0
2	0	1	0	0
3	1	1	0	0
4	0	0	1	0
5	1	0	1	0
6	0	1	1	0
▼				
Cont'd				

When unit U1 alone is on-line, the maximum power available is 210 MW, so that operating state 1 is not a possible solution to satisfy the load for any hour of the 8 h predicted load. The minimum and maximum possible powers for every combination of generator are computed and ordered by increasing power to find the possible operating states. The results of this calculation are presented in Table E6.3c.

TABLE E6.3c

Reordered States According to Increasing Power Capacity

Power Level	U1	U2	U3	U4	Plow	Phigh
0	0	0	0	0	0.000	0.000
1	1	0	0	0	30.000	210.000
2	0	0	1	0	40.000	300.000
3	0	1	0	0	60.000	500.000
4	1	0	1	0	70.000	510.000
5	0	0	0	1	80.000	600.000
6	1	1	0	0	90.000	710.000
7	0	1	1	0	100.000	800.000
8	1	0	0	1	110.000	810.000
9	0	0	1	1	120.000	900.000
10	1	1	1	0	130.000	1,010.000
11	0	1	0	1	140.000	1,100.000
12	1	0	1	1	150.000	1,110.000
13	1	1	0	1	170.000	1,310.000
14	0	1	1	1	180.000	1,400.000
15	1	1	1	1	210.000	1,610.000

Since the load varies between 300 and 1,200 MW, the load demand over the 8 h period can be satisfied by the reordered operating states between 2 and 11, inclusive. It is not necessary to consider other state combinations.

Solution (a)

Since generators U1 and U4 are on-line at the starting time, this corresponds to no. 8 in the reordered states shown in Table E6.3c. The operating cost for these units to supply 669 MW of power is a minimum when 600 MW of power at \$10/MWh is used from generator U4 and the remaining load from U1 at a constant incremental cost of \$14.50/MWh:

$$\begin{bmatrix} Cost \\ for \\ 669MW \end{bmatrix} = 600MW * (\$10/MWh) + (669MW - 600MW) * (\$14.5/MWh) = \$7000.50/h$$

Considering the load increase to 900 MW, the on-line generator capacity must be greater than the load, so state 9 is not viable. However, Table E6.3c shows that any new state from 10 to 15 has more capacity than required to supply the increased load of 900 MW. The transition to go from state 8 to 10 requires start-up costs for U2 and U3 and shutdown of U4.

As an example of the transition and operating calculation, let the transition be from state 8 to 12. The complete capacity of the lower incremental cost generation is used before the higher cost generators. Table E6.3a is used to obtain

$$\begin{bmatrix} Cost \\ for \\ 900MW \end{bmatrix} = \begin{aligned} & 600MW * (\$10/MWh) + 210MW * (\$14.50/MWh) \\ & + (900MW - 810MW) * (\$30/MWh) = \$11,745/h \end{aligned}$$

Operating cost for 900 MW	\$11,745
Cold start cost for U3	\$1,100
Offset U3 incremental cost	\$100
State transition + operating cost, state 8 to 12	\$12,945

The accumulated cost for the initial 2 h is \$7,000.5 + \$12,945 = \$19,945.50/h.

To go from state 8 to state 13, note the small difference in incremental costs between U2 and U3. The cost calculation for this transition is

Operating cost for 900 MW	\$11,754
Cold start cost for U2	\$400
Offset U2 incremental cost	\$500
State transition + operating cost, state 8 to 13	\$12,654

The accumulated cost for the initial 2 h is \$7,000.5 + \$12,654 = \$19,654.50/h. This process of transition and operating cost for every possible transition is shown below. The summary

Costs to Satisfy 669 MW for 1 h
and 900 MW for the Next Hour

Transition	Accumulated Cost
8 to 10	\$32,884.50
8 to 11	\$22,930.50
8 to 12	\$19,945.50
8 to 13	\$19,654.50
8 to 14	\$24,100.50
8 to 15	\$23,162.50

of costs shows that the minimum cost transition is from state 8 to 13 when there are only two load levels of power demand.

Solution (b)

When more than one state transition, such as eight transitions over an 8 h load cycle, the total cost is accumulated interval by interval over the cycle, and later transitions are affected by earlier changes. The computer program *ucomt.exe* is a dynamic programming UCP that retains the minimum accumulated cost path (state transitions and operating cost) from initial conditions to the final state. All possible transitions are tested and costs computed at each time interval such as demonstrated in part (a). The generators of Table E6.3a and the load cycle of E6.3b are input to the UCP to obtain the results presented in Table E6.3d.

TABLE E6.3d

Unit Commitment Program Results for Generators of Table E6.3a to Supply 8 h Load Cycle

Power Level	8	12	12	15	12	5	5	8
Hour	1	2	3	4	5	6	7	8
Unit 1	69	210	210	210	210	0	0	10
Unit 2	0	0	0	90	0	0	0	0
Unit 3	0	90	190	300	90	0	0	0
Unit 4	600	600	600	600	600	300	400	600
Total(MW) Load	669	900	1,000	1,200	900	300	400	610
Cum. $K	7.000	19.945	34.791	56.544	68.389	71.389	75.389	81.684

In Table E6.3d note the minimum accumulated cost considering hours 3 to 8 requires that from hour 1 to hour 2 the system switch from state 8 to 12, instead of the result computed in part (a). This is the effect of considering all state transitions in the dynamic programming algorithm. The accumulated cost is the row labeled "Cum. $K" in table E6.3d.

© The dynamic programming results of Example 6.3b may be verified using the computer program *ucomt.exe*. Copy the file *example63B.sss* and paste it into *ucomt.sss*, then execute *ucomt.exe* in order to duplicate the results. After execution, the solution is appended to the *ucomt.sss* file. The constraints on generators U1, ..., U4, such as minimum uptime or downtime, may be modified in the *ucomt.sss* file to observe their impact after again running the program.

Utilities that have a large number of generators to schedule often have UCPs employ power flows to check for transmission line constraint violations at the dispatch point. The UCPs may also employ parabolic or piecewise linear approximations to generator cost curves. These UCPs often minimize 7-day total costs considering 5 days of workdays and 2 days of weekend load cycle in their computation.

6.5 Power Transmission Line Losses for Economic Dispatch

Fuel costs for a 10,000 MW peak load utility are billions of dollars per year (in 2007), so a very small percentage in savings by considering transmission line losses (penalty factors) when dispatching generators would result in great revenue to the utility. In the past, the most widely used method to calculate the penalty factors and total transmission losses for generator economic allocation was by the loss coefficient, or B coefficient, method. This analysis was initiated by Kron [8], popularized by Kirchmayer [2], and had numerous extensions [9, 10].

The loss coefficient formulation assumes that the total transmission losses in the network may be expressed as a quadratic function of the power output of the M generators and K tie lines as

$$P_L = \sum_{\phi=1}^{lines} |i_\phi|^2 R_\phi = \sum_{n=1}^{M+K} \sum_{m=1}^{M+K} P_n B_{mn} P_m + \sum_{i=1}^{M+K} B_{i0} P_i + B_{00} \tag{6.16}$$

where i_ϕ is the current flow in line ϕ, which has the positive-sequence resistance R_ϕ. The tie-line flow may be either into or out of the system and transmission losses are assumed to be bilateral such that $B_{mn} = B_{nm}$. When numerical values are known for the B_{mn} coefficients of Equation 6.16, the incremental loss partial derivatives are easily calculated as

$$\frac{\partial P_L}{\partial P_i} = 2 \sum_{m=1}^{M+K} B_{mi} P_m + B_{i0} \quad (i = 1, 2, 3, \ldots, M) \tag{6.17}$$

Hence, the penalty factors, Equation 6.11, are determined for each of the M generators. The simultaneous solution of the $M + 1$ coordination equations to obtain values $P_1, P_2, \ldots, P_m, \lambda$ for economic dispatch is facilitated if the B coefficients are known.

There are many ways to calculate the B_{mn}, B_{i0}, coefficients, some of which reduce a large network to an equivalent consisting of generation buses, tie buses, and a single load, which is required for loss coefficient calculations. The network reduction technique to be described here, called the remote-equivalent-independent (REI) method [11, 12], is direct and easily adapted for on-line use with changing network conditions (generators in and out of service, transmission lines out of service, compensation added, etc.). The REI method to obtain loss coefficients uses the following eight steps:

Step 1: Execute a power flow or state estimator calculation to obtain the voltage magnitude and phase angle at each bus of the system. At this point all MW and Mvar loads are converted to lumped L, C, R elements to *linearize* the system.

Step 2: All network shunt elements from bus to ground are either combined with load lumped equivalents or converted into an injection for generator and ties. All autotransformers are converted to π-equivalents such that shunt currents to ground may be combined at adjoining buses.

Step 3: The lumped circuit elements for the N load buses are connected to a common node, called the g node. The potential of the g node is arbitrarily set to zero, which is the most convenient value. The network now appears as shown in Figure 6.8. Only the generators are connected to ground.

Step 4: The numerical value of all currents injected into the network as well as the bus voltages are known. Next, an artificial node, r, is defined whose injection current and power are equivalent to those of the N load buses.

$$i_r = \sum_{n=1}^{N} i_n \tag{6.18}$$

$$S_r = \sum_{n=1}^{N} S_n = \sum_{n=1}^{N} e_n i_n^* \tag{6.19}$$

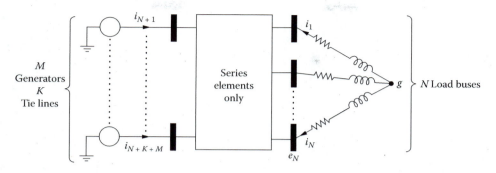

FIGURE 6.8
Lumped element loads connected to a common bus, bus *g*.

From these two quantities the equivalent voltage at node *r* may be calculated as

$$e_r = \frac{S_r}{i_r^*} \qquad (6.20)$$

An equivalent admittance from the *r* node to a zero potential *g* node is calculated as

$$Y_{rg} = \frac{-S_r^*}{e_r e_r^*} \qquad (6.21)$$

such that the *r* node has the same injection as the sum of the *N* load buses. The network now appears as shown in Figure 6.9.

Step 5: By connecting the loads to the *g* bus, buses 1, 2, 3, …, *N* and *g* have zero current injection, such that the system nodal equation is

$$
\mathbf{I}_{BUS} =
\begin{bmatrix}
i_1 \\
\vdots \\
i_N \\
i_g \\
i_{N+1} \\
\vdots \\
i_{N+M+K} \\
i_r
\end{bmatrix}
=
\begin{bmatrix}
0 \\
0 \\
\vdots \\
0 \\
i_{N+1} \\
\vdots \\
i_{N+M+K} \\
i_r
\end{bmatrix}
=
\begin{bmatrix}
\mathbf{Y}_{11} & \mathbf{Y}_{12} \\
\mathbf{Y}_{21} & \mathbf{Y}_{22}
\end{bmatrix}
\mathbf{E}_{BUS} = \mathbf{Y}_{BUS}\,\mathbf{E}_{BUS} \qquad (6.22)
$$

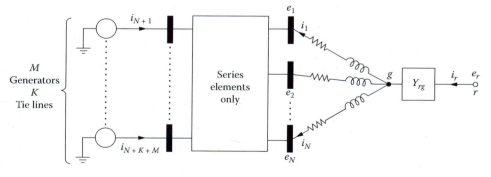

FIGURE 6.9
REI equivalent for a network.

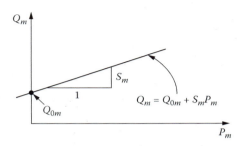

FIGURE 6.10
Linearly related real and reactive power.

Here $\mathbf{I}_{BUS}$ is the bus injection current vector (known numerical values) and $\mathbf{E}_{BUS}$ is the vector of bus voltages, including a zero for the g bus and the calculated potential for the REI bus, r. The $\mathbf{Y}_{11}$ partition of the matrix includes the N loads and the g node. The $\mathbf{Y}_{11}$ admittance matrix retains the sparsity of the original network except for the g node, such that sparse matrix methods may be used to invert $\mathbf{Y}_{11}$ (because its current injections are zero). To obtain the REI network equivalent, the buses with zero injection currents are eliminated to obtain

$$\begin{bmatrix} \mathbf{I}_3 \\ i_r \end{bmatrix} = \mathbf{I} = \begin{bmatrix} i_{N+1} \\ \vdots \\ i_{N+M+K} \\ i_r \end{bmatrix} = \left[\mathbf{Y}_{22} - \mathbf{Y}_{21}\mathbf{Y}_{11}^{-1}\mathbf{Y}_{12} \right] \mathbf{E} = \mathbf{Y}_r \mathbf{E} = \mathbf{Y}_r \begin{bmatrix} \mathbf{E}_3 \\ e_r \end{bmatrix} \tag{6.23}$$

The resulting $M + K + 1$ admittance matrix, $\mathbf{Y}_r$, is dense, usually with connecting elements between all generators, ties, and the single equivalent load of the r bus.

Step 6: A partition, $\mathbf{Y}_3$, using bus r as the reference, is taken from the $\mathbf{Y}_r$ matrix. The inverse of this partition is the matrix $\mathbf{Z}_3$ (with zero current to neutral):

$$\mathbf{Y}_r = \left[\begin{array}{c|c} \mathbf{Y}_3 & \mathbf{Y}_{1r} \\ \hline \mathbf{Y}_{r1} & \mathbf{Y}_{rr} \end{array} \right] \tag{6.24a}$$

$$\mathbf{Z}_3 = \mathbf{Y}_3^{-1} \tag{6.24b}$$

All transmission lines in $\mathbf{Z}_3$ have no elements to ground, and the impedances connected to the g node sum to zero. Therefore, the line real and reactive power losses are

$$P_L + jQ_L = (\mathbf{I}_3^*)^t \mathbf{E}_3 = (\mathbf{I}_3^*)^t \mathbf{Z}_3 \mathbf{I}_3 \tag{6.25}$$

The load bus has been eliminated, but the injected bus currents at the generators and tie lines remain the same.

Step 7: Next, the assumption is made that the real and reactive power at each generator bus and tie line are related linearly as shown in Figure 6.10. In addition, the voltage at each of the tie lines and generators is assumed fixed in magnitude plus phase angle regardless of load. Hence, the power injected at the tie buses and generator buses is

$$e_m^* i_m = P_m - jQ_m = (1 - jS_m)P_m - jQ_{0m} \tag{6.26}$$

and the current may be expressed as

$$i_m = \frac{(1 - jS_m)P_m - jQ_{0m}}{e_m^*} = \alpha_m P_m + \beta_m \tag{6.27}$$

In this equation, P_m is variable. The numerical value of the voltage is known from the power flow or state estimator, while at least two sets of such data must be used to determine the offset Q_{0m} and slope S_m. If only one set of power flow or state estimator data is known, P_m and Q_m are assumed to be proportional. When the currents are expressed in the form of Equation 6.27, P_m is implicit, α_m and β_m are numerical complex values, and the current injection vector becomes

$$\mathbf{I}_3 = \begin{bmatrix} \alpha_1 P_1 + \beta_1 \\ \vdots \\ \alpha_{K+M} P_{K+M} + \beta_{K+M} \end{bmatrix} \tag{6.28}$$

Step 8: The last step is to introduce Equation 6.28 into 6.25 to obtain the transmission losses, then equate the B coefficients of Equation 6.16 with the real part of the numerical values from Equation 6.25:

$$P_L = \mathbf{Re} \left\{ \begin{bmatrix} \alpha_1^* P_1 + \beta_1^* \\ \vdots \\ \alpha_{K+M}^* P_{K+M} + \beta_{K+M}^* \end{bmatrix}^t Z_3 \begin{bmatrix} \alpha_1 P_1 + \beta_1 \\ \vdots \\ \alpha_{K+M} P_{K+M} + \beta_{K+M} \end{bmatrix} \right\}$$

$$= \sum_{m=1}^{K+M} \sum_{n=1}^{K+M} P_m B_{mn} P_n + \sum_{i=1}^{K+M} B_{i0} P_i + B_{00} \tag{6.29}$$

where α_i, β_i, and Z_3 are numerical values. The B_{mn} are numerical combinations of α_i and β_i, which multiply $P_m P_n$, and the B_{i0} are the numerical coefficients that multiply the P_i.

In the case when the reactive power offset $S_m = 0$, and $i_m = \alpha_m P_m = (a_m + jb_m)P_m$, the expression for transmission losses simplifies to the following quadratic expression:

$$P_L = \sum_{m=1}^{K+M} \sum_{n=1}^{K+M} P_m \mathbf{Re}(Z_{3,mn})(a_m a_n + b_m b_n) P_n \quad \text{(no } Q_m \text{ offset)} \tag{6.30a}$$

Expressed as a matrix, the last equation is

$$P_L = [P_1\ P_2\ \dots\dots P_{M+K}] \mathbf{B} \begin{bmatrix} P_1 \\ P_2 \\ \vdots \\ P_{M+K} \end{bmatrix} \tag{6.30b}$$

The real power loss matrix, B, is not symmetric and may contain negative off-diagonal terms. In comparison to the impedance matrix Z_3, the alternative approach of expressing transmission losses in terms of an admittance formulation,

$$P_L = \mathbf{Re}\{(E_3^*)^t Y_3 E_3\} = \mathbf{Re}\ \{(P/I)^* Y_3 (P/I)\}$$

leads to expressing the generator voltages as

$$E_k^* = \frac{P_k + jQ_k}{i_k} = \frac{(1 + jS_k)P_k}{i_k}$$

where the steady-state power flow numerical value is used for i_k. This is a poor approximation of the loss gradient because the current does not remain fixed with power change:

$$\left. \frac{\partial P_L}{\partial P_k} \right|_{i_k} \neq \text{constant}$$

whereas the voltage at the generation bus remains comparatively constant. Because of this property, methods that depend on bus admittance matrices ultimately employ an inverse to compute the dense matrix $Z_3 = [Y_3]^{-1}$.

The calculation of loss coefficients by the REI method is demonstrated in Example 6.4.

Example 6.4

Use the REI method of network reduction and utilize the resulting admittance matrices for loss coefficient calculations of the resistive network shown in Figure E6.4.1. Steady-state values are shown.

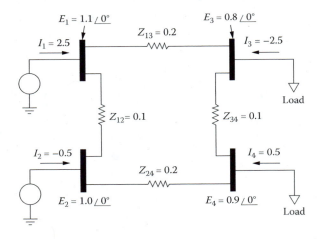

FIGURE E6.4.1

Solution

Loads at buses 3 and 4 converted to lumped admittances are (step 1) calculated as

$$Y_{3g} = \frac{2.5}{0.8} = 3.125 \quad Y_{4g} = \frac{-0.5}{0.9} = -0.555$$

The current injection and power flow at the equivalent r bus are (step 4)

$$i_r = i_3 + i_4 = -2.0$$

$$S_r = e_3 i_3 + e_4 i_4 = -1.55$$

The voltages at the r bus and Y_{rg} are calculated to be

$$e_r = \frac{-1.55}{-2} = 0.7775 \quad Y_{rg} = -2.581$$

The resulting network is shown in Figure E6.4.2 (step 4). The admittance formulation including the r and g nodes is the following (step 5):

$$I = \begin{bmatrix} 0 \\ 0 \\ 0 \\ i_1 \\ i_2 \\ i_r \end{bmatrix} = YE = \begin{bmatrix} 18.125 & -10 & -3.125 & -5 & 0 & 0 \\ -10 & 14.45 & 0.555 & 0 & -5 & 0 \\ -3.125 & 0.555 & -0.011 & 0 & 0 & 2.581 \\ -5 & 0 & 0 & 15 & -10 & 0 \\ 0 & -5 & 0 & -10 & 15 & 0 \\ 0 & 0 & 2.581 & 0 & 0 & -2.581 \end{bmatrix} \begin{bmatrix} e_3 \\ e_4 \\ e_g \\ e_1 \\ e_2 \\ e_r \end{bmatrix}$$

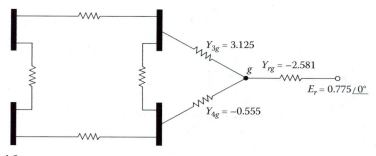

FIGURE E6.4.2

The inverse of the Y_{11} partition is used to obtain Y_r.

$$\begin{bmatrix} i_1 \\ i_2 \\ i_r \end{bmatrix} = [Y_{22} - Y_{21}Y_{11}^{-1}Y_{12}]E = \left\{ \begin{bmatrix} 15 & -10 & 0 \\ -10 & 15 & 0 \\ 0 & 0 & -2.581 \end{bmatrix} - \begin{bmatrix} 0.1027 & 0.4053 & 4.4922 \\ 0.4053 & 2.1893 & 2.4029 \\ 4.4922 & 2.4029 & -9.4787 \end{bmatrix} \right\} E$$

$$= Y_r E = \begin{bmatrix} 14.8973 & -10.4053 & -4.4922 \\ -10.4053 & 12.8107 & -2.4029 \\ -4.4922 & -2.4029 & 6.8977 \end{bmatrix} \begin{bmatrix} e_1 \\ e_2 \\ e_r \end{bmatrix}$$

This last equation represents the system admittance equations for generators 1 and 2 and the equivalent load bus r, with ground as a reference. Matrix Y_r is singular since it has no shunt elements to ground. The partition of Y_r corresponding to generators and ties using the load bus as a reference is the 2×2 matrix for buses 1 and 2. The inverse of this partition is the bus impedance matrix of Section 4.3.

$$\mathbf{Z}_3 = \mathbf{Y}_3^{-1} = \begin{bmatrix} 14.8972 & -10.4053 \\ -10.4053 & 12.8107 \end{bmatrix}^{-1} = \begin{bmatrix} 0.1551 & 0.1260 \\ 0.1260 & 0.1804 \end{bmatrix}$$

In this example there is only real power flow, so loss coefficients may be found by means of equating like coefficients on both sides of the equation for P_L:

$$P_L = B_{11}P_1^2 + B_{12}P_1P_2 + B_{22}P_2^2 = \begin{bmatrix} \dfrac{P_1}{1.1} & \dfrac{P_2}{1.0} \end{bmatrix} \begin{bmatrix} 0.1551 & 0.1260 \\ 0.1260 & 0.1804 \end{bmatrix} \begin{bmatrix} \dfrac{P_1}{1.1} \\ \dfrac{P_2}{1.0} \end{bmatrix}$$

$$= 0.1282P_1^2 + 0.2291P_1P_2 + 0.1804P_2^2$$

Total transmission losses are given by introducing generator base powers into the loss expression.

$$P_L = 0.1282(2.75)^2 + 0.2291(2.75)(-0.5) + 10804\,(0.5)^2$$

$$= \sum_{i=1}^{4} P_i = (2.75) + (-0.5) + (-2.0) + (0.45) = 0.70$$

© The loss coefficients calculated in Example 6.4 may be duplicated by copying the file *example64.sss* into *rei.sss*, then executing the *rei.exe* program. To match the loss coefficients, it is necessary to specify bus 1 and bus 2 as the generation buses, even though bus 2 has a negative power injection. Otherwise, *rei.exe* identifies generators at buses with real power injection into the network, and would assign bus 1 and bus 4 for loss coefficient calculation. The *rei.exe* calculates a power flow solution for the input network, then the reduced network Z_3, and finally the B loss coefficient matrix. The *rei.exe* program accepts data for general ac networks with complex injections and power flow.

As stated previously, the matrix $\mathbf{Y}_3$ is dense. However, an advantage of using the sparse matrix programs in step 6 is that one column of $\mathbf{Z}_3$ is obtained from the **LDU** decomposition for every right-hand-side vector. Each successive column of $\mathbf{Z}_3$ requires one less element, so there is some computational advantage. It is also simpler to match coefficients in Equation 6.30a,b on a column-by-column basis.

6.6 Utilizing the Load-Flow Jacobian for Economic Dispatch

Power systems short-circuit programs, contingencies, and load flow all utilize sparse matrix techniques and the bus admittance matrix, $\mathbf{Y}_{BUS}$, to compute desired quantities. It is therefore not surprising that programs multiplying or inverting $\mathbf{Y}_{BUS}$ are also employed in economic dispatch of generators [13]. Consider the network power flow equations with the slack generator as reference and bus injections as shown in Figure 6.11. All transformers

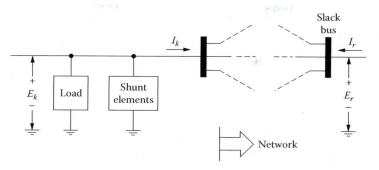

FIGURE 6.11
Bus impedance formulation for load flow.

are converted to π-equivalents, and line-charging elements are included in the shunt element to ground. Resistive power losses in these shunt elements are *assumed* to be negligible in comparison to series transmission line losses.

For each of the $n-1$ system buses, after execution of the load flow, the vector, $\mathbf{E}'_{\text{BUS}}$, and injection current, $\mathbf{I}'_{\text{BUS}}$, are known and related by

$$\mathbf{E}'_{\text{BUS}} = \mathbf{Z}_{\text{BUS}}\,\mathbf{I}'_{\text{BUS}} = \begin{bmatrix} E_1 - E_r \\ \vdots \\ E_{n-1} - E_r \end{bmatrix} \tag{6.31}$$

where $\mathbf{E}'_{\text{BUS}}$ is an $n-1$ voltage difference vector with respect to the slack bus. A new slack bus, in particular a *load bus*, is now selected as the reference bus. It is not necessary to calculate $\mathbf{Z}_{\text{BUS}}$ as in Equation 6.31, for only $\mathbf{Y}_{\text{BUS}}$ with this new reference is used in the subsequent derivation.

For any load bus as the reference, the series transmission system real and reactive losses are the residue of *all* the power injected into the network. With $\mathbf{E}_{\text{BUS}}$ as the n vector of bus voltages, the losses are

$$P_L + jQ_L = \sum_{i=1}^{n} E_i I_i^* = \mathbf{E}^t_{\text{BUS}}\,\mathbf{I}^*_{\text{BUS}}$$

$$= (\mathbf{E}'_{\text{BUS}})^t\,(\mathbf{I}'_{\text{BUS}})^* + E_r \sum_{i=1}^{n-1} I_i^* - E_r I_r \tag{6.32}$$

$$= (\mathbf{E}'_{\text{BUS}})^t (\mathbf{I}'_{\text{BUS}})^* = (\mathbf{I}'_{\text{BUS}})^t (\mathbf{Z}^t_{\text{BUS}})(\mathbf{I}'_{\text{BUS}})^* = (\mathbf{E}'_{\text{BUS}})^t \mathbf{Y}^*_{\text{BUS}}\,(\mathbf{E}'_{\text{BUS}})^*$$

In this equation, because the slack bus is the reference, its injection current is related to the other buses.

$$I_r = -\sum_{i=1}^{n-1} I_i \tag{6.33}$$

The incremental change in transmission line losses with a change in bus phase angle is obtained from Equation 6.32 as

$$\frac{\partial P_L}{\partial \delta_i} = \mathbf{Re}\left\{\frac{\partial}{\partial \delta_i}[(\mathbf{E}'_{\text{BUS}})^t \mathbf{Y}^*_{\text{BUS}}(\mathbf{E}'_{\text{BUS}})^*]\right\}$$

$$= \mathbf{Re}\left\{\frac{\partial}{\partial \delta_i}\begin{bmatrix} V_1\cos\delta_1 + jV_1\sin\delta_1 - E_r \\ \vdots \\ V_{n-1}\cos\delta_{n-1} + j_{n-1}V_{n-1}\sin\delta_{n-1} - E_r \end{bmatrix}^t \mathbf{Y}^*_{\text{BUS}} \times \right.$$

$$\left. \begin{bmatrix} V_1\cos\delta_1 - jV_1\sin\delta_1 - E_r \\ \vdots \\ V_{n-1}\cos\delta_{n-1} - jV_{n-1}\sin\delta_{n-1} - E_r \end{bmatrix}\right\}$$

$$= 2V_i\sum_{k=1}^{n-1}(V_k\sin\delta_k\cos\delta_i - V_k\cos\delta_k\sin\delta_i + V_r\sin\delta_i)\,\mathbf{Re}\{Y_{ik}\} \qquad (6.34)$$

where V_r is the magnitude of the slack bus voltage whose phase angle is taken as zero reference.

In a similar manner, the gradient of transmission line losses with respect to voltage magnitude V_i is obtained as

$$\frac{\partial P_L}{\partial V_i} = 2\sum_{k=1}^{n-1}(V_k\cos\delta_k\cos\delta_i - V_r\cos\delta_i + V_k\sin\delta_k\sin\delta_i)\,\mathbf{Re}\{Y_{ik}\} \qquad (6.35)$$

The numerical values for Equation 6.34 and Equation 6.35 are obtained by means of substituting load-flow numerical values into the right-hand sides. A symbolic form for the right-hand sides is found by using the chain rule for partial derivatives:

$$\frac{\partial P_L}{\partial \delta_i} = \sum_{k=1}^{n-1}\left\{\frac{\partial P_k}{\partial \delta_i}\left(\frac{\partial P_L}{\partial P_k}\right) + \frac{\partial Q_k}{\partial \delta_i}\left(\frac{\partial P_L}{\partial Q_k}\right)\right\} \qquad (6.36)$$

$$V_i\frac{\partial P_L}{\partial V_i} = \sum_{k=1}^{n-1}\left\{V_i\frac{\partial P_k}{\partial V_i}\left(\frac{\partial P_L}{\partial P_k}\right) + V_i\frac{\partial Q_k}{\partial V_i}\left(\frac{\partial P_L}{\partial Q_k}\right)\right\} \qquad (6.37)$$

where a V_i factor has been introduced into Equation 6.37 to normalize the voltage increment. In matrix form, Equation 6.36 and Equation 6.37 are written as

$$\begin{bmatrix} \dfrac{\partial P_L}{\partial \delta} \\[2ex] \mathbf{V}\dfrac{\partial P_L}{\partial \mathbf{V}} \end{bmatrix} = \begin{bmatrix} \dfrac{\partial \mathbf{P}}{\partial \delta} & \dfrac{\partial \mathbf{Q}}{\partial \delta} \\[2ex] \mathbf{V}\dfrac{\partial \mathbf{P}}{\partial \mathbf{V}} & \mathbf{V}\dfrac{\partial \mathbf{Q}}{\partial \mathbf{V}} \end{bmatrix}\begin{bmatrix} \dfrac{\partial P_L}{\partial \mathbf{P}} \\[2ex] \dfrac{\partial P_L}{\partial \mathbf{Q}} \end{bmatrix} = \mathbf{J}^t_v\begin{bmatrix} \dfrac{\partial P_L}{\partial \mathbf{P}} \\[2ex] \dfrac{\partial P_L}{\partial \mathbf{Q}} \end{bmatrix} \qquad (6.38)$$

The $(2n-2) \times (2n-2)$ matrix, $\mathbf{J}_v^t$ in Equation 6.38, is the transpose of the Jacobian from the *Newton-Raphson load flow* with rows in two quadrants multiplied by voltages. The transmission line gradient with respect to $n-1$ bus is

$$\begin{bmatrix} \dfrac{\partial P_L}{\partial \mathbf{P}} \\[2ex] \dfrac{\partial P_L}{\partial \mathbf{Q}} \end{bmatrix} = [\mathbf{J}_v^t]^{-1} \begin{bmatrix} \dfrac{\partial P_L}{\partial \boldsymbol{\delta}} \\[2ex] \mathbf{V} \dfrac{\partial P_L}{\partial \mathbf{V}} \end{bmatrix}$$

(6.39)

where the Jacobian from the load flow is used to calculate $\mathbf{J}_v$. The load-flow solution is used in Equation 6.34 and Equation 6.35 to calculate $\partial P_L/\partial \delta$ and $\mathbf{V}\,\partial P_L/\partial \mathbf{V}$. Only the generator buses $\partial P_L/\partial \mathbf{P}$ are required, the other buses are disregarded.

If a generator is retained as the slack bus, it must be treated separately starting with the power balance equation

$$\sum_{m=1}^{M} P_m - P_L - P_R = 0$$

(6.40)

Let generator M correspond to the slack bus, so the gradient is found from Equation 6.40 as

$$\frac{\partial P_L}{\partial P_M} = 1 + \sum_{m=1}^{M-1} \frac{\partial P_m}{\partial P_M} = 1 + \sum_{m=1}^{M-1} \left\{ \frac{\partial P_m}{\partial \delta_m} \left(\frac{\partial \delta_m}{\partial P_M} \right) + \frac{\partial P_m}{\partial V_m} \left(\frac{\partial V_m}{\partial P_M} \right) \right\}$$

(6.41)

Numerical values of the partial derivatives $\partial P_m/\partial \delta_m$ and $\partial P_m/\partial V_m$ are obtained from the Jacobian of the Newton-Raphson load flow. The real power flow from the slack bus is

$$P_M = \mathbf{Re}\{V_r I_r^*\} = V_r \,\mathbf{Re} \left\{ \sum_{k=1}^{n-1} Y_{rk}^* (V_r - V_k e^{-j\delta_k}) \right\}$$

$$= V_r \sum_{k=1}^{n-1} [V_r - V_k \cos \delta_k)\, \mathbf{Re}\,\{Y_{rk}\} - V_k \sin \delta_k \mathbf{Im}\,\{Y_{rk}^*\}]$$

(6.42)

Only if the slack generator is directly connected to another generator $m = 1, 2, 3, \ldots, M-1$ are the partial derivatives

$$\frac{\partial P_M}{\partial \delta_m} = \frac{\partial P_r}{\partial \delta_m} \qquad \frac{\partial P_M}{\partial V_m} = \frac{\partial P_r}{\partial V_m}$$

evaluated for use in Equation 6.40. If the numerical value is zero, a second derivative should be used. However, it is easier to switch the slack bus to a load after the solution as a new inverse is needed for the modified Jacobian.

Example 6.5 demonstrates the application of Equation 6.31 to Equation 6.39.

Example 6.5

A solved load flow and the topology of a three-bus direct-current system are given in Figure E6.5. Calculate the gradient of the transmission line losses with respect to the generator power and compare it with the classical loss coefficient method.

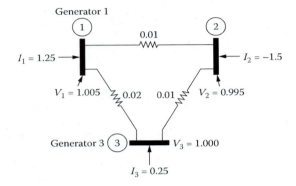

FIGURE E6.5

Solution

The bus admittance matrix with the load as reference is

$$\mathbf{Y}_{\text{BUS}} = \begin{matrix} & 1 & 3 \\ & \begin{bmatrix} 150 & -50 \\ -50 & 150 \end{bmatrix} \end{matrix} = [Y_{mn}]$$

Substituting numerical values into Equation 6.35, we obtain

$$\frac{\partial P_L}{\partial V_1} = 2 \sum_{k=1,3} (V_k - V_r)\mathbf{Re}\{Y_{1k}\}$$
$$= 2[(V_1 - V_2)Y_{11} + (V_3 - V_2)Y_{13}]$$
$$= 2[0.010(150) + 0.005(-50)] = 2.5$$
$$\frac{\partial P_L}{\partial V_3} = 2[(V_1 - V_2) + Y_{31} + (V_3 - V_2)Y_{33}]$$
$$= 2[0.010(-50) + 0.005(150)] = 0.5$$

The power flow equations for the two generators are

$$P_1 = V_1[100(V_1 - V_2) + 50(V_1 - V_3)]$$
$$P_3 = V_3[50(V_3 - V_1) + 100(V_3 - V_2)]$$

The Jacobian, $\partial \mathbf{P}/\partial \mathbf{V}$, for these equations is

$$\frac{\partial \mathbf{P}}{\partial \mathbf{V}} = \begin{bmatrix} 300V_1 - 100V_2 - 50V_3 & -50V_1 \\ -50V_3 & 300V_3 - 100V_2 - 50V_1 \end{bmatrix}$$

Multiplying the first row of the Jacobian and $\partial P_L / \partial V_1$ by V_1 and similar operations with respect to V_3 yields

$$
\begin{bmatrix} \dfrac{\partial P_L}{\partial P_1} \\[2mm] \dfrac{\partial P_L}{\partial P_3} \end{bmatrix} = \left[\mathbf{V} \dfrac{\partial \mathbf{P}^t}{\partial \mathbf{V}} \right]^{-1} \begin{bmatrix} V_1 \dfrac{\partial P_L}{\partial V_1} \\[2mm] V_3 \dfrac{\partial P_L}{\partial V_3} \end{bmatrix} = \begin{bmatrix} 152.76 & -50.50125 \\ -50.50125 & 150.25 \end{bmatrix}^{-1} \begin{bmatrix} 2.5125 \\ 0.5 \end{bmatrix}
$$

$$
= 10^{-2} \begin{bmatrix} 0.7364541 & 0.2475331 \\ 0.2475331 & 0.748757 \end{bmatrix} \begin{bmatrix} 2.5125 \\ 0.5 \end{bmatrix}
$$

$$
= \begin{bmatrix} 0.019741075 \\ 0.009963054 \end{bmatrix}
$$

These are the incremental transmission line losses with respect to the generators. Next, the classical approach to calculate the loss coefficients begins with $\mathbf{Z}_{\mathrm{BUS}}$ with respect to bus 2:

$$
\mathbf{Z}_{\mathrm{BUS}} = \begin{matrix} & 1 & 3 \\ \begin{bmatrix} 0.075 & 0.0025 \\ 0.0025 & 0.0075 \end{bmatrix} \end{matrix}
$$

The total transmission line losses are given by

$$
P_L = \mathbf{I}' \mathbf{Z}_{\mathrm{BUS}} \mathbf{I} = \begin{bmatrix} \dfrac{P_1}{V_1} & \dfrac{P_3}{V_3} \end{bmatrix} \mathbf{Z}_{\mathrm{BUS}} \begin{bmatrix} \dfrac{P_1}{V_1} \\[2mm] \dfrac{P_3}{V_3} \end{bmatrix}
$$

The gradient of P_L with respect to the generators is

$$
\frac{\partial P_L}{\partial P_1} = \frac{2Z_{11}P_1}{V_1^2} + \frac{2Z_{13}P_3}{V_1 V_3} = \frac{2Z_{11}I_1}{V_1} + \frac{2Z_{13}I_3}{V_1} = 0.0199005
$$

$$
\frac{\partial P_L}{\partial P_3} = \frac{2Z_{13}P_1}{V_1 V_3} + \frac{2Z_{33}P_3}{V_3^2} = \frac{2Z_{13}I_1}{V_3} + \frac{2Z_{33}I_3}{V_3} = 0.0100
$$

The numerical values of the gradient from the classical method and the modified Jacobian compare very closely.

Thus, by changing the slack bus and modifying the Jacobian, a Newton-Raphson load-flow program may be used to calculate the transmission loss gradient. Only the real power for generators portion of Equation 6.39 is needed, even though it is possible to allocate network reactive power using the terms from $\partial P_L / \partial \mathbf{Q}$.

Using the digital Newton-Raphson load flow as a computational element, the gradient $\partial P_L / \partial \mathbf{P}$ may be numerically calculated for the penalty factors, Equation 6.11, and an economic dispatch performed. But if the load flow is used, it is also available for execution in order to "track" the changing system load demand and avoid the mismatch between high or low sets of penalty factors and the actual system load. This was the impetus for the iterative scheme [13] to find the optimum economic dispatch point, whose algorithm is shown in Figure 6.12. The convergence factor α was found by trial and error and may be

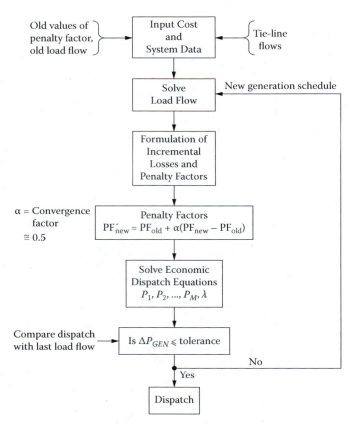

FIGURE 6.12
Iterative method to determine power system optimum economic dispatch.

different for other power systems. However, the algorithm strongly converges in six iterations around the major loop for the IEEE standard 118 bus test case. It is probably sufficient for hourly execution of this type of program, with the gradient used for 5 min periodic economic adjustments.

6.7   Economic Exchange of Power between Areas

Members of power pools often buy and sell blocks of energy to each other, or to tie-line-connected nonneighbors. This is done for economic reasons or to assist each other in situations such as forced generator outages. A scheduled interchange sale of power is easily implemented by offsetting the ACE signal with P_s as shown in Figure 6.13. The exchange of power between members of a pool is shown schematically in Figure 1.9 and often involves all utilities in the interconnection.

Interchange schedules are normally implemented via 20 min ramp times. Some utilities start 10 min before the hour and reach the full amount 10 min after the hour, as shown in Figure 6.13, while others begin the up and down ramps on the hour mark. The ramps and constant power periods comprise a block of energy, MWh.

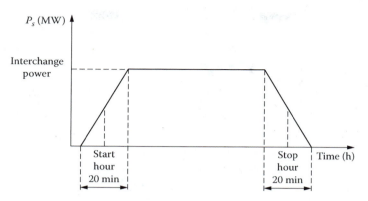

FIGURE 6.13
Block of interchange energy.

The purpose of rigidly scheduled start and stop times is that all members of the pool with knowledge of the exchange can closely monitor their transmission facilities for possible overloads. All lines of the interconnection are affected by the exchange. In the case of "wheeling" power flow through an area, the net power appears as losses on its transmission facilities. The area subject to wheeling power knows what to expect for tie flows because of predetermined, pool-wide power flow calculations. Thus, even if an affected area does not buy or sell the power, it adjusts its K tie-line flows in Equation 6.16 to account for the new distribution, which disturbs its own economic dispatch.

Areas buying and selling power have precalculated tie-line factors from power flow calculations to use in Equation 6.16 for their respective transmission line losses. The flow factors are called tie-line distribution factors (TLDFs) and take the form of incremental power flow on tie line K for incremental change in power of its internal generator m:

$$\text{TLDF}(k, m) = \frac{\Delta P_{Tk}}{\Delta P_m} \quad \begin{array}{l} k = 1, 2, 3, \ldots, K \\ m = 1, 2, 3, 4, \ldots, M \end{array} \tag{6.43}$$

These factors are also found by successively using each generator as the slack bus to supply an increment of power to another area of the power pool. The TLDFs eliminate tie-line flows in Equation 6.16, so economic dispatch is based on M generators delivering power:

$$\text{Generation} = \sum_{i=1}^{M} P_i = P_L + P_R + P_{ex} \tag{6.44}$$

where P_{ex} is the exchange power. The other symbols were defined previously. Let a power system A be requested to sell a block of P_{ex} megawatts for H hours at a price of b dollars. The average selling price, c, is

$$c = \frac{b}{P_{ex}H} \quad \$/(\text{MWh}) \tag{6.45}$$

The start and stop time for the exchange is specified, so the question posed to power system A is as follows: With known generation incremental costs, $\$/(\text{MWh})$, known transmission line loss coefficients, and historical system A load data, is it economically advantageous for system A to sell a block of energy at b dollars?

A computer program that provides the answer is often called an economy A interchange program or cost-of-interchange program. It performs the following tasks:

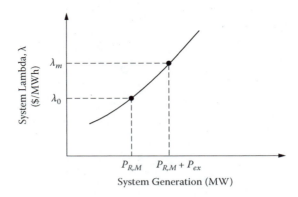

FIGURE 6.14
System generation costs ($MW/h).

6.7.1   Economy A Program

1. Divide the proposed power exchange time period into N equal-duration segments, t_n. The number or segments depends on how rapidly the historical load data vary for a time of day similar to the proposed interchange period.

2. Perform an economic dispatch calculation for each time segment, computing λ_n with and without the exchange as shown in Figure 6.14. $P_{R,n}$ is the average system load recorded in segment n for the system. The present set of on-line generators is used.

3. Sell the block of energy if the average incremental cost is less than the sale price:

$$\frac{1}{N}\sum_{m=1}^{N}\lambda_m \le \frac{b}{p_{ex}H} \quad (\$/\mathrm{MWh})$$

and do not sell the block otherwise. Compare the production cost ($/h) with and without the transaction.

A number of operational features are often found in the economy A programs that are usually restricted to the generating units presently on-line. Economy B programs can use a variable number of generating units on-line during the transaction. The historical records indicate which units were on-line for a similar time period or else a unit commitment program can be called by an economy B. If the spinning reserve is inadequate for the proposed transaction,

$$\begin{bmatrix}\text{spinning}\\\text{reserve}\end{bmatrix}=\begin{bmatrix}\text{total}\\\text{generating}\\\text{capacity}\\\text{online}\end{bmatrix}-\begin{bmatrix}\text{system}\\\text{load}\end{bmatrix}-\begin{bmatrix}\text{exchange}\\\text{power}\end{bmatrix}\le\begin{bmatrix}\text{margin}\\\text{of}\\\text{safety}\end{bmatrix}$$

it may be necessary to specify additional units to be on-line. The economy A and economy B programs often have as input data a required spinning reserve margin. Multiple and overlapping interchange schedules are common, with the programs taking in account the prior schedules to compute a cost advantage. A composite P_s versus time, which is the sum of all the schedules, is then used in the ACE signal.

Example 6.6 demonstrates the use of an economy A program or cost-of-interchange program.

Example 6.6

Power areas *A* and *B* are interconnected by tie lines. Initially, there is zero tie-line power flow. Area *A* contains generators 1 and 2, both on-line in economic dispatch and satisfying a 500 MW load demand. Area *B* contains generators 3 and 4, both on-line in economic dispatch delivering a total of 1,000 MW. A schematic of the interconnection is shown in Figure E6.6. The quadratic operation cost curves and power limits for the generators are as follows:

Unit	Max. Cap. (MW)	Min. Cap. (MW)	Cost Curve Parameters $C + BP + AP^2$		
			C ($/h)	*B* ($/MWh)	*A* ($/MW²h)
1	600	25	580	9.023	0.00173
2	700	130	400	7.654	0.0016
3	600	165	600	8.752	0.00127
4	420	130	420	8.431	0.0016

The transmission line losses are neglected in both areas for dispatching and interchanging power. A utility question and a pool question are the following:

1. Area *B* telephones the dispatcher of area *A* that it desires to purchase a 50 MWh block for the next hour and offers $467.50 for the energy. Assuming that the area *A* load demand does not change during the next hour, how much of this block should area *A* supply? Compute the incremental and production costs for area *A* to sell 0, 25, 50, 75, 100, 125, and 150 MWh of energy. Compute area *B* costs without and with the import power.
2. Compute the pool cost for a 50 MWh exchange of energy. Is this the best amount of exchange energy to minimize pool costs?

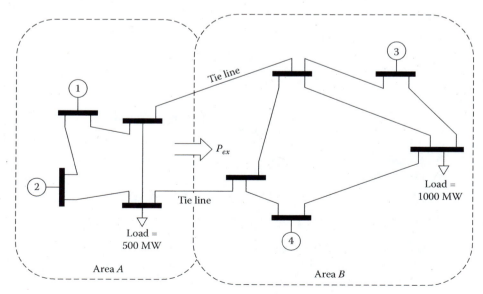

FIGURE E6.6

Solution

For question 1, the performance index for system A is

$$J = F_1 + F_2 - \lambda_A (P_1 + P_2 - P_{RA}) \, (\$/h)$$

where F_1 and F_2 are the generator production costs and P_{RA} is the load plus exchange power. When minimized with respect to the variables λ_A, P_1, P_2, and numerical values are used from the table above, the variables as related to P_{RA} are

$$\left. \begin{array}{l} \lambda_A = 8.311 + 0.0016625 P_{RA} \\ P_1 = -205.56 + 0.4805 P_{RA} \\ P_2 = 205.56 + 0.5195 P_{RA} \end{array} \right\} \text{ system } A \text{ minimum cost } (\$/MWh)$$

The production cost for system A is found by introducing values of P_1 and P_2 in the quadratic cost curves. For the range of block sizes the costs computed and presented to the dispatcher are

Block Size to Sell (MW)	Foreign Bid ($/MWh)	System A Cost, λ ($/MWh)	System A Gen. (MW)	Production Cost ($/h)	Spinning Reserve (MW)
0	—	9.142	500	5202.99	500
25	9.350	9.184	525	5432.09	475
50	9.350	9.225	550	5662.22	450
75	9.350	9.267	575	5893.40	425
100	9.350	9.309	600	6125.61	400
125	9.350	9.350	625	6358.86	375
150	9.350	9.391	650	6593.15	350

The incremental cost of supplying system A base load plus 50 MW is \$9.225/MWh, which is less than the sale price rate of \$9.350/MWh, so system A should *sell* the entire block. System A could sell up to 125 MW of power before it reaches the break-even incremental cost point. However, its net return on the sale is

$$\begin{bmatrix} \text{area } A \\ \text{net revenue} \end{bmatrix} = \begin{bmatrix} \text{sale} \\ \text{price} \end{bmatrix} - \begin{bmatrix} \text{increased} \\ \text{production cost} \end{bmatrix}$$

$$= \$467.50 - (\$5662.22 - \$5202.99)$$
$$= \$8.27/h$$

If an additional generator has to be brought on-line for a sale, its start-up cost is added to the production cost at the appropriate power level.

When system B is supplying its own load in economic dispatch:

$$\begin{array}{l} \text{system } B \\ \text{base case} \\ P_{RB} = 1000 \end{array} \left\{ \begin{array}{l} P_3 = 483.57 \text{ MW} \\ P_4 = 516.43 \text{ MW} \\ \lambda_B = 9.981 \$/MWh \\ \text{Generation cost} = \$10,303.35/h \\ \text{Operating cost} = \$10,303.35/h \end{array} \right.$$

When system B is receiving 50 MW of tie-line power for \$9.35/MWh and is economic dispatch,

$$\text{system } B \text{ importing 50 MW} \begin{cases} P_3 = 456.50 \text{ MW} \\ P_4 = 493.50 \text{ MW} \\ \lambda_B = 9.911 \text{ \$/MWh} \\ \text{Generation cost} = \$9806.06/h \\ \text{Operating cost} = \text{gen.} + \$467.50 = \$10,273.56/h \end{cases}$$

At 50 MW exchange, the lambda for system B is considerably higher than for system A, so that more power could be exchanged. However, there is a net savings of \$29.79/h to system B *when it purchases 50 MW of power, so it realizes more benefit than system A.* Very often the exchanging utilities agree to share benefits from the exchange.

When the pool controls the dispatch point of all the generators, the minimum-cost operating point is calculated according to the methods of Section 6.1 because there are no line losses. The optimum dispatch point is

$$\text{system } A \text{ and system } B \text{ dispatch together} \begin{cases} \left.\begin{array}{l} P_1 = 167.11 \\ P_2 = 608.50 \end{array}\right\} P_{EXA} = P_1 + P_2 - 500 = 275.61 \\ \left.\begin{array}{l} P_3 = 334.33 \\ P_4 = 390.06 \end{array}\right\} P_{EXB} = P_3 + P_4 - 1000 = -275.61 \\ \lambda = 9.601 \end{cases}$$

There should be 275.61 MW of interchange power from system A to system B in order to minimize the power pool operating cost.

A summary of production costs for the three interchanges conditions is

No interchange:

Sys. *A* cost	\$5,202.99
Sys. *B* cost	10,303.35
Pool cost	\$15,506.34
($/h)	

50 MW interchange:

Sys. *A* cost	\$5,662.22
Sys. *B* cost	9,806.06
Pool cost	\$15,468.28
($/h)	

Pool dispatch:

Sys. *A* cost	\$7,785.99
Sys. *B* cost	7.604.85
Pool cost	\$15,390.84
($/h)	

There is a \$115.50/h savings if a pool dispatch allocates 275.61 MW of exchange power from system A to system B.

In Example 6.6, if the economy A program resident in the computer of system A has a scale of \$/MWh versus block size from the buyer (or seller), rather than a fixed price bid, the dispatcher of system A determines the interchange to equalize lambdas and achieve a more optimum cost. This additional information is often provided and updated periodically. The average incremental cost for two systems, A and B, exchanging power,

$$\lambda_{ex} = \frac{1}{2}(\lambda_A + \lambda_B) \quad (\$/\text{MWh}) \tag{6.46}$$

is called the *strike price*. The strike price (rate) multiplied by the exchange power P_{ex} is the buy/sell cost used by the participants to *share* benefits:

$$\text{cost of exchange} = \lambda_{ex} P_{ex} \quad (\$/h) \tag{6.47}$$

When a member of a power pool has several possible buyers/sellers of a block of energy, the dispatcher could compare costs with all other pool dispatches hourly and negotiate economy interchanges. This random approach is inefficient and not likely to achieve the full economic potential for the pool. An improved pool cooperative method, which preserves each member's autonomy, is called the *energy-broker* system [14].

In the energy-broker system the members exchange hourly quotations of prices at which each is willing to buy and sell energy. The energy-broker system may be manual or computerized. The buyers and sellers are matched by the broker on a high-to-low basis. The buyer with the highest decremental cost and the seller with the lowest incremental cost (both with adjustments for transmission charges and losses) are matched to negotiate an interchange transaction. The next-to-highest and next-to-lowest are matched, and so on, until all offerers are matched. The savings resulting from a specific transaction are then normally shared equally by the two utilities exchanging power. The energy-broker system is the next-best method to central pool dispatch of all members' generators (see Problem 6.8).

Problems

6.1. Three generators are supplying 274.4 MW of power to a common load with negligible transmission line losses. The operating costs and power level constraints for the three units are as follows:

$$F_1 = 240 + 2.45\, P_1 + \frac{P_1^2}{2} \quad (\$/h) \quad 50 \le P_1 \le 200$$

$$F_2 = 80 + 3.51\, P_2 + \frac{P_2^2}{2} \quad (\$/h) \quad 10 \le P_2 \le 200$$

$$F_3 = 80 + 3.89\, P_3 + \frac{P_3^2}{2} \quad (\$/h) \quad 10 \le P_3 \le 100$$

Find the base power settings P_1, P_2, P_3 for the generators that supply the load at minimum operating cost. What are the system lambda and the total operating cost, $/h?

© The economic dispatch program *econ.exe* can be used to calculate the solution for this problem. Edit the file *econ.sss* to enter the data for the three generators.

6.2. The fuel inputs to two generating units in MBtu/h are given as

$$\text{MBtu/h } F_1 = 80 + 8P_1 + 0.024P_1^2 \quad 10 \le P_1 \le 100\text{MW}$$

$$\text{MBtu/h } F_2 = 120 + 6P_2 + 0.04P_2^2 \quad 10 \le P_2 \le 100\text{MW}$$

The cost of fuel is approximately $2/mil ≡ $2 per million Btu.

(a) Assume that the system load on these two units varies from 50 to 200 MW. Plot the output of the two units as a function of system load when they are scheduled by equal incremental cost. Assume that both units are on-line at all times.

(b) If only one unit is in operation, when should the second unit be brought on-line for most economical operation as the load increases from 50 MW to 200 MW? What unit is on-line as the load begins to increase? Assume that start-up or shutdown costs are negligible.

(c) Consider a 24 h period from 6 A.M. one morning to 6 A.M. the next morning and assume that the daily load cycle is as shown in Figure P6.2. The cost to take a unit off-line and return it to service after 12 h is $250. Would it be more economical to keep both units in service for the 24 h period or to remove one of the units from 6 P.M. to 6 A.M. the next morning?

(d) What is the hour-by-hour operating schedule for the two units for 1 day?

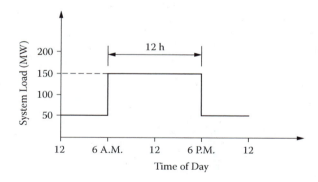

FIGURE P6.2

6.3. The yearly load duration curve for a generator is shown in Figure P6.3.

$$F = 160 + 16P + 0.048P^2 \quad \$/h$$

$$10 \le P \le 100$$

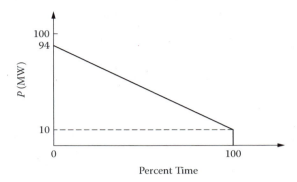

FIGURE P6.3

This is a cost for fuel of approximately $2/MBtu. What is the average cost of operating the generator over 1 year? Assume that the generator is never shut down for maintenance.

6.4. The transmission line resistances for a five-bus dc system are shown in Figure P6.4. The data for a solved power flow are:

Bus Voltages	Bus Injections
$e_1 = 1.10$	$i_1 = 0.115$
$e_2 = 1.05$	$i_2 = 0.0001$
$e_3 = 1.00$	$i_3 = -0.0167$
$e_4 = 0.93333$	$i_4 = -0.0555$
$e_5 = 1.00$	$i_5 = -0.0277$

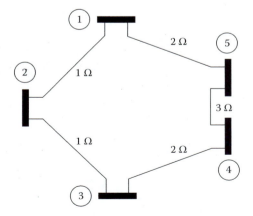

FIGURE P6.4

(a) Determine the B_{mn} loss coefficients at the operating point by the REI method.

$$P_L = \sum_{m=1}^{2} \sum_{m=1}^{2} P_m B_{mn} P_n$$

© Format the system data into *rei.sss* and then run *rei.exe*

(b) Verify that the loss coefficients describe transmission losses for the solved power flow case.

(c) The incremental cost curves for the two generators are at buses 1 and 2

$$\frac{dF_1}{dP_1} = P_1 \qquad \frac{dF_2}{dP_2} = 2P_2$$

Determine the optimum dispatch power set point P_1, P_2.

(d) Increase the system load by 50% and determine the new power set point P_1, P_2.

6.5. A solved power flow for a three-bus network is specified in the following table. The power base is 100 MVA.

	Bus Voltage	Power Injections
1	1.018 $\angle 1.687°$	$0.525 + j0.151$
2	1.018 $\angle 1.853°$	$1.196 + j0.222$
3	1.000 $\angle 0°$	$-1.701 + j0.313$

The topology is shown in Figure P6.5. The line admittances of the network are:

$$Y_{12} = 2.5 - j7.5 = 7.9 \angle{-71.°}6$$
$$Y_{23} = 10 - j30 = 31.6 \angle{-71.°}6$$
$$Y_{31} = 5 - j15 = 15.8 \angle{-71.°}6$$

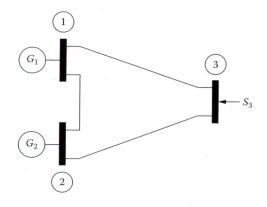

FIGURE P6.5

The line-charging capacitance is negligible for the network. The generators at buses 1 and 2 have the following per-unit incremental costs:

$$\text{Generator 1: } \frac{dF_1}{dP_1} = 1 + P_1$$

$$\text{Generator 2: } \frac{dF_2}{dP_2} = 0.5 + P_2$$

Assume that the reactive power is a fixed ratio of the real power, without offset, at each generator bus.

(a) Calculate the B_{mn} loss coefficients in

$$P_L = \sum_{m=1}^{2} \sum_{n=1}^{2} P_m B_{mn} P_n$$

for the steady-state power flow condition. © Use the *rei.exe* program.

(b) Determine the classical dispatch output of generators 1 and 2 to satisfy the load demand using the loss coefficients calculated in part (a) and the given incremental cost curves. Note the differences in P_1 and P_2.

(c) Use a Newton-Raphson power flow and the optimal dispatch method of Section 6.6 to calculate generator outputs P_1 and P_2 at the fixed-load condition.

Compare these results with part (b).

6.6. A base case solved power flow is shown in Figure P6.6 for a three-bus system. All generators are phase locked at 0°.

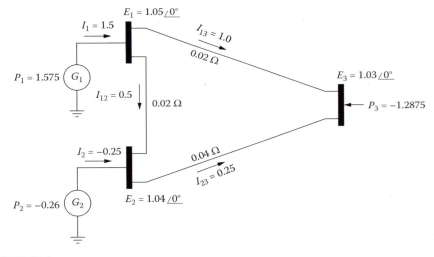

FIGURE P6.6

(a) Find $\mathbf{Z}_{BUS}$ with respect to bus 3 for the system.

(b) Find $\mathbf{Y}_{BUS}$ with respect to bus 3 for the system.

(c) Find the B_{mn} loss coefficients using $\mathbf{Z}_{BUS}$.

$$P_L = \sum_{m=1}^{2} \sum_{n=1}^{2} P_m P_{mn} P_n$$

(d) Find the B'_{mn} loss coefficients using $\mathbf{Y}_{BUS}$, expressing the voltages as implicit functions of power and steady currents, that is,

$$E_1 = \frac{P_1}{I_1} = \frac{P_1}{1.5} \quad E_2 = \frac{P_2}{-0.25}$$

$$P'_L = \sum_{m=1}^{2}\sum_{n=1}^{2} P_m B'_{mn} P_n + \sum_{k=1}^{2} P_k B'_{k0} + B'_{00}$$

(*Hint*: Introduce E_3 numerically in the calculations.)

(e) Compute $\partial P_L/\partial P_i$ and $\partial P'_L/\partial P_i$ for the two sets of coefficients $i = 1, 2$.

(f) *A difficult part.* Now check whether $\partial P_L/\partial P_i$ or $\partial P'_L/\partial P_i$ is a better approximation. There may be several ways to do this, but one hint is the following: To compute $\partial P_L/\partial P_1$, hold E_2 constant at 1.04 and increase I_{12} and I_{23} so that $I_{12} - I_{23} = 0.25$ = current into bus 2. The load voltage must decrease, but P_1 increases, which is the perturbation needed to compute $\Delta P_L/\Delta P_1$ from the *network* (as in load flow).

6.7. Power system A has two generators whose operating costs are

$$\text{Unit 1:} \quad A_1 = 5P_1 + 0.025P_1^2 \quad (\$/h) \quad 0 \le P_1 \le 125$$

$$\text{Unit 2:} \quad A_2 = 6.25P_2 + 0.0125P_2^2 \quad (\$/h) \quad 0 \le P_2 \le 150$$

Power system B has two generators with operating costs:

$$\text{Unit 1:} \quad B_1 = 8.644P_3 + 0.03P_3^2 \quad (\$/h) \quad -1 \le P_3 \le 75$$

$$\text{Unit 2:} \quad B_2 = 7.552P_4 + 0.015P_4^2 \quad (\$/h) \quad 0 \le P_4 \le 100$$

Assume that both power systems are compact, so they do not consider transmission line losses in dispatching generators. The two systems are connected by a single tie line with zero initial flow (see Figure P6.7). Both systems have a base load of 100 MW.

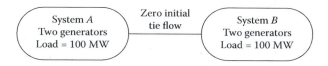

FIGURE P6.7

(a) Plot λ_A versus load and λ_B versus load. λ is the incremental cost curve $/MWh. What are both systems' operating costs at 100 MW?

(b) How much power (MW) can system A sell to system B to operate both systems at the same incremental cost, λ?

(c) What is the minimum price that system A should ask for the power that makes $\lambda_A = \lambda_B$? In other words, what is the break-even price for the power?

6.8. Four interconnected municipal power systems are organized in an energy-broker association. The interarea transmission losses are negligible. The four area production cost curves, present load demand, and MW offered sale/purchase are:

Area	Max. Gen. (MW)	Min. Gen. (MW)	Cost Curves ($/h) $C + BP + AP^2$			Present Load (MW)	Sell (+) Buy (−) (MW)
			C	B	A		
A	600	25	580	9.023	0.00173	200	−50
B	700	130	400	7.654	0.00160	400	+100
C	600	165	600	8.752	0.00127	500	−75
D	420	130	420	8.431	0.0015	300	+25

The utilities agree to share benefits between buyer and seller of a block of power according to Equation 6.46 and Equation 6.47. Find the blocks of power to be exchanged between A, B, C, and D and the $/h to be paid or received.

6.9. A network reduction method has combined all the power system loads in an equivalent at bus R while retaining the generators at buses 1 and 2, plus a tie bus 3, as shown in Figure P6.9. The bus voltages and injections on a 100 MVA, 100 kV base are

$$E_1 = 1.056\angle 10.892° \qquad S_1 = 0.456 - j0.300$$
$$E_2 = 1.073\angle 10.46° \qquad S_2 = 2.0 + j0.397$$
$$E_3 = 1.037\angle 2.65° \qquad S_3 = -0.94 + j0.082$$
$$E_R = 1.048\angle 0° \qquad S_R = -1.44 + j0.21$$

The equivalent bus impedance matrix with R as a reference is (on the same 100 MW base)

$$\mathbf{Z}_{BUS} = \begin{bmatrix} & 1 & 2 & 3 \\ 0.027 + j0.15 & 0.016 + j0.11 & 0.011 + j0.081 \\ \swarrow & 0.021 + 0.12 & 0.014 + j0.086 \\ \text{symmetry} & & 0.039 + j0.16 \end{bmatrix}$$

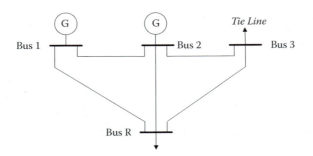

Bus 1   Bus 2   Tie Line   Bus 3

Bus R

FIGURE P6.9

(a) Determine the B_{ij} loss coefficients assuming that the generator real and reactive output remains proportional. That is, find the B_{ij} in $P_L = \sum \sum B_{ij} P_i P_j$ for the generator and tie.

(b) Use a power flow calculation to determine the tie-line distribution factors $\Delta P_3 / \Delta P_1 = K_{31}$, $\Delta P_3 / \Delta P_2 = K_{32}$.

(c) Let the incremental costs for generators 1 and 2 be

$$\frac{\partial F_1}{\partial P_1} = 864.4 + 2141.4 P_1 \quad \frac{\partial F_2}{\partial P_2} = 755.2 + 283.2 P_2$$

where the cost is in (\$/p.u. h). Determine the generator output powers to dispatch the same load and tie-line power considering transmission line losses expressed as P_L.

(d) What is the price that the utility should receive in \$/h for the 94 MW tie power? What price should be received for an additional 0.20 p.u. MW? Assume that $F_1(0) = 0$ and $F_2(0) = 0$.

6.10 The MBTU/h fuel-power costs for to generators at bus #1 and bus #2 of the IEEE 5 bus test case (file **stagg.sss**) are as follows:

$$\$h \quad F_1 = 160 + 16 P_1 + 0.048 P_1^2 \quad 10 \le P_1 \le 150 \text{ MW}$$

$$\$/h \quad F_2 = 240 + 12 P_2 + 0.08 P_2^2 \quad 10 \le P_2 \le 100 \text{ MW}$$

The cost of fuel is approximately $\$2/\text{Mil} \equiv \$2$ per million Btu.

(a) Use the parabolic cost curves to compute the cost of supplying fuel to the network is generator at bus #1 supplies 129.6 MW and the generator at bus #2 supplies 20 MW to the system.

(b) Compute the minimum cost operating power for the 2 generators if the total power to be dispatched is 149.6 MW for the generators. i.e., find new values of P1 and P2 to minimize total operating cost.

(c) Copy the file **stagg.sss** into **rei.sss** and schedule the generators to have the power output as computed in part (b). Execute **rei.exe** in order to obtain the B_{ij} loss coefficients for this operating point.

(d) Use the B_{ij} loss coefficients calculated in part (c) along with the cost curves in **econ.exe** to find the 'best' dispatch power considering the line losses. Compare these results with those from part (b).

References

1. Steinberg, M. J., and Smith, T. H. 1943. *Economy loading of power plants and electric systems.* New York: John Wiley & Sons.
2. Kirchmayer, L. K. 1959. *Economic control of interconnected systems.* New York: John Wiley & Sons.
3. Buck, C. R. 1956. *Advanced calculus.* New York: McGraw-Hill Book Company.
4. Ott, A. L. 2003. Experience with PJM market operation, system design and implementation. *IEEE Trans. Power Syst.* 18(2).

5. Gil, H. A. 2006. Nodal price control: A mechanism for transmission network cost allocation. *IEEE Trans. Power Syst.* 21(1).

6. Lew, A., and Mauch, H. 2007. *Dynamic programming: A computational tool.* New York: Springer.

7. Wood, A. J., and Wollenberg, B. F. 1984. *Power generation, operation, and control.* New York: John Wiley & Sons.

8. Kron, G. 1951. Tensorial analysis of integrated transmission systems. Part I. The six basic reference frames. *AIEE Trans.* 70:1239–48.

9. Happ, H. H. 1964. Direct calculation of transmission loss formula, II. *IEEE Trans. Power Appar. Syst.* 83:702–7.

10. Meyer, W. S., and Albertson, V. D. 1971. Improved loss formula computation by optionally ordered elimination techniques. *IEEE Trans. Power Appar. Syst.* 90(1).

11. Dimo, P. 1975. *Nodal analysis of power systems.* Turnbridge Wells, Kent, England: Abacus Press.

12. Tinney, W. F., and Powell, W. L. 1977. The REI approach to power network equivalents. Paper presented at the IEEE PICA Conference.

13. Happ, H. H. 1974. Optimal power dispatch. *IEEE Trans. Power Appar. Syst.* 93(3).

14. U. S. Department of Energy. 1980. *Power pooling: Issues and approaches.* Report DOE/ERA/6385-1.

7

State Estimation from Real-Time Measurements

In recent years, methods have been developed using measurements from the network to calculate the state of the network—voltage magnitude and phase angle at every bus. These methods are called state estimators because they are essentially weighted-least-squares techniques to find the best state vector to fit a scatter of data. The scatter of data is due to imperfect measurements of rapidly changing voltages or currents on the network in addition to errors in the transmission line linear models, line charging, and so on. The first source of errors, imperfect measurements due to signal noise, metering accuracy, and analog-to-digital conversion, is treated in this chapter. The network topology and parameters are assumed as known until they are fitted to the data in parameter estimation.

The power system is assumed to be operating in a steady-state condition with fixed voltages, currents, and power flow. The remote terminal units (RTUs) that sample network analog variables and convert the signals to digital form are periodically interrogated for the latest values of the signals. For example, in Figure 7.1 the RTUs are sequentially interrogated causing a "time skew" in the data from unit 1 compared to unit M, not only due to when the unit is scanned, but also due to the time when the actual analog signal was sampled. The set of M measurements is called a snapshot of the power system, even though the data may have as much as 2 sec of time skew.

The data collected by the RTUs are often redundant. There may be voltage sensing by step-down transformers on each phase—a, b, c—of the transmission line, whereas only one is needed for balanced operation. In addition, each transmission line voltage to the substation may be monitored on both sides of a circuit breaker, introducing redundancy when all lines are in service. There may be single-phase watt and var meters in addition to current measurements on all phases. The state estimator should incorporate all measurements to obtain the greatest possible accuracy.

Because the power system data are redundant, the state estimator may be used with statistical methods to detect bad or grossly incorrect data. For example, if all current transformers near a bus, except one, agree on the current flows into and out of the bus, and the exception shows 50% error, the 50% value is far beyond a 2% metering accuracy specified for the current transducers, so this measurement is considered bad. It is removed from subsequent state estimates until it can be physically checked. The operator of the system is usually alarmed of this action. The establishment of acceptable error limits based on the number, types, and accuracy of all measurements is an important design aspect of state estimators.

Another purpose of a state estimator is to detect changes in network configuration. If a transmission line has one phase abruptly open circuited, the average power flow on the intact phases is far less than a value given by the last state estimate. The operator is alerted as to this condition at the first data scan. At this time, corrective action by the control computer is not automatic, but may be implemented in the future through remedial-action programs.

Finally, another purpose of a state estimator is to complete a set of measurements in order to replace faulty or missing data. It is possible to estimate power flows and voltages at a bus whose measurements are lost due to a communication line failure or RTU failure.

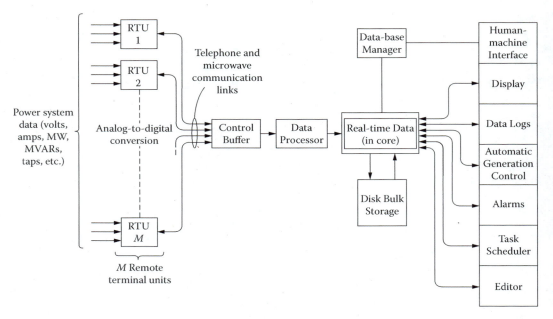

FIGURE 7.1
Data collection from the power system.

Significant problems in this regard are to determine the minimum number of measurements in order to calculate the state, often called *observability*, and how to improve state estimates by additional measurements.

The development of state estimators begins with the line-flow state estimator [1] as first introduced by J. F. Dopazo et al., then extended by many other power systems analysts.

7.1 The Line Power Flow State Estimator

Consider a generator directly connected to several transmission lines in a positive-sequence diagram as shown in Figure 7.2. The instrumentation around the bus can be divided into injections, bus metering, and line-flow measurements. Each of the measurements monitors all three phases in order to calculate positive-sequence values. The following assumptions are made:

1. Balanced three-phase flow conditions are present, and the system is in steady-state operation.

2. Accuracy of metering is known (i.e., the meter readings are accurate to 1%, 2%, etc., of the true physical value).

3. The full-scale range of each meter is known (e.g., 0 to 100 MW, 0 to 50 Mvar, etc.).

4. The errors in converting the analog quantities to digital signals for the data link to the central computer are known—0.25% error, 0.1% error, and so on—due to discretization.

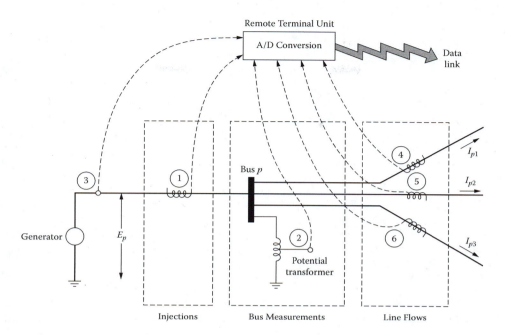

FIGURE 7.2
Positive-sequence measurements at a bus.

In Figure 7.2 measurements 4, 5, and 6 are for power flow in terms of the real part (MW) and reactive part (Mvar). The three phases are checked for balance, and then products of instantaneous voltages and currents are obtained for scaled, limited ranges and converted into serial digital data by the RTU. Each measurement of line flow, in complex form,

$$\begin{bmatrix} \text{measurement } i \\ \text{on line } m \\ \text{in MW} + j \text{ Mvar} \end{bmatrix} = S_{mi} \tag{7.1}$$

has a specified weighting factor inversely proportional to its accuracy.

$$W_i = \frac{50 \times 10^{-6}}{[c_1 \mid S_{mi} \mid + c_2 (\text{F.S.})]^2} \tag{7.2}$$

where c_1 is the accuracy, expressed as a decimal, of the real and reactive flow, typically 0.01 or 0.02; c_2 is the transducer and analog-to-digital converter accuracy, in decimal form, typically 0.0025 or 0.005; and F.S. is the full-scale range of both the MW and Mvar readings.

In Equation 7.2 the factor 50×10^{-6} is selected to normalize subsequent numerical quantities. The absolute value of the measurement is used in the denominator, so the weighting factors W_i are recomputed for each new set of measurements.

The weighted-least-squares state estimator is to find the best state vector, $\hat{\mathbf{X}}$, which minimizes the performance index of M measurements:

$$\boxed{\begin{array}{c} \text{Performance index for weighted measurements} \\[4pt] J = \sum_{i=1}^{M} W_i \mid S_{mi} - S_{ci} \mid^2 \end{array}} \tag{7.3}$$

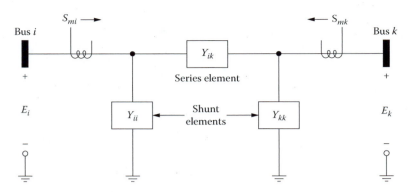

FIGURE 7.3
Power flow measurements at both ends of a line.

where S_{mi} is a power flow measurement and S_{ci} is a *calculated* value based on the state $\hat{\mathbf{X}}$. The state vector is an $n-1$ complex vector or $2(n-1)$ real vector for an n-bus power system. Since power constraints are again to be imposed on an n-bus system, a slack bus must be specified, just as in the case of load-flow calculations.

Each transmission line is assumed to be modeled by a π-section equivalent valid for lines less than 100 miles in length, so power flow measurements near the buses include power to the shunt elements. This is schematically shown in Figure 7.3. The measurements at buses i and k on the line are considered positive into the line as seen from the buses.

The power flow calculated at bus i is

$$
\begin{aligned}
S_{ci} &= E_i \{ Y_{ik}^* (E_i^* - E_k^*) + Y_{ii}^* E_i^* \} \\
&= E_i \left\{ \frac{E_i^* - E_k^*}{Z_{ik}^*} + Y_{ii}^* E_i^* \right\}
\end{aligned}
\tag{7.4}
$$

which is a combination of the series element flow and power to the shunt element. The line parameters are known numerical values, but as yet the bus voltages (part of the state) are *unknown*.

Several approximations are necessary to obtain an iterative scheme to solve Equation 7.3 for the voltages at $n-1$ buses. The approximations are equivalent to those applied to the Jacobian matrix in a load flow, in the sense that the final accuracy of the state estimator is not dependent on the approximation; only the rate of convergence is affected. The measured *numerical values*, S_{mi}, are equated to flow from bus voltages to be calculated:

$$
S_{mi} = E_i \left\{ \frac{(E_i^* - E_k^*)}{Z_{ik}^*} + Y_{ii}^* E_i^* \right\}
\tag{7.5}
$$

The power flow measurement depends primarily on the difference in voltage from bus i to bus k, so the equivalent measured differences are

$$
E_{im} - E_{km} = E_{ikm} = \frac{Z_{ik}}{E_i^*} S_{mi}^* - Z_{ik} Y_{ii} E_i
\tag{7.6}
$$

In matrix form this equation is

$$
\mathbf{E}_M = \begin{bmatrix} E_{1m} - E_{pm} \\ E_{2m} - E_{nm} \\ \vdots \end{bmatrix} = \mathbf{H}^{-1} \mathbf{S}_M^* - \mathbf{K}
\tag{7.7}
$$

FIGURE 7.4
Three-phase SF$_6$ insulated circuit breaker for 800 kV UHV application. Each arm of a U-shaped pole contains two interrupting gaps in the top of the arm. The cylindrical structure at the left of the breaker is a current transformer to measure line current and compute power flow.

where **H** is an $M \times M$ diagonal matrix, **K** is a vector, and $\mathbf{E}_M$ is a vector of measured differences associated with each power flow measurement. Even though the bus voltages on the right-hand side of Equation 7.7 are unknown, $\mathbf{E}_M$ can be considered to be derived from $\mathbf{S}_M$.

A matrix formulation of the state estimator performance index is

$$J = \sum_{i=1}^{M} W_i \, | \, S_{mi} - S_{ci} \, |^2 \tag{7.8}$$

$$= [\mathbf{S}_M - \mathbf{S}_c]^{*t} \, \mathbf{W} [\mathbf{S}_M - \mathbf{S}_c]$$

where $\mathbf{S}_M$ is an M vector of the measurements, **W** is a diagonal weighting matrix, and $\mathbf{S}_c$ is the calculated power flow vector.

Into Equation 7.8 substitute the measurement vector, Equation 7.7, and a calculated power flow

$$\mathbf{S}_c = (\mathbf{H}\mathbf{E}_c + \mathbf{H}\mathbf{K})^* \tag{7.9}$$

in order to obtain

$$J = [\mathbf{S}_M - \mathbf{S}_c]^{*t} \, \mathbf{W} [\mathbf{S}_M - \mathbf{S}_c]$$

$$= [\mathbf{H}\mathbf{E}_M + \mathbf{H}\mathbf{K} - (\mathbf{H}\mathbf{E}_c + \mathbf{H}\mathbf{K})]^{*t} \, \mathbf{W} [\mathbf{H}\mathbf{E}_M + \mathbf{H}\mathbf{K} - (\mathbf{H}\mathbf{E}_c + \mathbf{H}\mathbf{K})]$$

$$= [\mathbf{E}_M - \mathbf{E}_c]^{*t} \, \mathbf{H}^* \mathbf{W} \mathbf{H} [\mathbf{E}_M - \mathbf{E}_c] \tag{7.10}$$

The calculated voltage difference vector can be expressed in terms of the bus voltages, $\mathbf{E}_{\mathrm{BUS}}$, and the slack bus voltage, E_r:

$$\mathbf{E}_c = \hat{\mathbf{A}}\mathbf{E} = \mathbf{B}\mathbf{E}_{\mathrm{BUS}} + \mathbf{b}E_r \tag{7.11}$$

In Equation 7.11, $\hat{\mathbf{A}}$ is a double branch to bus incidence matrix, in other words, +1 and −1 in each row, corresponding to a measurement along a directed branch element:

Measurements   buses $\rightarrow$

$$\hat{\mathbf{A}} = \begin{bmatrix} 1 & -1 & 0 & 0 & . & . & . & . & | & 0 \\ 0 & 1 & 0 & 0 & 0 & . & . & -1 & | & 0 \\ -1 & 0 & 1 & 0 & . & . & . & . & | & . \\ & & & & & & & & | & 1 \\ & & & & & & & & | & -1 \end{bmatrix} = [\mathbf{B} \quad \mathbf{b}] \tag{7.12}$$

When line flows are measured on each end, the series element appears twice. The order of $\hat{\mathbf{A}}$ is $M \times n$, where n is the number of buses. $\hat{\mathbf{A}}$ is partitioned into a bus voltage part, $\mathbf{B}$, and a slack bus vector $\mathbf{b}$.

Substituting Equation 7.11 into 7.10 yields

$$J = [\mathbf{E}_M - \mathbf{B}\mathbf{E}_{\mathrm{BUS}} - \mathbf{b}E_r]^{*t}\mathbf{H}^*\mathbf{W}\mathbf{H}[\mathbf{E}_M - \mathbf{B}\mathbf{E}_{\mathrm{BUS}} - \mathbf{b}E_r] \tag{7.13}$$

The product $\mathbf{H}^*\mathbf{W}\mathbf{H}$ is assumed to be constant and Equation 7.13 is minimized with respect to the bus voltage vector

$$\frac{\partial J}{\partial \mathbf{E}_{\mathrm{BUS}}} = 0 = -\mathbf{B}^t\mathbf{H}^*\mathbf{W}\mathbf{H}[\mathbf{E}_M - \mathbf{B}\mathbf{E}_{\mathrm{BUS}} - \mathbf{b}E_r] \tag{7.14}$$

$$- [\mathbf{E}_M - \mathbf{B}\mathbf{E}_{\mathrm{BUS}} - \mathbf{b}E_r]^{*t}\mathbf{H}^*\mathbf{W}\mathbf{H}\mathbf{B} = 0$$

Either of the conjugate forms may be used to solve for $\mathbf{E}_{\mathrm{BUS}}$ in order to minimize the performance index:

$$[\mathbf{B}^t\mathbf{H}^*\mathbf{W}\mathbf{H}\mathbf{B}]\mathbf{E}_{\mathrm{BUS}} = \mathbf{B}^t\mathbf{H}^*\mathbf{W}\mathbf{H}[\mathbf{E}_M - \mathbf{b}E_r] \tag{7.15}$$

For $\mathbf{D} = \mathbf{H}^*\mathbf{W}\mathbf{H}$, a diagonal matrix of order M, the iterative line-flow state estimator is

$$\boxed{\begin{array}{l} \text{Iterative equation for line-flow state estimator} \\ \mathbf{E}_{\mathrm{BUS}}^{k+1} = [\mathbf{B}^t\mathbf{D}\mathbf{B}]^{-1}\mathbf{B}^t\mathbf{D}\left[\mathbf{E}_{\mathbf{M}}^k - \mathbf{b}E_r\right] \end{array}} \tag{7.16}$$

where $k + 1$ is the next iteration, and $\mathbf{E}_M^k$ is evaluated using the previous value of $\mathbf{E}_{\mathrm{BUS}}$. The algorithm to solve for the bus voltage vector in the state estimator is:

1. Guess an initial $\mathbf{E}_{\mathrm{BUS}}^0$. A flat start sets all bus voltages equal to the slack bus.
2. Use the measurement data to calculate the initial voltage difference:

$$\mathbf{E}_M^0 = \mathbf{H}^{-1}\mathbf{S}_M^* - \mathbf{K}$$

where both $\mathbf{H}$ and $\mathbf{K}$ depend on $\mathbf{E}_{\mathrm{BUS}}^0$, as in Equation 7.6.

3. Use the state estimator iterative Equation 7.16 to solve for $\mathbf{E}_{\mathrm{BUS}}^1$.

4. Replace $\mathbf{E}_{\mathrm{BUS}}^0$ by $\mathbf{E}_{\mathrm{BUS}}^1$. Repeat steps 2 and 3 until convergence to within a desired ε error tolerance. The tolerance is usually expressed in terms of change in bus voltages between iterations:

$$|E_i^{k+1} - E_i^k| < \varepsilon$$

Observe that the approximation of constant $\mathbf{H}^*\mathbf{WH} = \mathbf{D}$ does not affect the accuracy of converged results. The matrix $\mathbf{D}$ appears as an approximation to a gradient multiplication on the right-hand side of Equation 7.16.

Example 7.1 clarifies the use of the line-flow state estimator algorithm.

Example 7.1

Real and reactive line flows are measured at both ends of three transmission lines on the three-bus, three-line power transmission system shown in Figure E7.1. The set of power flow measurements and the transmission line impedances, using a 100 MVA floating voltage base, are:

$$S_1 = 0.41 - j0.11$$
$$S_2 = -0.40 - j0.10$$
$$S_3 = -0.105 + j0.11$$
$$S_4 = 0.14 - j0.14$$
$$S_5 = 0.72 - j0.37$$
$$S_6 = -0.70 + j0.35$$

$$Z_{12} = 0.08 + j0.24 \qquad Z_{23} = 0.06 + j0.18 \qquad Z_{31} = 0.02 + j0.06$$

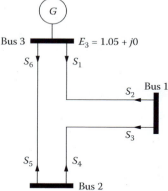

FIGURE E7.1

The line-charging and shunt terms are negligible. Consider each line-flow measurement to be performed with $c_1 = 0.02$, or 2% accurate meters with 100 MW, 100 Mvar full-scale range, and $0.005 = c_2$ transducer errors. The weighting factor for measurement 1 computed using Equation 7.2 is

$$W_1 = \frac{50 \times 10^{-6}}{\{0.02 \times 100 \,|\, 0.41 - j.11 \,|\, + 0.005 \times 100\}^2}$$

$$= 27.4755 \times 10^{-6}$$

In a similar manner, the other weighting factors are found:

$$W_2 = 28.4962 \times 10^{-6} \qquad W_3 = 77.323 \times 10^{-6} \qquad W_4 = 62.2836 \times 10^{-6}$$
$$W_5 = 11.1353 \times 10^{-6} \qquad W_6 = 11.7226 \times 10^{-6}$$

Find the state vector, $\mathbf{E}_{BUS}$, so the absolute magnitude of the error in voltage at any bus between successive iterations is less than 0.0001 p.u.

Solution

Let all voltages be 1.0 in the product $\mathbf{H}^t\mathbf{W}\mathbf{H}$, so the diagonal weighting matrix is

$$\mathbf{D} = \text{diag}\,[d_1, d_2, d_3, d_4, d_5, d_6]$$

$$= \text{diag}\left[\frac{W_1}{|Z_{13}|^2} \quad \frac{W_2}{|Z_{13}|^2} \quad \frac{W_3}{|Z_{12}|^2} \quad \frac{W_4}{|Z_{12}|^2} \quad \frac{W_5}{|Z_{23}|^2} \quad \frac{W_6}{|Z_{23}|^2} \right]$$

$$= \text{diag}\,[6.869 \times 10^{-3}, 7.124 \times 10^{-3}, 1.208 \times 10^{-3}, 0.9732 \times 10^{-3}, 0.3093 \times 10^{-3}, 0.3256 \times 10^{-3}]$$

Measurements $S_1, S_2, \ldots, S_6$ are considered positive into the line from the bus, so that Equation 7.11 is

measurements   buses $\rightarrow$

$\downarrow$

$$\mathbf{E}_c = \hat{\mathbf{A}}\mathbf{E} = \mathbf{B}\mathbf{E}_{BUS} + \mathbf{b}E_r = \begin{bmatrix} -1 & 0 & 1 \\ 1 & 0 & -1 \\ 1 & -1 & 0 \\ -1 & 1 & 0 \\ 0 & 1 & -1 \\ 0 & -1 & 1 \end{bmatrix} \begin{bmatrix} E_1 \\ E_2 \\ E_3 \end{bmatrix} = \begin{bmatrix} -1 & 0 \\ 1 & 0 \\ 1 & -1 \\ -1 & 1 \\ 0 & 1 \\ 0 & -1 \end{bmatrix} \mathbf{E}_{BUS} + \begin{bmatrix} 1 \\ -1 \\ 0 \\ 0 \\ -1 \\ 1 \end{bmatrix} E_r$$

The matrix product $\mathbf{B}^t\mathbf{D}$ is

$$\mathbf{B}^t\mathbf{D} = \begin{bmatrix} -1 & 1 & 1 & -1 & 0 & 0 \\ 0 & 0 & -1 & 1 & 1 & -1 \end{bmatrix} \mathbf{D} = \begin{bmatrix} -d_1 & d_2 & d_3 & -d_4 & 0 & 0 \\ 0 & 0 & -d_3 & d_4 & d_5 & -d_6 \end{bmatrix}$$

The 2×2 constant coefficient matrix that must be inverted is

$$\mathbf{B}^t\mathbf{D}\mathbf{B} = \begin{bmatrix} d_1 + d_2 + d_3 + d_4 & -d_3 - d_4 \\ -d_3 - d_4 & d_3 + d_4 + d_5 + d_6 \end{bmatrix} = \begin{bmatrix} 0.0161743 & -0.00218135 \\ -0.00218135 & d_3 + d_4 + d_5 + d_6 \end{bmatrix}$$

$$[\mathbf{B}^t\mathbf{D}\mathbf{B}]^{-1} = \begin{bmatrix} 69.038 & 53.473 \\ 53.473 & 396.49 \end{bmatrix}$$

The state estimator iterative equation is

$$\mathbf{E}_{BUS}^{k+1} = \begin{bmatrix} E_1^{k+1} \\ E_2^{k+1} \end{bmatrix} = [\mathbf{B}^t\mathbf{D}\mathbf{B}]^{-1}\mathbf{B}^t\mathbf{D}\left[\mathbf{E}_M^k - \mathbf{b}E_r \right]$$

$$= \begin{bmatrix} -0.47422 & 0.49183 & 0.01881 & -0.01515 & 0.01654 & -0.01741 \\ -0.36730 & 0.38095 & -0.41443 & 0.33382 & 0.12264 & -0.12911 \end{bmatrix} \left[\mathbf{E}_m^k - \mathbf{b}E_r \right]$$

Using a flat start with $E_1 = E_2 = E_r = 1.05 + j0$, the results of alternately using the measurements

$$\mathbf{E}_M^k = [\mathbf{H}^k]^{-1}\mathbf{S}_M^* - \mathbf{K}^k$$

and the state estimator iterative equation are as follows:

Iteration	E_1	E_2
0	$1.05 + j0$	$1.05 + j0$
1	$1.048269 - j0.020709$	$1.029173 + j0.044660$
2	$1.047938 - j0.020683$	$1.027649 + j0.044970$
3	$1.047937 - j0.020685$	$1.027618 + j0.045015$

The absolute magnitude of the change in voltage is less than 1×10^{-4} for the next iteration, so the algorithm converges in three iterations. The voltages from the last iteration are used to calculate the best estimate for the line flows. The calculated values are

$$\text{Best Estimates} \begin{cases} \hat{S}_1 = 0.337 - j0.076 & \hat{S}_3 = -0.228 + j0.170 & \hat{S}_5 = 0.201 - j0.184 \\ \hat{S}_2 = -0.334 + j0.083 & \hat{S}_4 = 0.234 - j0.153 & \hat{S}_6 = -0.197 - j0.196 \end{cases}$$

Observe that considerable error exists between the measured line flows and the best estimate for the flows. One method to detect snapshot errors is subsequently presented.

© The software program *stat_est.exe* may be used to compute the best estimate for Example 7.1. The measurement default weights are $R_i^{-1} = W_i = 1.0$. Since relative weights affect the results, the user can enter various values to examine the effect of weights on the best estimate for measurements. Copy the file *example71.sss* into *p_flow.sss*, then execute *stat_est.exe* to compare the results with different weighting factors. Use the weighting factor option to change values.

Example 7.1 demonstrated properties of the line-flow state estimation algorithm. The matrix $\mathbf{B}^t\mathbf{DB}$ has on the diagonal, for any bus i, *positive* weighting factors for all line measurements connected to bus i. The off-diagonal terms, from bus i to bus k, for example, are negatives of the weighting factors on the line between bus i and bus k.

$$\begin{bmatrix} ik \text{ element of} \\ \mathbf{B}^t\mathbf{DB} \end{bmatrix} = \begin{cases} \displaystyle\sum_{m=1}^{M} d_m & \text{for all lines to bus } i \text{ if } i = k \\ -\displaystyle\sum_{m=1}^{M} d_m & \text{for all lines from bus } i \text{ to } k \text{ if } i \neq k \end{cases} \tag{7.17}$$

If a measurement is not present, the d_i weighting factor is set to zero. The matrix $\mathbf{B}^t\mathbf{DB}$ is of order $(n-1) \times (n-1)$ and has, at most, the same nonzero pattern as $\mathbf{Y}_{\text{BUS}}$. Hence, for large networks $\mathbf{B}^t\mathbf{DB}$ is a *sparse* matrix with real coefficients, so its inverse may be computed using the ordered Gaussian elimination method of Chapter 3 or other sparse matrix methods. The matrix can be written by inspection of the measurements on a topological graph of the power system. If measurements are not performed on either end of a transmission line from bus i to bus k, the (ik)th entry of the matrix is zero.

The nonzero terms of the matrix $\mathbf{B}^t\mathbf{D}$ can also be written by inspection. The matrix has $(n-1)$ rows corresponding to each bus and M columns corresponding to the set of measurements. For row i corresponding to bus i, if a measurement is at the near end of a line connected to bus i, the weighting factor d_m is positive, and negative for a measurement at the far end of a line to this bus.

It is worth emphasizing that the sparse matrix inverse $[\mathbf{B}^t\mathbf{D}\mathbf{B}]^{-1}$ is triangularized only one time. Repeated solutions are calculated for updated right hand side vectors, $\mathbf{R}$, which are formed from $\mathbf{B}^t\mathbf{D}$ and the constants of Equation 7.7:

$$\mathbf{R}\left(\mathbf{E}_{\mathrm{BUS}}^k\right) = \mathbf{B}^t\mathbf{D}[\mathbf{H}^{-1}\mathbf{S}_M^* - \mathbf{K} - b E_r] \tag{7.18}$$

where only $\mathbf{H}$ and $\mathbf{K}$ depend on the bus voltages at the kth iteration.

A limitation of the line-flow state estimator is that it does not accept other available measurements, such as bus power injections, and partial information such as bus voltage magnitudes or current magnitudes that do not have phase-angle data. A more general formulation is subsequently derived in terms of in-phase and out-of-phase measurements.

7.2 State Estimation and Noisy Measurements

Another approach to obtain a state estimate [2] is to consider the measurements performed on voltages, currents, and power of the network to be mixed or "corrupted" by noise. Let the state vector be $2(n-1)$ real-valued phase angles and voltage magnitudes at all network buses except the referenced bus:

$$\mathbf{X} = \begin{bmatrix} \boldsymbol{\delta} \\ \mathbf{V} \end{bmatrix} = \begin{bmatrix} \delta_1 \\ \vdots \\ \delta_{n-1} \\ V_1 \\ \vdots \\ V_{n-1} \end{bmatrix} \tag{7.19}$$

It is possible to measure the reference bus voltage directly and increase the state vector to $2n-1$, but Equation 7.19 is used to be consistent with prior notation. The real-valued set of M measurements, $\mathbf{Z}_M$, is nonlinearly dependent on the state of the power system, and is corrupted by noise:

$$\mathbf{Z}_M = \mathbf{h}(\mathbf{X}) + \zeta \tag{7.20}$$

where $\mathbf{h}(\mathbf{X})$ is a nonlinear vector function of the state $\mathbf{X}$, or equivalently of $\mathbf{E}_{\mathrm{BUS}}$. The measurement vector, $\mathbf{Z}_M$, is comprised of Mvar, amperes, volts, and so on, which are obtained from the network and would be accurate except for noise ζ. If the network is in steady state and the measurements performed a large number of times, the noise would average to zero and $\mathbf{Z}_M$ would be a true, or unbiased measurement. More formally, the noise is assumed to be zero-mean, Gaussian white noise whose probability of magnitude ζ_i is given by

$$p(\zeta_i) = \frac{e^{-\zeta_i^2/2\sigma_i^2}}{\sqrt{2\pi\sigma_i^2}} \tag{7.21}$$

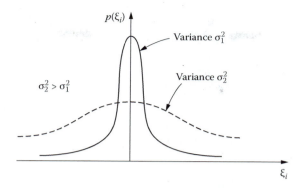

FIGURE 7.5
Zero-mean, Gaussian white noise probability density function for two values of variance.

The standard deviation of this noise source is σ_i. A large value of variance, σ_i^2, implies there is a high probability that the noise ζ_i takes on large values. A sketch of the probability density function, Equation 7.21, is shown in Figure 7.5 for two values of variance $\sigma_2^2 > \sigma_1^2$.

When a periodic measurement, Z_i, is performed a number of times, and an attempt is made to correlate the noise from a sample at time t_k with the noise from samples made at earlier or later times by means of averaging, the result is

$$R_i = \sum_{m=-\infty}^{m+\infty} \zeta_i(t_k)\zeta_i(t_k - m\Delta t) = \begin{cases} 0 & \text{if } m \neq 0 \\ r_{ii} & \text{if } m = 0 \end{cases} \tag{7.22}$$

In other words, the noise is uncorrelated so the error introduced at one measurement time does not influence the error at other times. This white noise assumption is further extended to assume that noise in measurement Z_{mi} is not correlated with noise in another measurement, Z_{mk}. Both assumptions are more for mathematical convenience than physically true. For example, an error in a voltage measurement (via a step-down potential transformer) may also affect Mvar and MW, which require the voltage as an input to multiply with current signals. It is furthermore very difficult to measure noise statistics, as in Equation 7.22, or to obtain the probability density, Equation 7.21, so investigators [2] use metering equipment accuracy to define the variance, such as

$$\frac{1}{\sigma_i^2} = \frac{1}{r_{ii}} = \frac{3}{\{c_1 \mid Z_i \mid + c_2(\text{F.S.})\}^2} \tag{7.23}$$

This definition agrees with Equation 7.2 to within a multiplying factor.

The multiplying factor 50×10^{-6} cancels out of Equation 7.16, just as it will for state estimation in the presence of noise, which is now formulated.

With this brief description of noise-corrupted measurements on a power system, the performance index to be minimized by calculating the best state estimate for a mixture of $\mathbf{Z}_M$ measurements is

$$J(\mathbf{X}) = [\mathbf{Z}_M - \mathbf{h}(\mathbf{X})]^t \mathbf{R}^{-1} [\mathbf{Z}_M - \mathbf{h}(\mathbf{X})] \tag{7.24}$$

where $\mathbf{R}^{-1}$ is a diagonal weighting matrix of the variances for each of the M measurements (Equation 7.23). The best estimate $\hat{\mathbf{X}}$ is the state that makes the gradient of Equation 7.24 go to zero:

$$\frac{\partial J}{\partial \mathbf{X}}(\hat{\mathbf{X}}) = 0 = \frac{\partial \mathbf{h}^t}{\partial \mathbf{X}}(\hat{\mathbf{X}})(\mathbf{R})^{-1}[\mathbf{Z}_M - \mathbf{h}(\hat{\mathbf{X}})] \tag{7.25a}$$

This is a set of M nonlinear equations to be solved by an iterative method. The Newton-Raphson method is selected and the Jacobian is approximated as

$$\mathbf{J}(\mathbf{X}) = \frac{\partial^2 J}{\partial \mathbf{X}^2}(\mathbf{X}) \cong -\frac{\partial \mathbf{h}^t}{\partial \mathbf{X}} \mathbf{R}^{-1} \frac{\partial \mathbf{h}}{\partial \mathbf{X}} \tag{7.25b}$$

where higher-order derivatives in $\mathbf{X}$ are neglected. Equation 7.25b is often called the *information matrix*. Using Equation 7.25b to solve Equation 7.25a, by means of a Newton-Raphson iterative scheme to find the best state estimate for noisy measurements is

$$\boxed{\begin{array}{c} \text{Iterative equation for state estimator using} \\ \text{general measurements} \\ \mathbf{X}^{k+1} = \mathbf{X}^k + \left\{\frac{\partial \mathbf{h}^t}{\partial \mathbf{X}} \mathbf{R}^{-1} \frac{\partial \mathbf{h}}{\partial \mathbf{X}}\right\}^{-1} \frac{\partial \mathbf{h}^t}{\partial \mathbf{X}} \mathbf{R}^{-1}[\mathbf{Z}_M - \mathbf{h}(\mathbf{X})] \end{array}} \tag{7.26}$$

The iterations are carried out until the calculated difference in state for successive iterations is less than a convergence tolerance.

Since the Jacobian (information matrix) is already approximated, almost all applications use constants for the elements of this matrix and calculate the inverse one time only. Let the physical measurements, $\mathbf{Z}_M$, and the nonlinear function of state, $\mathbf{h}$, be separated into real and reactive parts such that

$$\mathbf{h} = \begin{bmatrix} \mathbf{f} \\ \hline \mathbf{g} \end{bmatrix} = \begin{bmatrix} \text{real measurements} \\ \hline \text{reactive measurements} \end{bmatrix} \tag{7.27}$$

so that partial derivatives can be obtained as follows:

$$\frac{\partial \mathbf{h}}{\partial \mathbf{X}}(\mathbf{X}) = \begin{bmatrix} \dfrac{\partial \mathbf{f}}{\partial \boldsymbol{\delta}} & \dfrac{\partial \mathbf{f}}{\partial \mathbf{V}} \\ \dfrac{\partial \mathbf{g}}{\partial \boldsymbol{\delta}} & \dfrac{\partial \mathbf{g}}{\partial \mathbf{V}} \end{bmatrix} \tag{7.28}$$

The diagonal weighting matrix, $\mathbf{R}^{-1}$, is accordingly separated into active and reactive diagonal parts:

$$\mathbf{R}^{-1} = \begin{bmatrix} \mathbf{R}_a^{-1} & 0 \\ 0 & \mathbf{R}_r^{-1} \end{bmatrix} \tag{7.29}$$

With Equation 7.27 to Equation 7.29 the Jacobian (or information matrix) is

$$
\mathbf{J}(\delta, \mathbf{V}) =
\begin{bmatrix}
\dfrac{\partial \mathbf{f}^t}{\partial \boldsymbol{\delta}} \mathbf{R}_a^{-1} \dfrac{\partial \mathbf{f}}{\partial \boldsymbol{\delta}} + \dfrac{\partial \mathbf{g}^t}{\partial \boldsymbol{\delta}} \mathbf{R}_r^{-1} \dfrac{\partial \mathbf{g}}{\partial \boldsymbol{\delta}} & \vdots & \dfrac{\partial \mathbf{f}^t}{\partial \boldsymbol{\delta}} \mathbf{R}_a^{-1} \dfrac{\partial \mathbf{f}}{\partial \mathbf{V}} + \dfrac{\partial \mathbf{g}^t}{\partial \boldsymbol{\delta}} \mathbf{R}_r^{-1} \dfrac{\partial \mathbf{g}}{\partial \mathbf{V}} \\
\cdots & \vdots & \cdots \\
\dfrac{\partial \mathbf{f}^t}{\partial \mathbf{V}} \mathbf{R}_a^{-1} \dfrac{\partial \mathbf{f}}{\partial \boldsymbol{\delta}} + \dfrac{\partial \mathbf{g}^t}{\partial \mathbf{V}} \mathbf{R}_r^{-1} \dfrac{\partial \mathbf{g}}{\partial \boldsymbol{\delta}} & \vdots & \dfrac{\partial \mathbf{f}^t}{\partial \mathbf{V}} \mathbf{R}_a^{-1} \dfrac{\partial \mathbf{f}}{\partial \mathbf{V}} + \dfrac{\partial \mathbf{g}^t}{\partial \mathbf{V}} \mathbf{R}_r^{-1} \dfrac{\partial \mathbf{g}}{\partial \mathbf{V}}
\end{bmatrix}
\tag{7.30}
$$

When the phase-angle differences between directly connected buses are approximated as zero and all bus voltages are approximated as 1.0 p.u., the Jacobian is a *constant matrix*, not changing with iterations in Equation 7.26. Finally, for power systems with transmission line reactances much larger than the resistive component, and comparatively large shunt reactances from bus to ground, the Jacobian is approximated using reactances and autocorrelation terms as

$$
\mathbf{J}_{FD} =
\begin{bmatrix}
\dfrac{\partial \mathbf{f}^t}{\partial \boldsymbol{\delta}} \mathbf{R}_a^{-1} \dfrac{\partial \mathbf{f}}{\partial \boldsymbol{\delta}} & \vdots & 0 \\
\cdots & \vdots & \cdots \\
0 & \vdots & \dfrac{\partial \mathbf{g}^t}{\partial \mathbf{V}} \mathbf{R}_r^{-1} \dfrac{\partial \mathbf{g}}{\partial \mathbf{V}}
\end{bmatrix}_{\text{constant}}
\tag{7.31}
$$

The approximations result in neglecting the partial derivatives of real (in-phase) flows with voltages and neglecting reactive flows depending on phase-angle changes. Equation 7.31 is called a fast-decoupled approximation to the information matrix. The gradient $\partial \mathbf{h}/\partial \mathbf{X}$ in Equation 7.26 is also approximated by constants. When the decoupling of Equation 7.31 is used, the fast-decoupled state estimator requires updating the $\mathbf{g}(\mathbf{X})$ vector with the last δ^k before $\mathbf{V}^k$ is calculated [2] in the same manner as the fast-decoupled load flow (FDLF) (see the flowchart of Figure 5.6):

$$
\delta^{k+1} = \delta^k + \left[\dfrac{\partial \mathbf{f}^t}{\partial \boldsymbol{\delta}} \mathbf{R}_a^{-1} \dfrac{\partial \mathbf{f}}{\partial \boldsymbol{\delta}} \right]^{-1} \dfrac{\partial \mathbf{f}^t}{\partial \boldsymbol{\delta}} \mathbf{R}_a^{-1} [\mathbf{f}_m - \mathbf{f}(\mathbf{X}^k)]
\tag{7.32a}
$$

$$
\mathbf{V}^{k+1} = \mathbf{V}^k + \left[\dfrac{\partial \mathbf{g}^t}{\partial \mathbf{V}} \mathbf{R}_r^{-1} \dfrac{\partial \mathbf{g}}{\partial \mathbf{V}} \right]^{-1} \dfrac{\partial \mathbf{g}^t}{\partial \mathbf{V}} \mathbf{R}_r^{-1} [\mathbf{g}_m - \mathbf{g}(\mathbf{X}^k)]
\tag{7.32b}
$$

where the in-phase portions of the measurement are $\mathbf{f}_m$, and the reactive measurements are $\mathbf{g}_m$:

$$
\mathbf{Z}_M =
\begin{bmatrix}
\mathbf{f}_m \\
\cdots \\
\mathbf{g}_m
\end{bmatrix}
\tag{7.33}
$$

Example 7.2 demonstrates separation of the measurements into real and reactive parts and implementing the fast decoupled state estimator (FDSE).

Example 7.2

The power injections (MW, Mvar) are measured on the three-bus, three-line power system shown in Figure E7.2. The line parameters are on a 100 MVA base. Bus 1 is taken as the reference bus. The accuracy of the metering, or equivalently, the autocorrelation matrix, and the measurements are given below:

$$S_1 = 12 - j24 \text{ (MW, Mvars)}$$

$$S_2 = 21 - j24 \text{ (MW, Mvars)}$$

$$S_3 = -30 + j50 \text{ (MW, Mvars)}$$

$$R_1^{-1} = 3 = W_1$$

$$R_2^{-1} = 5 = W_2$$

$$R_3^{-1} = 2 = W_3$$

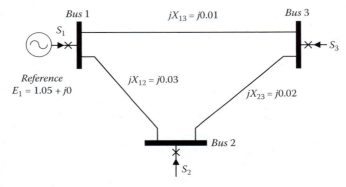

FIGURE E7.2

Use a flat start and iterate until the absolute magnitude of the voltage at any bus does not change more than 0.0001 p.u. between iterations. Find the state estimate.

Solution

The performance index using the 3 complex, or 6 real measurement, is

$$J(\mathbf{X}) = [\mathbf{Z}_M - \mathbf{h}(\mathbf{X})^t] \mathbf{R}^{-1} [\mathbf{Z}_M - \mathbf{h}(\mathbf{X})]$$

$$= \begin{bmatrix} \mathbf{Re}\,(S_1) - f_1 \\ \mathbf{Re}(S_2) - f_2 \\ \mathbf{Re}(S_3) - f_3 \\ \mathbf{Im}(S_1) - g_1 \\ \mathbf{Im}(S_2) - g_2 \\ \mathbf{Im}(S_3) - g_3 \end{bmatrix}^t \begin{bmatrix} R_1^{-1} & 0 & 0 & 0 & 0 & 0 \\ 0 & R_2^{-1} & 0 & 0 & 0 & 0 \\ 0 & 0 & R_3^{-1} & 0 & 0 & 0 \\ 0 & 0 & 0 & R_1^{-1} & 0 & 0 \\ 0 & 0 & 0 & 0 & R_2^{-1} & 0 \\ 0 & 0 & 0 & 0 & 0 & R_3^{-1} \end{bmatrix} \begin{bmatrix} \mathbf{Re}\,(S_1) - f_1 \\ \mathbf{Re}(S_2) - f_2 \\ \mathbf{Re}(S_3) - f_3 \\ \mathbf{Im}(S_1) - g_1 \\ \mathbf{Im}(S_2) - g_2 \\ \mathbf{Im}(S_3) - g_3 \end{bmatrix}$$

The state vector for this power system is

$$\mathbf{X} = [\delta_2 \ \ \delta_3 \ \ V_2 \ \ V_3]^t$$

Consider the calculated injection at bus 1, which is the reference bus at known potential V_1:

$$f_1 + jg_1 = E_1 I_1^* = V_1 [I_{12}^* + I_{13}^*] = V_1 \left[\frac{V_1 - V_2 e^{-j\delta_2}}{-j0.03} + \frac{V_1 - V_3 e^{-j\delta_3}}{-j0.01} \right]$$

The real and reactive parts, their partial derivatives, and decoupled approximations are:

$$f_1 = -\frac{V_1 V_2 \sin \delta_2}{X_{12}} - \frac{V_1 V_3 \sin \delta_3}{X_{13}}$$

$$\frac{\partial f_1}{\partial \delta_2} = -\frac{V_1 V_2 \cos \delta_2}{X_{12}} \cong -\frac{1}{X_{12}} \qquad \frac{\partial f_1}{\partial \delta_3} = -\frac{V_1 V_3 \cos \delta_3}{X_{13}} \cong -\frac{1}{X_{13}}$$

$$\frac{\partial f_1}{\partial V_2} = \frac{-V_1 \sin \delta_2}{X_{12}} \cong 0 \qquad \frac{\partial f_1}{\partial V_3} = \frac{-V_1 \sin \delta_3}{X_{13}} \cong 0$$

$$g_1 = \frac{V_1^2}{X_{12}} - \frac{V_1 V_2}{X_{12}} \cos \delta_2 + \frac{V_1^2}{X_{13}} - \frac{V_1 V_3 \cos \delta_3}{X_{13}}$$

$$\frac{\partial g_1}{\partial \delta_2} = \frac{V_1 V_2}{X_{12}} \sin \delta_2 \cong 0 \qquad \frac{\partial g_1}{\partial \delta_3} = \frac{V_1 V_3 \sin \delta_3}{X_{13}} \cong 0$$

$$\frac{\partial g_1}{\partial V_2} = \frac{-V_1 \cos \delta_2}{X_{12}} \cong -\frac{1}{X_{12}} \qquad \frac{\partial g_1}{\partial V_3} = \frac{-V_1 \cos \delta_3}{X_{13}} \cong -\frac{1}{X_{13}}$$

The power injections at buses 2 and 3 are

$$f_2 + jg_2 = V_2 e^{j\delta_2} \left[\frac{V_2 e^{-j\delta_2} - V_1}{-jX_{12}} + \frac{V_2 e^{-j\delta_2} - V_3 e^{-j\delta_3}}{-jX_{23}} \right]$$

$$f_3 + jg_3 = V_3 e^{j\delta_3} \left[\frac{V_3 e^{-j\delta_3} - V_2 e^{-j\delta_2}}{-jX_{23}} + \frac{V_3 e^{-j\delta_3} - V_1}{-jX_{13}} \right]$$

It is easily verified that the fast-decoupled approximation for the calculated gradient is the 6×4 matrix.

$$\frac{\partial \mathbf{h}}{\partial \mathbf{X}} \cong \begin{bmatrix} \dfrac{\partial \mathbf{f}}{\partial \boldsymbol{\delta}} & 0 \\ 0 & \dfrac{\partial \mathbf{g}}{\partial \mathbf{V}} \end{bmatrix} \cong \left[\begin{array}{cc:cc} \dfrac{-1}{X_{12}} & \dfrac{-1}{X_{13}} & 0 & 0 \\[2mm] \dfrac{1}{X_{12}} + \dfrac{1}{X_{23}} & \dfrac{-1}{X_{23}} & 0 & 0 \\[2mm] \dfrac{-1}{X_{23}} & \dfrac{1}{X_{13}} + \dfrac{1}{X_{23}} & 0 & 0 \\ \hdashline 0 & 0 & \dfrac{-1}{X_{12}} & \dfrac{-1}{X_{13}} \\[2mm] 0 & 0 & \dfrac{1}{X_{12}} + \dfrac{1}{X_{23}} & \dfrac{-1}{X_{23}} \\[2mm] 0 & 0 & \dfrac{-1}{X_{23}} & \dfrac{1}{X_{13}} + \dfrac{1}{X_{23}} \end{array} \right]$$

Either W_i or R_i^{-1} may be used to compute the Jacobian, or information matrix, as

$$
\mathbf{J} = \frac{\partial \mathbf{h}'}{\partial \mathbf{X}} \mathbf{R}^{-1} \frac{\partial \mathbf{h}}{\partial \mathbf{X}} \cong
\begin{bmatrix}
\mathbf{A}_1 & \begin{matrix} 0 & 0 \\ 0 & 0 \end{matrix} \\
\begin{matrix} 0 & 0 \\ 0 & 0 \end{matrix} & \mathbf{A}_2
\end{bmatrix}
=
\begin{bmatrix}
\dfrac{\partial \mathbf{f}'}{\partial \boldsymbol{\delta}} \mathbf{R}_a^{-1} \dfrac{\partial \mathbf{f}}{\partial \boldsymbol{\delta}} & \begin{matrix} 0 & 0 \\ 0 & 0 \end{matrix} \\
\begin{matrix} 0 & 0 \end{matrix} & \dfrac{\partial \mathbf{g}'}{\partial \mathbf{V}} \mathbf{R}_r^{-1} \dfrac{\partial \mathbf{g}}{\partial \mathbf{V}} \\
\begin{matrix} 0 & 0 \end{matrix} & \begin{matrix} 0 & 0 \end{matrix}
\end{bmatrix}
$$

The $\mathbf{A}_1$ and $\mathbf{A}_2$ matrices in terms of line parameters and weighting factors are

$$
\mathbf{A}_1 = \mathbf{A}_2 =
\begin{bmatrix}
\dfrac{W_1}{X_{12}^2} + W_2\left(\dfrac{1}{X_{12}} + \dfrac{1}{X_{23}}\right)^2 + \dfrac{W_3}{X_{23}^2} & \dfrac{W_1}{X_{12}X_{13}} - \dfrac{W_2}{X_{23}}\left(\dfrac{1}{X_{12}} + \dfrac{1}{X_{23}}\right) - \dfrac{W_3}{X_{23}}\left(\dfrac{1}{X_{13}} + \dfrac{1}{X_{23}}\right) \\[4mm]
\dfrac{W_1}{X_{12}X_{13}} - \dfrac{W_2}{X_{23}}\left(\dfrac{1}{X_{12}} + \dfrac{1}{X_{23}}\right) - \dfrac{W_3}{X_{23}}\left(\dfrac{1}{X_{13}} + \dfrac{1}{X_{23}}\right) & \dfrac{W_1}{X_{13}^2} + \dfrac{W_2}{X_{23}^2} + W_3\left(\dfrac{1}{X_{13}} + \dfrac{1}{X_{23}}\right)^2
\end{bmatrix}
$$

The measurements separated into real and reactive parts using the 100 MVA base are

$$
\mathbf{f}_m = [0.12 \quad 0.21 \quad -0.30]^t \quad \text{and} \quad \mathbf{g}_m = [-0.24 \quad -0.24 \quad 0.50]^t
$$

The iterative expression for the phase angles is then

$$
\begin{bmatrix} \delta_2 \\ \delta_3 \end{bmatrix} = \begin{bmatrix} \delta_2 \\ \delta_3 \end{bmatrix}^k + [\mathbf{A}_1]^{-1} \frac{\partial \mathbf{f}^t}{\partial \boldsymbol{\delta}} \mathbf{R}_a^{-1}
\begin{bmatrix}
0.12 - f_1(\mathbf{X}^k) \\
0.21 - f_2(\mathbf{X}^k) \\
-0.30 - f_3(\mathbf{X}^k)
\end{bmatrix}
$$

and the voltage iterative equation is

$$
\begin{bmatrix} V_2 \\ V_3 \end{bmatrix}^{k+1} = \begin{bmatrix} V_2 \\ V_3 \end{bmatrix}^k + [\mathbf{A}_1]^{-1} \frac{\partial \mathbf{g}^t}{\partial \mathbf{V}} \mathbf{R}_r^{-1}
\begin{bmatrix}
-0.24 - g_1(\mathbf{X}^k) \\
-0.24 - g_2(\mathbf{X}^k) \\
0.50 - g_3(\mathbf{X}^k)
\end{bmatrix}
$$

The results of iterations from a flat start are given as follows:

Iteration, k	δ_2 (rad)	δ_3 (rad)	V_2	V_3	Largest Voltage Error Absolute Magnitude
1	0.001490	−0.001600	1.048792	1.052866	0.002866
2	0.001338	−0.001431	1.048850	1.052718	0.0001479
3	0.001354	−0.001449	1.048847	1.052776	0.0000079

Using the best estimate for the state, the best estimate for the p.u. injections are
$\hat{S}_1 = 0.11051 - j0.24572$, $\hat{S}_2 = 0.20439 - j0.24346$, $\hat{S}_3 = -0.31490 + j0.4925$.

© The software program *stat_est* is a fast-decoupled state estimation program that can be used to compute the best estimate. Copy the file ***example72.sss*** into ***p_flow.sss***, and in this copied file, eliminate line flows. (This is done by setting the numerical value of all line flows to 0.0.) The weights $R_i^{-1} = W_i = 1$ closely match these results, but the program permits the exact weights to be entered. Use the measured injections at buses 1 to 3 to execute ***stat_est.exe*** and compare results.

In Example 7.2 there is a great deal of similarity between a load-flow calculation and state estimation using weighted injection measurements. The state estimator incorporates slack bus injection measurements in computing a weighted best estimate, so the state estimation should be more accurate than a load-flow calculation to determine the power system state.

An additional benefit of this more general state estimator is that it can include voltage and current measurements. The measured voltage can be directly compared to a calculated magnitude, but the measurements do not have phase information unless there are special devices such as GPS timing. Example 7.3 demonstrates the use of current and voltage measurements that do not have phase information.

Example 7.3

The three-bus system whose topology is given Figure E7.3 is instrumented for a mixture of line power flows, voltages, and current. A data scan yields the following set of measurements:

$$S_1 = 0.70 + j1.30 \text{ p.u.}$$

$$P_2 = -0.65 \text{ p.u.}$$

$$jQ_3 = -j0.04 \text{ p.u.}$$

$$S_4 = -0.065 + j0.035 \text{ p.u.}$$

$$S_5 = -0.28 - j0.40 \text{ p.u.}$$

$$jQ_6 = j0.42 \text{ p.u.}$$

$$|I_m| = -0.07049 \text{ p.u.}$$

$$|V_m| = 1.0383$$

$$R_i^{-1} = W_i = 5 \quad \text{for } i = 1, 2, 3, 4, 5, 6$$

$$W_c = R_7^{-1} = 7.0 \qquad W_V = R_8^{-1} = 6.0$$

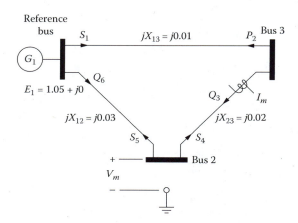

FIGURE E7.3

The accuracy of the metering, or equivalently, the autocorrelation matrix, is also specified. Calculate the state estimate of the system, using a flat start, to within an error tolerance of 0.00001 p.u. between successive iterations. Perform the state estimation first without voltage and current measurements, then repeat using these measurements. The power base is 100 MW.

Solution

There are four real power flow measurements, five reactive, one current, and one voltage measurement. The calculated line current from bus 3 toward bus 2 is

$$I_c = \frac{E_3 - E_2}{jX_{23}} = f_c + jg_c = \frac{V_3 e^{j\delta_3} - V_2 e^{j\delta_2}}{jX_{23}} = |I_c| \angle \gamma$$

$$= \frac{V_3 \sin \delta_3 - V_2 \sin \delta_2}{X_{23}} - \frac{j}{X_{23}} [V_3 \cos \delta_3 - V_2 \cos \delta_2]$$

The angle γ is defined as $\gamma = \tan^{-1}(g_c/f_c)$. The decoupled matrix A_2 is

$$\frac{\partial \mathbf{g}^t}{\partial \mathbf{V}} \mathbf{R}_r^{-1} \frac{\partial \mathbf{g}}{\partial \mathbf{V}} = \begin{bmatrix} \dfrac{W_3}{X_{23}^2} + \dfrac{W_4}{X_{23}^2} + \dfrac{W_5}{X_{12}^2} + \dfrac{W_6}{X_{12}^2} + \dfrac{W_c}{X_{23}^2} + W_V & \vdots & -\dfrac{W_3}{X_{23}^2} - \dfrac{W_4}{X_{23}^2} - \dfrac{W_c}{X_{23}^2} \\ \cdots & \vdots & \cdots \\ -\dfrac{W_3}{X_{23}^2} - \dfrac{W_4}{X_{23}^2} - \dfrac{W_c}{X_{23}^2} & \vdots & \dfrac{W_1}{X_{13}^2} + \dfrac{W_3}{W_{23}^2} + \dfrac{W_4}{X_{23}^2} + \dfrac{W_c}{X_{23}^2} \end{bmatrix} = A_2$$

The A_1 matrix does not include the W_V term. The current and voltage measurements use the phase angles calculated in the previous iteration. The sequential iterative scheme for phase angles and voltages is

$$\begin{bmatrix} \delta_2 \\ \delta_3 \end{bmatrix}^{k+1} = \begin{bmatrix} \delta_2 \\ \delta_3 \end{bmatrix}^k + [\mathbf{A}_1]^{-1} \frac{\partial f^t}{\partial \boldsymbol{\delta}} \begin{bmatrix} W_1 & 0 & 0 & 0 & 0 \\ 0 & W_2 & 0 & 0 & 0 \\ 0 & 0 & W_4 & 0 & 0 \\ 0 & 0 & 0 & W_5 & 0 \\ 0 & 0 & 0 & 0 & W_c \end{bmatrix} \begin{bmatrix} 0.70 - f_1(\mathbf{X}^k) \\ -0.65 - f_2(\mathbf{X}^k) \\ -0.065 - f_4(\mathbf{X}^k) \\ -0.28 - f_5(\mathbf{X}^k) \\ -0.07 \cos \gamma_\delta^k - f_c(\mathbf{X}^k) \end{bmatrix}$$

$$\begin{bmatrix} V_2 \\ V_3 \end{bmatrix}^{k+1} = \begin{bmatrix} V_2 \\ V_3 \end{bmatrix}^k + [\mathbf{A}_2]^{-1} \frac{\partial \mathbf{g}^t}{\partial \mathbf{V}} \begin{bmatrix} W_1 & 0 & 0 & 0 & 0 & 0 & 0 \\ 0 & W_3 & 0 & 0 & 0 & 0 & 0 \\ 0 & 0 & W_4 & 0 & 0 & 0 & 0 \\ 0 & 0 & 0 & W_5 & 0 & 0 & 0 \\ 0 & 0 & 0 & 0 & W_6 & 0 & 0 \\ 0 & 0 & 0 & 0 & 0 & W_c & 0 \\ 0 & 0 & 0 & 0 & 0 & 0 & W_V \end{bmatrix} \begin{bmatrix} 1.3 - g_1(\mathbf{X}^k) \\ -0.04 - g_3(\mathbf{X}^k) \\ 0.035 - g_4(\mathbf{X}^k) \\ -0.40 - g_5(\mathbf{X}^k) \\ 0.42 - g_6(\mathbf{X}^k) \\ -0.070 \sin \gamma^k - g_c(\mathbf{X}^k) \\ (1.0383 - V_2^k) \sin \delta_2^k \end{bmatrix}$$

The voltage and current measurements are eliminated from the measurement set by using $W_c = W_v = 0.0$ in the sequential iterative scheme. This is often done in operational state estimators as more efficient than a reformulation and recompiling of computer code.

The results of iterating from a flat start without voltage and current measurements are as follows.

Case 1: Without Current and Voltage Measurements

Iteration, k	δ_2 (rad)	δ_3 (rad)	V_2	V_3	Largest-Magnitude Voltage Change
1	−0.008167	−0.006763	1.037745	1.037015	0.012984
2	−0.007451	−0.006161	1.038331	1.037649	0.000634
3	−0.007508	−0.006211	1.038304	1.037619	0.000030
4	−0.007503	−0.006206	1.038305	1.037620	0.000001

Using the fourth iteration state variables, the best estimates for the line flows are as follows:

$$
\text{best estimate}\atop\text{using only}\atop\text{flow measurements}
\begin{cases}
\hat{S}_1 = 0.6762 + j1.302 \\[4pt]
\hat{S}_2 = -0.6762 - j1.282 \\[4pt]
\hat{S}_3 = 0.06987 - j0.03547 \\[4pt]
\hat{S}_4 = -0.06987 + j0.03559 \\[4pt]
\hat{S}_5 = -0.2727 - j0.4037 \\[4pt]
\hat{S}_6 = 0.2727 + j0.4103
\end{cases}
$$

Each of the calculated line flows is close to the measured values. However, without quantitative methods to analyze the differences, it is not possible to detect probable errors. Section 7.3 presents one error detection method.

© The software program *stat_est* is a fast-decoupled state estimation program that can be used to compute case 1 results. Copy the file **example73.sss** into **p_flow.sss**, then execute **stat_est.exe** to compare results. The angle γ and δ_2 after each iteration are used to update the right hand side for current and voltage respectively.

When the voltage and current measurements are included in the measurement set, the results of iterating from a flat start are as given below.

Case 2: Including Current and Voltage Measurements

Iteration, k	δ_2 (rad)	δ_3 (rad)	V_2	V_3	Largest-Magnitude Voltage Change
1	−0.006801	−0.006839	1.037545	1.037062	0.012938
2	−0.006195	−0.006231	1.038444	1.037624	0.000898
3	−0.006243	−0.006281	1.038584	1.037557	0.000140
4	−0.006238	−0.006277	1.038692	1.037535	0.000108
5	−0.006238	−0.006278	1.038760	1.037521	0.000068
6	−0.006238	−0006278	1.038804	1.037511	0.000044
7	−0.006238	−0.006278	1.038832	1.037505	0.000028
8	−0.006238	−0.006278	1.038850	1.037501	0.000018
9	−0.006237	−0.006278	1.038861	1.037498	0.000012
10	−0.006237	−0.006278	1.038869	1.037497	0.000007

The best estimates for the line flows using the final iterated state values are

$$
\text{best estimate using current and voltage measurements}
\begin{cases}
\hat{S}_1 = 0.6839 + j1.315 \\[4pt]
\hat{S}_2 = -0.6839 - j1.295 \\[4pt]
\hat{S}_3 = -0.0022 - j0.0712 \\[4pt]
\hat{S}_4 = 0.0022 + j0.0713 \\[4pt]
\hat{S}_5 = -0.2268 - j0.3848 \\[4pt]
\hat{S}_6 = 0.2268 + j0.3903
\end{cases}
$$

In comparing the results of the state estimates without and with the current and voltages, it is seen that the number of iterations increases when they are included. The power flow in line 2–3 is affected primarily by including these measurements, with the real flow forced to zero by the additional current measurement.

In general, voltage and current measurements decrease the rate of convergence of a state estimator because they do not have phase-angle information. As a result, on-line state estimation programs make little use of these data, and employ small weighting factors when they are included.

The Jacobian (information matrix) of Example 7.3 has the same nonzero pattern as $\mathbf{Y}_{BUS}$ because only bus voltage and flow measurements are given. It is more difficult to discern from a case where all buses are connected, but using the injections the state estimation of Example 7.2 has the same nonzero pattern as $\mathbf{Y}_{BUS}$. However, there are more weighting terms per matrix element because an injection measurement affects each line connected to the bus. As a result, in the matrix

$$
\mathbf{A}_1 = \frac{\partial \mathbf{f}^t}{\partial \boldsymbol{\delta}} \, \mathbf{R}_a^{-1} \, \frac{\partial \mathbf{f}}{\partial \boldsymbol{\delta}}
$$

the injection measurement at bus i, given by terms multiplied by W_i, affects every bus *directly* connected by lines to bus i. The sparsity of the Jacobian is that of $\mathbf{Y}_{BUS}$, so the ordered Gaussian elimination methods are also applicable here.

7.3   Monitoring the Power System

Data are collected by RTU units, such as shown in Figure 7.6, on each scan cycle of the central digital computer. Using new data sets, the performance index for a state estimator serves as a continuous and simple monitor for the power system. The best state estimate, $\hat{\mathbf{E}}_{BUS}$, minimizes the performance index:

$$
J(\hat{\mathbf{E}}_{BUS}) = \sum_{i=1}^{M} \left| Z_i - h_i(\hat{\mathbf{E}}_{BUS}) \right|^2 R_i^{-1} = \text{minimum} \tag{7.34}
$$

FIGURE 7.6
Remote terminal unit (RTU) installed in a substation. The unit has its door open. The RTU provides the interface for analog measurements of voltages, currents, power, and status information to be digitally encoded and transmitted to the central computer via telemetry, dedicated telephone line, or line carrier. The RTU also implements commands from the central computer to the equipment. (Photograph courtesy of Advanced Control Systems, Atlanta, GA)

and has a residual value after the state estimation calculation. When new data scans are performed on the power system, the new measurements $Z_{i,\text{new}}$ and old $\widehat{\text{E}}_{\text{BUS}}$ are substituted into the performance index to calculate a new residual value:

$$J_{\text{new}}(\widehat{\text{E}}_{\text{BUS}}) = \sum_{i=1}^{M} \left| Z_{i,\text{new}} - h_i(\widehat{\text{E}}_{\text{BUS}}) \right|^2 R_i^{-1} \tag{7.35}$$

Whenever the new residual value exceeds a threshold level, or confidence level, the power system state has significantly changed to the point that contingency outage studies should be recalculated. A time plot of the performance index is shown in Figure 7.7.

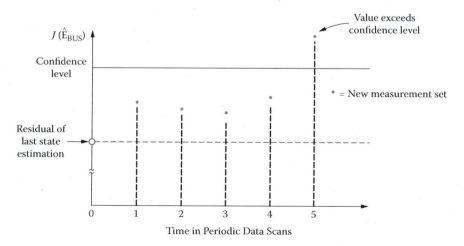

FIGURE 7.7
Monitoring the power system with state estimate performance index.

Because the power system changes slowly, it is sufficient to evaluate the performance index once per minute to test if the power system change exceeds the confidence level. In many energy management computers, a new state calculation is periodically executed, even if the threshold is not exceeded, because the contribution of a large change at a bus (as reflected by a few measurements) is not easily detected in the performance index.

A difficult problem is to detect gross errors in the measurements when they are compared to values calculated from the state. This is known as *bad data detection* and is used to identify errors in metering, analog-to-digital conversion, and the telemetry system. To develop the basic ideas for bad data detection, let $\hat{\mathbf{Z}}_c$ be the vector of values calculated from the best state estimate:

$$\hat{\mathbf{Z}}_c = \mathbf{h}(\hat{\mathbf{E}}_{\text{BUS}}) \tag{7.36}$$

The basic problem is to quantify the difference between the measured and calculated values:

$$\zeta = \mathbf{Z}_M - \hat{\mathbf{Z}}_c \tag{7.37}$$

in order to establish an acceptable error limit.

It was initially assumed every measurement Z_{mi} has noise (or errors), ζ_i, independent of all other measurements. The network electrical flows couple data at one point to all other data points. To compare differences between measured and calculated values, it is necessary to have a common basis. This is done by defining a zero-mean, Gaussian probability distribution, random variable Y_i that is normalized dependent on the network coupling:

$$Y_i = \frac{Z_{mi} - \hat{Z}_{ci}}{\Sigma_i} \tag{7.38}$$

The new random variable Y_i has zero mean if there is no bias (static error) between the calculated and measured values. The standard deviation, Σ_i, required to normalize measurement i is due to network coupling and is subsequently derived. The Gaussian probability function for the random variable Y_i is plotted in Figure 7.8. It has unity variance.

When the standard deviation Σ_i is obtained for each measurement, differences between measured and calculated values may be compared using the probability density function of Figure 7.8.

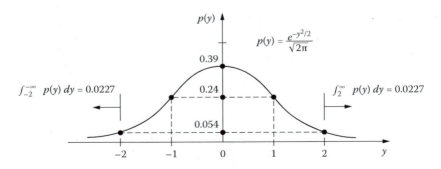

FIGURE 7.8
Probability density function for normalized measurement.

TABLE 7.1

Good versus Bad Data Based on Unit Variance
(Gaussian) Probability Function

$\left\| \dfrac{Z_{mi} - \hat{Z}_{ci}}{\Sigma_i} \right\| = t_i$	Probability That the Measurement Is Good	Probability That the Measurement Is Bad
0.10	0.9204	0.0796
0.25	0.8026	0.1974
0.5	0.6170	0.3830
1.0	0.3174	0.6826
2.0	0.0454	0.9546
2.57	0.01	0.99
3.0	0.0026	0.9974

The sum of all probable errors is 1.0, which is the area under the probability density function curve:

$$\int_{-\infty}^{+\infty} p(y)\, dy = 1.0 \tag{7.39}$$

Therefore, given a measurement datum, a calculated value, and Σ_i, the probability that the measurement is valid can be computed. For example, if the normalized difference $Y_i > 2$, there is <0.0454 probability that the measurement is good compared to >0.9546 probability that the measurement is bad. These probable values are obtained by integrating the curve of Figure 7.8 for the region $-2 \le y \le 2$ (the integral is called a cumulative distribution function). Therefore, given a value $|t_{imax}|$ that represents an acceptable limit, a measurement is ruled invalid or bad data if Y_i exceeds this limit. Table 7.1 is based on the normalized Gaussian random variable.

It is necessary to set an acceptable error limit, $|t_{imax}|$, for the measurements. Measurements with normalized error differences more than t_{imax} are called bad data. Such bad measurements may immediately be removed form subsequent sets of measurements and the power system operator alarmed. Alternatively, the bad measurements may be identified, and only if they repeatedly exceed the limit in the next few state estimations are they removed from the data scan until corrected in the field.

7.4 Determination of Variance Σ^2 to Normalize Measurements

Power flow in all parts of a network is dependent to some extent on voltages and currents at every bus. In the same manner, a measurement error at one point in the power system affects all calculated states in the system to some extent. Consequently, the standard deviation Σ_i, used to normalize measurement difference $Z_{mi} - \hat{Z}_{ci}$, depends on the topology of the network and the accuracy of all measurements (W_i in Equation 7.2 or r_{ii} in Equation 7.23).

Consider a linearization of the calculated measurement vector

$$\mathbf{H} = \frac{\partial \mathbf{h}}{\partial \mathbf{X}} \cong \begin{bmatrix} \dfrac{\partial \mathbf{h}}{\partial \boldsymbol{\delta}} \\ \dfrac{\partial \mathbf{h}}{\partial \mathbf{V}} \end{bmatrix}_{\hat{\boldsymbol{\delta}}, \hat{\mathbf{v}}}$$

(7.40)

With this definition, the best set of linearized, calculated measurements are approximately

$$\hat{\mathbf{Z}}_c = \mathbf{H}\hat{\mathbf{X}}_c = \mathbf{H} \begin{bmatrix} \hat{\boldsymbol{\delta}} \\ \hat{\mathbf{V}} \end{bmatrix}$$

(7.41)

where $\hat{\mathbf{X}}$ is the magnitude and phase angle from $\hat{\mathbf{E}}_{\mathrm{BUS}}$. The size of matrix $\mathbf{H}$ is $M \times (2n - 2)$. It is necessary that the number of measurements is $M \geq 2n - 2$.

If the rank of $\mathbf{H}$ is $2n - 2$, then $2n - 2$ nonsingular measurements may be ordered first, the matrix $\mathbf{H}$ partitioned, and Equation 7.41 written in terms of measured values as

$$\mathbf{Z}_M = \begin{bmatrix} Z_{M1} \\ \vdots \\ Z_{M,2n-2} \\ \vdots \\ Z_{M,M} \end{bmatrix} = \mathbf{H}\mathbf{X}_M = \begin{bmatrix} \mathbf{C}_1 \\ \mathbf{C}_2 \end{bmatrix} \mathbf{X}_M$$

(7.42)

Because $\mathbf{C}_1$ is nonsingular, a measured state of the system is found directly from

$$\mathbf{X}_M = [\mathbf{C}_1]^{-1} \begin{bmatrix} Z_{M1} \\ \vdots \\ Z_{M,2n-2} \end{bmatrix}$$

(7.43)

Expressed in another way, the state of the power system is *observable* (measurable) if the rank of the matrix $\partial \mathbf{h}/\partial \mathbf{X}$ is $2n - 2$ (the number of states). This is a significant result because it is the first available test to determine if the state can be calculated from a set of measurements. It does not ensure that the state estimator iterative calculations will converge because the set of measurements could result in ill-conditioned equations to solve for the state vector.

When the linearized gradient gradient matrix $\partial \mathbf{h}/\partial \mathbf{X}$ is substituted into the performance index, this yields a quadratic form

$$J' = [\mathbf{Z}_M - \mathbf{Z}_c]^t \mathbf{R}^{-1} [\mathbf{Z}_M - \mathbf{Z}_c]$$
$$= [\mathbf{Z}_M - \mathbf{H}\mathbf{X}]^t \mathbf{R}^{-1} [\mathbf{Z}_M - \mathbf{H}\mathbf{X}]$$

(7.44)

which is easily minimized for the best value of $\mathbf{X}$ in terms of the measured data:

$$\hat{\mathbf{X}} = [\mathbf{H}^t \mathbf{R}^{-1} \mathbf{H}]^{-1} \mathbf{H}^t \mathbf{R}^{-1} \mathbf{Z}_M = \mathbf{C}\mathbf{Z}_M$$

(7.45)

Using Equation 7.45, the best calculated set of values is

$$\hat{\mathbf{Z}}_c = \mathbf{H}\hat{\mathbf{X}} = \mathbf{H}[\mathbf{H}^t\mathbf{R}^{-1}\mathbf{H}]^{-1}\mathbf{H}^t\mathbf{R}^{-1}\mathbf{Z}_M \qquad (7.46)$$

It is now necessary to borrow several concepts from the field of probability and statistics. Let μ_i be the mean or average value, expected value, or most probable value of the random variable Y_i based on its probability density function (as in Figure 7.8, for example), It is calculated as

> Mean Value
>
> $$\mu_i = \Xi\{Y_i\} = \int_{-\infty}^{+\infty} Y_i p(Y_i) dY_i$$

where $\Xi\{\cdot\}$ is defined as an expectation operator. Note that the probability density function of Figure 7.8 has zero mean. In the same manner, the variance is defined as

> Variance
>
> $$\sigma_i^2 = \Xi\{(Y_i - \mu_i)^2\} = \int_{-\infty}^{+\infty} (Y_i - \mu_i)^2 p(Y_i) dY_i$$

The random variable shown in Figure 7.8 has a variance $\sigma^2 = 1$.

Using our definition of mean value, the linearized mean differences between measured and best calculated values, called *residues*, are

$$\Xi\{(\mathbf{Z}_M - \hat{\mathbf{Z}}_c)\} = \Xi\{\mathbf{HX} + \zeta - \hat{\mathbf{Z}}_c\} = \Xi\{HX + \zeta - \mathbf{H}\hat{\mathbf{X}}\}$$

$$= \mathbf{H}\Xi\{\mathbf{X}\} - \mathbf{H}\Xi\{\hat{\mathbf{X}}\} + \mathbf{J}\{\zeta\zeta\} = 0 \qquad (7.47)$$

where the noise (or error) associated with every measurement has zero mean according to Equation 7.21. Therefore, the calculated values are unbiased (no static error) estimates of the measurements.

The variance of the difference between measured and calculated values, in other words, the covariance of the residues, is

$$\Sigma^2 = \Xi\{(\mathbf{Z}_M - \hat{\mathbf{Z}}_c)(\mathbf{Z}_M - \hat{\mathbf{Z}}_c)^t\} = \Xi\{(\mathbf{HX} + \zeta - \hat{\mathbf{Z}}_c)(\mathbf{HX} + \zeta - \hat{\mathbf{Z}}_c)^t\}$$

$$= \Xi\{(\mathbf{HX} + \zeta - \mathbf{HC}(\mathbf{X} + \zeta))(\mathbf{HX} + \zeta - \mathbf{HC}(\mathbf{X} + \zeta))^t\} \qquad (7.48)$$

$$= \Xi\{\zeta\zeta^t - \zeta\zeta^t\mathbf{C}^t\mathbf{H}^t - \mathbf{HC}\zeta\zeta^t + \mathbf{HC}\zeta\zeta^t\mathbf{C}^t\mathbf{H}^t\}$$

$$= \mathbf{R} - \mathbf{H}[\mathbf{H}^t\mathbf{R}^{-1}\mathbf{H}]^{-1}\mathbf{H}^t$$

Because each measurement Z_{mi} is in error by zero-mean, Gaussian random noise, the residues are also a zero-mean, Gaussian joint probability density, the variances of which are *diagonal* elements [3] of Equation 7.48. Thus, the elements are the squared value of the Σ_i standard deviation required in Equation 7.38 for normalization. It may be shown [3,4] that the measurement with the *highest* normalized residue,

$$\underset{i}{\text{Max}} \left| \frac{Z_{mi} - \hat{Z}_{ci}}{\Sigma_i} \right| = \begin{bmatrix} \text{Most probable} \\ \text{bad measurement} \end{bmatrix}$$

is most likely the erroneous measurement, and it is considered bad data. An example will be used to demonstrate use of the normalized residue.

Example 7.4

Consider a dc network with three parallel 1.0 Ω lines as shown in Figure E7.4. A set of two potentials and three currents are measured with meters of known accuracy. The measurements and their covariances are as follows:

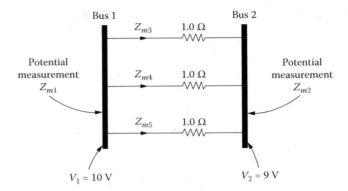

FIGURE E7.4

Potential measurements:

$$Z_{m1} = 10.5$$

$$Z_{m2} = 8.5$$

Current measurements:

$$Z_{m3} = 1.1$$

$$Z_{m4} = 0.9$$

$$Z_{m5} = 5.0$$

Covariances:

$$\Xi\{\zeta_1\zeta_1\} = r_{11} = r_{22} = r_{33} = r_{44} = 1.0$$

$$\Xi\{\zeta_5\zeta_5\} = 0.001$$

A measurement is considered to be bad data if its normalized residual is more than $t_i = 5$. Detect any bad data, remove them, then recalculate the state of the system. In comparison to the others, measurement 5 is extremely accurate.

Solution

There are no phase angles required for steady-state power flow on a dc network, and because only potentials and currents are used, the measurements are linear functions of the system state, $\mathbf{X} = [V_1 \ V_2]^t$. The calculated values are

$$\mathbf{Z}_c = \mathbf{h}(\mathbf{X}) = \mathbf{H}\mathbf{X} = \begin{bmatrix} 1 & 0 \\ 0 & 1 \\ 1 & -1 \\ 1 & -1 \\ 1 & -1 \end{bmatrix} \begin{bmatrix} V_1 \\ V_2 \end{bmatrix}$$

The best estimate for the state is given by

$$\hat{X} = [H^t R^{-1} H]^{-1} H^t R^{-1} Z_M$$

$$= \left\{ \begin{bmatrix} 1 & 0 & 1 & 1 & 1 \\ 0 & 1 & -1 & -1 & -1 \end{bmatrix} \begin{bmatrix} 1 & 0 & 0 & 0 & 0 \\ 0 & 1 & 0 & 0 & 0 \\ 0 & 0 & 1 & 0 & 0 \\ 0 & 0 & 0 & 1 & 0 \\ 0 & 0 & 0 & 0 & 1000 \end{bmatrix} \begin{bmatrix} 1 & 0 \\ 0 & 1 \\ 1 & -1 \\ 1 & -1 \\ 1 & -1 \end{bmatrix} \right\}^{-1} H^t R^{-1} \begin{bmatrix} 10.5 \\ 8.5 \\ 1.10 \\ 0.90 \\ 5.00 \end{bmatrix}$$

$$= \left\{ \begin{bmatrix} 1003 & -1002 \\ -1002 & 1003 \end{bmatrix} \right\}^{-1} \begin{bmatrix} 5012.5 \\ -4993.5 \end{bmatrix}$$

$$= \begin{bmatrix} 0.5002494 & 0.4997506 \\ 0.5997506 & 0.5002494 \end{bmatrix} \begin{bmatrix} 5012.5 \\ -4993.5 \end{bmatrix}$$

$$= \begin{bmatrix} 11.9955 \\ 7.0045 \end{bmatrix}$$

The best estimate for the measurements is

$$\hat{Z}_c = H\hat{X} = [11.9955, 7.0045, 4.991, 4.991, 4.991]^t$$

The covariance matrix for the residues using $a = 0.5002494$ and $b = 0.4997506$ is

$$\Sigma^2 = R - H[H^t R^{-1} H]^{-1} H^t = R - H \begin{bmatrix} a & b \\ b & a \end{bmatrix} H^t$$

$$= \begin{bmatrix} 1 & 0 & 0 & 0 & 0 \\ 0 & 1 & 0 & 0 & 0 \\ 0 & 0 & 1 & 0 & 0 \\ 0 & 0 & 0 & 1 & 0 \\ 0 & 0 & 0 & 0 & 0.001 \end{bmatrix} - \begin{bmatrix} a & b & a-b & a-b & a-b \\ b & a & b-a & b-a & b-a \\ a-b & b-a & 2(a-b) & 2(a-b) & 2(a-b) \\ a-b & b-a & 2(a-b) & 2(a-b) & 2(a-b) \\ a-b & b-a & 2(a-b) & 2(a-b) & 2(a-b) \end{bmatrix}$$

$$= \begin{bmatrix} 0.4997 & -0.4997 & -4.988 \times 10^{-5} & -4.998 \times 10^{-5} & -4.998 \times 10^{-5} \\ -0.4997 & 0.4997 & 4.988 \times 10^{-5} & 4.998 \times 10^{-5} & 4.998 \times 10^{-5} \\ -4.998 \times 10^{-5} & 0.4998 \times 10^{-5} & 1 & -1 \times 10^{-4} & -1 \times 10^{-4} \\ -4.998 \times 10^{-5} & 0.4998 \times 10^{-5} & -1 \times 10^{-4} & 1 & -1 \times 10^{-4} \\ -4.998 \times 10^{-5} & 0.4998 \times 10^{-5} & -1 \times 10^{-4} & -1 \times 10^{-4} & 2.4 \times 10^{-6} \end{bmatrix}$$

The normalized residuals $(Z_{mi} - \hat{Z}_{ci})/\Sigma_i$, where Σ_i is the square root of the ith diagonal element of the above matrix, are $[Y_1, Y_2, Y_3, Y_4, Y_5] = [-2.257, -2.116, -3.891, -4.091, 5.809]$. It is clear that the measurement Z_{M5} is bad data, as it is greater than $t_i = 5$ standard deviations. It is then eliminated from the set of measurements. Using only measurements 1 through 4, it may be verified that the best estimate for the state is

$$\hat{X} = \begin{bmatrix} 10.1 \\ 8.9 \end{bmatrix} \quad \text{(measurements 1–4)}$$

The best estimate calculated for measurements, using measurements 1 through 4, is

$$\hat{\mathbf{Z}}_c = \begin{bmatrix} 10.1 \\ 8.9 \\ 1.2 \\ 1.2 \end{bmatrix}$$

With only four measurements the covariance matrix is

$$\Sigma^2 = \mathbf{R} - \mathbf{H}[\mathbf{H}^t\mathbf{R}^{-1}\mathbf{H}]^{-1}\mathbf{H}^t = \begin{bmatrix} 0.6 & 0.4 & 0.2 & 0.2 \\ 0.4 & 0.6 & -0.2 & -0.2 \\ 0.2 & -0.2 & 0.4 & 0.4 \\ 0.2 & -0.2 & 0.4 & 0.4 \end{bmatrix}$$

The normalized residuals are then calculated as

$$[Y_1, Y_2, Y_3, Y_4 = [0.6666. - 0.66660, -0.2500, -0.7500]$$

Each residual is considerably less without the bad data point. All are less than one standard deviation, much less than the $t_i = 5$ limit.

A significant reduction in computation is possible using a method to calculate diagonal elements of the residual covariance matrix, Equation 7.48, without using a full inverse. The diagonal elements are given by

$$\sum\nolimits_{\ell}^2 = \mathbf{R}_{\ell\ell} - \sum_{j=1}^{2(M-1)} \sum_{k=1}^{2(M-1)} \mathbf{H}(\ell, j)\mathbf{N}(k, j)\mathbf{H}^t(k, \ell) \tag{7.49}$$

In this method [3], only the nonzero elements of the product $H(\ell, j)H^t(k, \ell)$, which are passed *before* inversion, are required for **N**:

$$\mathbf{N} \cong [\mathbf{H}^t\mathbf{R}^{-1}\mathbf{H}]^{-1} \tag{7.50}$$

That is, the diagonal elements of the residual covariance matrix may be calculated using elements that correspond to the nonzero terms of $\mathbf{H}^t\mathbf{R}^{-1}\mathbf{H}$ before inversion. The matrix $\mathbf{H}^t\mathbf{H}$ is symmetric and sparse, so it is readily adapted to this "sparse inverse matrix" technique to calculate specific elements of an otherwise full inverse. The method is fast and requires only the pointer storage space of $\mathbf{H}^t\mathbf{H}$. Figure 7.9 shows typical real time data—MW, MVAR, MV—that are displayed on an LED screen to a dispatcher and used in the state estimator program.

This chapter has presented the basic concepts of collecting real-time measurements from the power system and estimating the state of the power system according to the weighted accuracy of the measurements. It is one of the most modern and useful operating techniques in power systems. It has become prominent because of simultaneous technical advances in the remote terminal data acquisition units (RTUs), parallel or preprocessing of the data in the central computer, very fast central processing units of the main computer, efficient sparse matrix techniques to perform the calculations, and methods to analyze the results.

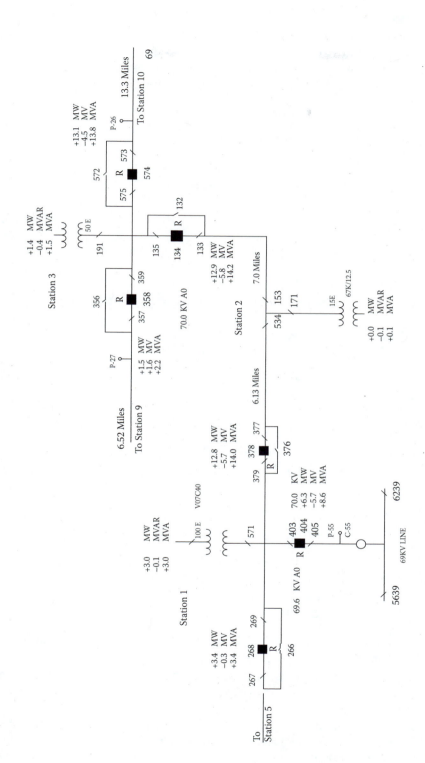

FIGURE 7.9

Presentation of line flow data and voltages on a dispatcher LED screen. Data is in real time and periodically updated. (Display courtesy of South Texas Electric Cooperative, Inc. All rights reserved)

Problems

7.1. Line-flow measurements are available on two of the lines of the three-bus system shown in Figure P7.1 The measurements at a 100 MVA base are

$$S_{m1} = -0.22 - j0.075$$

$$S_{m2} = 0.11 - j0.056$$

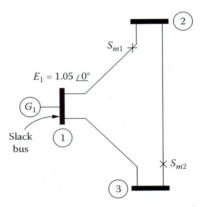

FIGURE P7.1

The line data are

$$\text{Line 1-2}: Z_{12} = j0.24$$
$$\text{Line 2-3}: Z_{23} = j0.18$$
$$\text{Line 1-3}: Z_{13} = j0.06$$

The metering accuracy is given by

$$W_1 = R_1^{-1} = 3$$
$$W_2 = R_2^{-1} = 2$$

(a) Using a fast-decoupled state estimator, derive the information matrix (Schweppe's method), or equivalently the **B'DB** matrix (Dopazo). Is this matrix singular?

(b) If something should cause the Mvar reading from measurement 2 to fail, would it still be possible to measure the state of the system? Clearly explain your reason.

7.2. By virtue of their location near a bus selected as the reference voltage, the phase angle of current measurements as well as magnitude are known. It is desired to use these current measurements in a state estimator calculation along with several other line-flow and injection measurements. The three-bus system is shown in Figure P7.2. The measurement data at a 100 MW base are as follows:

$$I_{m1} = 0.4 - j0.4$$
$$I_{m2} = -0.3 - j0.2$$
$$S_{m3} = -0.22 - j0.075$$
$$S_{m4} = -0.11 - j0.056$$
$$S_{m5} = 0.15 - j0.05$$

The line data (100 MW base) are

$$Line\,1-2:\ Z_{12} = j0.24$$
$$Line\,2-3:\ Z_{23} = j0.18$$
$$Line\,1-3:\ Z_{13} = j0.06$$

The metering accuracy is given by

$$W_1 = W_2 = R_1^{-1} = R_2^{-1} = 3$$
$$W_3 = R_3^{-1} = 2$$
$$W_4 = R_4^{-1} = 3$$
$$W_5 = R_5^{-1} = 1$$

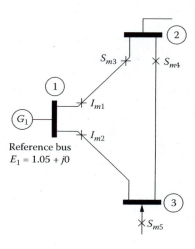

FIGURE P7.2

(a) Use a fast-decoupled state estimator to calculate the voltages at buses 2 and 3. Iterate until the voltages do not change more than 1×10^{-5} in absolute magnitude between successive iterations.

(b) If injection measurement 5 should fail, could the state of the system be calculated?

7.3 Define the performance index for a line-flow state estimator as

$$J(\mathbf{E}_{BUS}) = [\mathbf{S}_M - \mathbf{S}_c]^{*t} \mathbf{W}[\mathbf{S}_M - \mathbf{S}_c] + [\mathbf{E}_M - \mathbf{E}_c]^{*t} \mathbf{U}[\mathbf{E}_M - \mathbf{E}_c]$$

and incorporate the voltage measurement and weighting factor

$$|E_{2,m}| = 1.03 \qquad U_2 = 10^{-4}$$

into the calculation of Example 7.1. Start from a flat start and iterate until the largest voltage error between successive iterations is less than 0.0001.

7.4. For a six-bus power system, the transmission line parameters on a 100 MVA base are as follows:

Line i–k	R	X	ω C/2 or Transformer Tap, Bus i to Bus k
1–2	0.02	0.035	0.025
1–3	0.03	0.08	0.030
3–4	0	0.06	$\alpha = 1.02$
5–4	0	0.08	$\alpha = 1.01$
2–5	0.04	0.15	0.035
2–6	0.01	0.03	0.03
5–6	0.05	0.18	0.01

The real and reactive power flow on both ends of each line is measured with 50 MW, 50 Mvar meters. The accuracy of the metering is 2% and the A/D conversion error is 0.35%. The p.u. line-flow measurements are as follows:

Line	Real Power Measurement	Reactive Power Measurement
1–2	−0.15	−0.08
2–1	0.15	0.03
1–3	−0.43	−0.02
3–1	0.44	−0.04
3–4	0.36	0.14
4–3	−0.36	−0.13
5–4	−0.11	0.02
4–5	0.11	−0.02
2–5	0.02	0.02
5–2	−0.02	−0.10
2–6	−0.27	−0.10
6–2	0.27	0.04
5–6	−0.06	−0.07
6–5	0.06	0.05

Bus 6 is the reference at $E_r = 1.06 + j0$. Find the state of the power system using a line-flow state estimator. Use a flat start. Terminate iterations when $|E_i^{k+1} - E_i^k| \le 0.00001$ for each bus $i = 1, 2, 3, 4, 5$, where k is the iteration number.

© The computer program *stat_est* can compute an approximate solution using the input file *prob7-4.sss*. The program does not consider the off-nominal tap ratios for transformers.

7.5. A state estimator program may be verified by means of using data calculated in a power flow. The power flow program *p_flow.exe* writes results into the file *p_flow.sss*, which may be used as the input data to *stat_est.exe*. Use the power flow results calculated in Example 5.5, part (b), to do the following:

(a) © Use the state estimator program *stat_est.exe* with seven line flow measurements, no injections, to calculate the state that matches results of Example 5.5. The reference bus 4 is $E_r = 1.05 + j0.0$.

(b) Reverse the algebraic signs of a measurement and observe bad data detection and a change in the best estimate for the state.

(c) Set any measurement to a very small numerical value (e.g., 0.0001) to observe bad data detection of *stat_est.exe*. A numerical data value of 0.0 is read as "no measurement." The software program computes the standard deviation Σ_i for each measurement to normalize the difference between the measurements and the best estimate. The test for bad data is a normalized residual greater than a standard deviation of 0.1. The program permits the user to specify any threshold for bad data.

References

1. Dopazo, J. F., Kliton, O. A., and Van Slyck, L. S. 1970. State calculations of power systems from line flow measurements. Part II. *IEEE Trans. Power Appar. Syst.* 89:1698–708.
2. Horisberger, H. P., Richard, J. C., and Rossier, C. 1976. Fast decoupled state estimation. *IEEE Trans. Power Appar. Syst.* 95.
3. Broussolle, F. 1978. State estimation in power systems detecting bad data through the sparse inverse matrix method. *IEEE Trans. Power Appar. Syst.* 97.
4. Handschin, E., Schweppe, F. C., et al. 1975. Bad data analysis for power state estimation. *IEEE Trans. Power Appar. Syst.* 94(20).

Appendix A

Conductor Resistance and Rating

Power systems employ a variety of copper and aluminum conductors for current-carrying capability. The conductors are often reinforced with steel, titanium, or mechanically harder grades of conductor material. Low-voltage, low-power circuits tend to employ more copper conductors, while distribution lines and high/extra-high-voltage (EHV)/ultra-high-voltage (UHV) lines employ reinforced aluminum. Generally, utilities are replacing older copper installations by aluminum conductors for both cost and weight considerations, even though the aluminum has greater resistive losses.

The dc resistance of a conductor is found from the cross-sectional area, intrinsic resistivity of the material, and length of the conductor. Basic electrical and mechanical properties of conductor materials are given in Table A.1. The aluminum used for electrical conductors is mostly grade Commercial 1350. The table entries are taken from the American National Standards Institute (ANSI) and American Society for Testing and Materials (ASTM) standards [1].

Since transmission lines usually employ aluminum-composite or aluminum-reinforced conductors, a few of these are tabulated and described below.

Homogeneous designs:

AAC: All-aluminum conductor; given flower code names; concentrically standard alloy 1350-H19.

AAAC-5005-H19: All-aluminum alloy conductor; given city code names; with grade 5005-H19 aluminum.

AAAC-6201-T81: All-aluminum alloy conductor; given city code names; with higher strength than obtainable with H19-grade aluminum.

AAC/COMP: All-aluminum conductors (compact round).

ACAR: aluminum conductor alloy reinforced, conductor grade 1350-H19 stranded concentrically over 6201 core; no code names given.

Composite designs:

ACSR: Aluminum conductor steel reinforced; given code names of birds; coated steel core, 1350-H19-grade aluminum.

ACSR/AW: Same as above with aluminum-clad steel core.

ACSR/SD: Trapezoidal aluminum conductors and one layer of round 135-H19 wires over a high-strength stranded or solid steel core; the SD suffix is for self-damping; given code names of birds.

In addition to the foregoing types of conductors, there are compact and expanded grades of ACSR conductors available. A side view of an ACSR concentrically stranded bare conductor for overhead use is shown in Figure A.1a. ACSR typical cross sections are shown in Figure A.1b. The axial length of one complete turn, or helix, of a wire in a stranded conductor is sometimes called the *lay* or *pitch*. The "class" rating of all-aluminum or composites is defined as follows:

TABLE A.1

Physical and Electrical Properties

| | Conductivity Percent IACS Min. Percent 20°C | Resistivity, Ohms | | | | Temperature Coefficients of Resistance/Degree C | | Density at 20°C | | | Coefficient of Linear Expansion per Degree | | Modulus of Elasticity | |
| | | Circular Mil/Ft | | Sq. Millimeter/Meter | | | | Grams/Cubic Centimeter | Lb/Cubic Inch | Lb/Million Circular Mils/1,000 Ft | | | Lb/Sq. Inch | Kg/Sq. Millimeter |
		20°C	25°C	20°C	25°C	20°C	25°C				F	C		
Commercial 1350 aluminum wire	61.0	17.002	17.345	0.028264	0.028834	0.00403	0.00395	2.705	0.0975	920.3	0.0000128	0.000023	10×10^6	7,030
Aluminum alloy wire 6201	52.5	19.754	20.097	0.032840	0.033373	0.00347	0.00340	2.69	0.0969	915.2	0.0000128	0.000023	10×10^6	7,030
Commercial hard-drawn copper wire	97.0	10.692	10.895	0.017774	0.018113	0.00381	0.00374	8.89	0.321	3027	0.0000094	0.0000169	17×10^6	11,950
Standard annealed copper wire	100.0	10.371	10.575	0.017241	0.017579	0.00393	0.00385	8.89	0.321	3027	0.0000094	0.0000169	17×10^6	11,950
Aluminum-coated steel-core wire	9.0[a]	115.23[a]	—	0.19157[a]	—	—	—	7.78	0.281	2649	0.0000064	0.0000115	29×10^6	20,400
Zinc-coated steel-core wire	9.0[a]	115.23[a]	—	0.19157	—	—	—	7.78	0.281	2649	0.0000064	0.0000115	29×10^6	20,400
Aluminum-clad steel-core wire	20.33	51.01	51.52	0.0848	0.08563	0.0036	0.00356	6.59	0.2380	2243	0.0000072	0.0000130	23.5×10^6	16,500

[a] Typical.

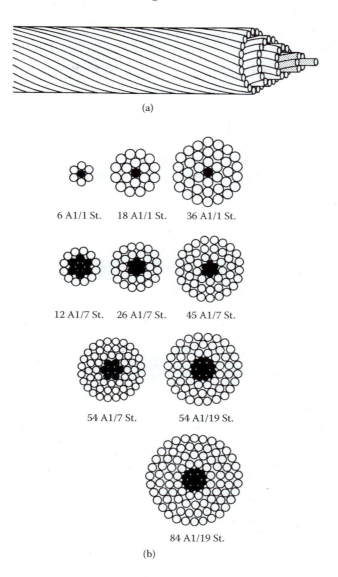

(a)

6 A1/1 St.   18 A1/1 St.   36 A1/1 St.

12 A1/7 St.   26 A1/7 St.   45 A1/7 St.

54 A1/7 St.   54 A1/19 St.

84 A1/19 St.

(b)

FIGURE A.1
a) ACSR conductor side viewing lay and standing
b) typical cross-sections of ACSR (aluminum is open circle, steel is a dark circle)

Classifications for Relative Conductor Flexibility

Class AA: For bare conductors usually used in overhead lines.

Class A: Conductors to be covered with weather-resistant materials and bare conductors where greater flexibility is required than that afforded by class AA.

Class B: For conductors to be insulated with various materials, such as rubber, paper, cloth, and so on, and for conductors indicated under class A where greater flexibility is desired.

Class C: Greater flexibility than class B.

For any single layer of conductor or reinforcements as shown in Figure A.1, if the length of lay is known, the dc resistance of the stranded layer is calculated at a temperature T_0 as

$$R_0 = \frac{\rho_0(5280)}{A}\sqrt{1+\left(\frac{2\pi r}{S}\right)^2} \tag{A.1}$$

where R_0 is ohms per mile per layer at temperature T_0, °C; ρ_0 is resistivity, ohm-circular mil/foot at temperature T_0; S is the length of lay, ft (axial length of one complete turn); r is the radius to the center of the layer, ft; and A is the cross section of conductors in the layer, circular mils.

For small ranges in temperature, $T_1 \neq T_0$, the resistance of the layer may be calculated as

$$R_1 = R_0[1 + \alpha_1(T_1 - T_0)] \tag{A.2}$$

where T_1 is in degrees celsius and α_1 is the temperature coefficient of resistance (e.g., Table A.1).

For N layers of reinforcement and conductors, the total resistance is the parallel combination of layers:

$$R_T = \frac{1}{\sum_{K=1}^{N}(1/R_k)} \tag{A.3}$$

The resistance for ac current flow is increased compared to the dc value because of skin effect. For a homogeneous cylindical conductor, the ac resistance compared to dc is

$$R' = K_r R_0 \tag{A.4}$$

The multiplier K_r is a function of the parameter x:

$$x = 2\pi r\sqrt{\frac{f\mu}{\rho \times 10^9}} \tag{A.5}$$

where r is the radius of the conductor (centimeters), f the frequency in hertz, μ the magnetic permeability, and ρ is in units of ohm-centimeters. Interdependenty, the magnetic fields in the wire cause the ac current distribution to be more concentrated at the conductor surface, which increases the voltage drop per length of conductor. Graphical plots and numerical tabulations of K_r versus x are available [2] to determine the ac resistance of a conductor.

Skin effect also causes a conductor's internal inductance to be lower than calculated using a uniform current distribution. A multiplier, K_L, to correct the internal inductance for skin effect,is also given in reference [2].

For overhead conductors, the temperature of the conductor strands is generally unknown, although surface temperature can be remotely monitored. For a specified current flow in the phase wire, the interior strands are at a higher temperature than the surface because heat transfer is strand to strand to surrounding air. Voids or air between strands inhibits heat flow. Heat transfer at the cable surface is dependent on:

Ambient air temperature

Winds perpendicular to the cable

Cable surface emissivity factor

Radiant energy (direct and diffuse solar energy)

Percipitation and humidity

As a consequence, an equivalent operating temperature is assumed for the conductor, and the appropriate ac resistance is read from the tables in Chapter 2.

The lifetime of service for the conductor is limited by aging phenomena. Two aging phenomena common to conductors under tension, such as overhead lines, are annealing and creep. Both lead to the loss of strength of the conductor with time. The rate that a conductor anneals and creeps increases as the temperature of the conductor increases. Therefore, the operating current plus the climate with daily and seasonal variations determines the conductor temperature rise, and consequently the expected lifetime. No method has been generally accepted by the industry to use climatological data for predicting lifetime at a current rating. Samples of long-term ratings by a manufacturer [1] are given in Table A.2.

As the temperature of an overhead conductor increases, its resistance increases. If the current through the conductor remains constant, a *nonuniform* part of the conductor could experience a self-reinforcing cycle of increasing temperature, increasing resistance, and decreasing cross-sectional area due to elongation from cable weight. This reinforced cycle could continue until the cable ruptures or sags to create a short circuit at some span. As a result, transmission lines have "shutdown" and protective device current limits that are dependent on the tensile strength of the materials, surface properties, and conductivity. Before the shutdown current limit is reached, several emergency ratings are established for time durations that allow the utility to take corrective action. The emergency current ratings are higher than normal, resulting in higher conductor temperatures that decrease the lifetime of the overhead lines. Two classes of emergency rating used by the New York Power Pool [3] are

Long-term emergency (LTE): 3 h

Short-term emergency (STE): 15 min

Example emergency ratings for a Drake ACSR conductor are shown in Table A.3. Observe in Table A.3 that the normal rating is based on a 45°C temperature rise over ambient in the year 1971, while a 40°C rise is specified in Table A.2 for the Drake conductor. This accounts for the difference in normal ratings.

The problem of transmission line rating and lifetime prediction is very extensive. New analytical methods, instrumentation techniques, statistical records, and on-line temperature monitoring devices are continually being developed for transmission lines. The reader is referred to the Institute of Electrical Engineers' (IEEE) *Transactions of Power Apparatus and Systems* for more details on ratings for overhead transmission lines.

For insulated conductors, copper or aluminum, both in air and enclosed in ducts, the properties of the insulating material determine the ratings. The current-carrying capabilities of

TABLE A.2

Current-Carrying Capability for Several Overhead Conductors[a]

	Type	Aluminum Cross Section (kcmi 1 or AWG)	Current-Carrying Capacity (rms A)
1. ACSR			
	Code name Bluebird	2,516	1,615
	Code name Chukar	1,780	1,435
	Code name Drake	795	900
	Code name Kingbird	636	750
	Code name Linnet	336.4	510
	Code name Waxwing	266.8	430
	Code name Raven	1/0	230
2. ACAR		Equiv. EC	
	3000 kcmi 1	2,913.7	1,870
	200 kcmi 1	1,885.4	1,475
	1000 kcmi 1	928.4	965
	500 kcmi 1	474.2	625
	1/0	99.25	225
3. AAC 1350-H19			
	Code name Trillium, A	3,000	1,885
	Code name Cowslip, A	2,000	1,500
	Code name Camelia, A	1,000	990
	Code name Hyacinth, A	500	635
	Code name Poppy, A, AA	1/0	235
4. AAAC 6201-T81		Equiv. EC	
	Code name Greeley (927.2 kcmi 1)	795	890
	Code name Darien (559.5 kcmi 1)	477	645
	Code name Azuza (123.3 kcmi 1)	1/0	240

[a] Ampacity for conductor temperature rise of 40°C over 40°C ambient with 2 ft/sec crosswind and an emissivity factor of 0.5 without sun.

TABLE A.3

New York Power Pool Ratings for Drake 795 kcmi 1 26/7 ACSR

Air Temperature (°C)	Wind (ft/sec)	Normal Rating (95°C) (A)	LTE (115°C, 3 h) (A)	STE (125°C, 15 min) (A)
1971 rating (assumed life 25 years)				
Summer 40	2	970	1,140	1,310
Winter 10	2	1,240	1,370	1,520
1981 revised rating (assumed life 40 years)				
Summer 35	3	1,101	1,270	1,430
Winter 10	3	1,347	1,476	1,616

these lines follow the recommendations of the Insulated Power Cable Engineers Association (IPCEA) [1].

References

1. Fink, D. G., and Beaty, H. W. 2000. Standard handbook for electrical engineers, Table 4–10, New York: M'Graw-Hill Book Company.
2. Fink, D. G., and Carroll, J. M. 1968. *Standard handbook for electrical engineers: Electric and magnetic circuits*, section 2. Boast and Hale. New York: McGraw-Hill Book Company.
3. Foss, S. D., Lin, S. H., and Fernandes, R. A. 1983. Dynamic thermal line ratings. Part I. Dynamic ampacity rating algorithm. *IEEE Trans. Power Appar. Syst.* 102(6).

Index

A

Accelerated Gaussian iterative method
    equation, 240
Accelerated Gauss-Seidel method, 184
Accelerated Gauss-Seidel method equation, 241
ACE, *see* Area control error (ACE)
Adjustments
    day-ahead economic dispatch,
      303
    network operating conditions, 262–275
Admittance, phasor notation, 44, *see also* Bus
    admittance matrix
Aerial transmission line parameters, 87–96
AGC, *see* Automatic generation control (AGC)
Alarms, *see also* Emergency states
    contingency analysis, 213
    control centers, 3
    dispatcher, control features, 5
Aluminum-conductor-steel-reinforced (ACSR) cable
    aerial transmission line parameters, 93
    balanced three-phase lines, 67, 70
    capacitive reactance, 81
    characteristics of multilayer, 82–84
    code names, 67
    conductor resistance and rating, 369–374
    inductive reactance, 68
    transmission line capacitance, 80
Amortisseur windings, 122
Ampere's law, 59
Annealing, 373
Approximations, short-circuits, 206
Arc furnace, 264
Area control error (ACE)
    adjustments, 255
    CPS1, 1 minute average, 12
    CPS2, 10 minute average, 12
    daily load factor, 295–296
    disturbance conditions, 14
    economy A interchange program, 322
    exchange of power, 320
    fundamentals, 7–9
    iteration methods, 242
    power flow, 231
    power flow programs, 277
Area lumped dynamic model, 30–34
Autocorrelation matrix, 352
Automatic generation control (AGC)
    area control error, 9–14
    area lumped dynamic model, 32–33
    CPS1, 1 minute average, 12
    CPS2, 10 minute average, 12
    daily load factor, 296

data entry, 4
dispatcher, control features, 4
disturbance conditions, 14
fundamentals, 7–9
minimum requirements, 11
Automatic phase-shifting transformers, 242
Autotransformers, 100

B

Back substitution, 172–175
Bad data detection, 356, 357
Balanced case, contingency analysis, 219–223
Balanced currents, 47
Balanced networks, 55
Balanced three-phase properties
    fault calculations, 199
    network connection, 39
    overhead transmission line representation, 65–73
    phasor notation, 41–42
    symmetrical component transformation, 47
Balanced voltages, 46
Base load deviation (BLD), 296
Base power, *see* Generator base power setting
Benefit sharing, 325–326
Bias, 231
Bisection method, 298
BLD, *see* Base load deviation (BLD)
Blondel's two-reaction theory, 111
Bolzano method, 298
Branches
    bus impedance matrix, mutual uncoupling, 152
    current reference direction, 153
    power flow programs, 277
Btu term expression, 286
Bus admittance matrix
    bus reference frame, 160–169
    defined, 146
    linear network injections and loads, 145, 147,
      148–151
Bus charging capacitance, 39
Bus current injection matrix, 163
Buses
    defined, 39, 142
    power flow programs, 276
Bus impedance formulation and matrix
    bus reference frame, 151–160, 165–169
    defined, 146
    linear network injections and loads, 145
    slack bus, 237
Bus reference frame
    bus admittance matrix, 160–169
    bus impedance matrix, 151–160, 165–169